Nissan 300ZX Automotive Repair Manual

by Homer Eubanks and John H Haynes

Member of the Guild of Motoring Writers

Models covered:

All Nissan 300ZX models - 1984 through 1989
Including Turbo, 2-seater and 2 + 2 with 3.0L V6 engine

(72010-8U9)

Haynes Group Limited
Haynes North America, Inc.
www.haynes.com

Acknowledgements

We are grateful for the help and cooperation of the Nissan Motor Co., Ltd. for their assistance with technical information, certain illustrations and vehicle photos.

© **Haynes North America, Inc. 1991**

With permission from Haynes Group Limited

A book in the Haynes Automotive Repair Manual Series

ISBN-10: 1-85010-563-4

ISBN-13: 978-1-85010-563-3

Library of Congress Control Number 91-70710

While every attempt is made to ensure that the information in this manual is correct, no liability can be accepted by the authors or publishers for loss, damage or injury caused by any errors in, or omissions from, the information given.

Contents

1985 Nissan 300ZX Turbo

About this manual

Its purpose

The purpose of this manual is to help you get the best value from your vehicle. It can do so in several ways. It can help you decide what work must be done, even if you choose to have it done by a dealer service department or a repair shop; it provides information and procedures for routine maintenance and servicing; and it offers diagnostic and repair procedures to follow when trouble occurs.

It is hoped that you will use the manual to tackle the work yourself. For many simpler jobs, doing it yourself may be quicker than arranging an appointment to get the vehicle into a shop and making the trips to leave it and pick it up. More importantly, a lot of money can be saved by avoiding the expense the shop must pass on to you to cover its labor and overhead costs. An added benefit is the sense of satisfaction and accomplishment that you feel after having done the job yourself.

Using the manual

The manual is divided into Chapters. Each Chapter is divided into numbered Sections, which are headed in bold type between horizontal lines. Each Section consists of consecutively numbered paragraphs.

At the beginning of each numbered section you will be referred to any illustrations which apply to the procedures in that section. The reference numbers used in illustration captions pinpoint the pertinent Section and the Step within that section. That is, illustration 3.2 means the illustration refers to Section 3 and Step (or paragraph) 2 within that Section.

Procedures, once described in the text, are not normally repeated. When it is necessary to refer to another Chapter, the reference will be given as Chapter and Section number i.e. Chapter 1/16). Cross references given without use of the word ''Chapter'' apply to Sections and/or paragraphs in the same Chapter. For example, ''see Section 8'' means in the same Chapter.

Reference to the left or right side of the vehicle is based on the assumption that one is sitting in the driver's seat, facing forward.

Even though extreme care has been taken during the preparation of this manual, neither the publisher nor the author can accept responsibility for any errors in, or omissions from, the information given.

NOTE

A Note provides information necessary to properly complete a procedure or information which will make the steps to be followed easier to understand.

CAUTION

A Caution indicates a special procedure or special steps which must be taken in the course of completing the procedure in which the **Caution** is found which are necessary to avoid damage to the assembly being worked on.

WARNING

A Warning indicates a special procedure or special steps which must be taken in the course of completing the procedure in which the **Warning** is found which are necessary to avoid injury to the person performing the procedure.

Introduction to the Nissan 300 ZX

The Nissan 300 ZX is a highly redesigned version of the popular Z series cars. While the Z cars were orginally designed as sports cars, the ZX series is intended to fit a more luxurious grand touring image.

Some ZX's come with a turbocharged engine, which uses an exhaust gas powered turbine to force more fuel/air mixture into the cylinders for additional power.

All 300 ZX cars utilize a V6, overhead-cam engine with a computer-controlled electronic fuel injection system, and are available with either a 5-speed manual transmission or a 3-speed automatic transmission. Power is transmitted through a rear drive differential carrier which transmits power to the independently suspended rear wheels by driveaxles which incorporate the use of Constant Velocity (CV) joints to join them to the carrier.

Two body styles are available; the 2-seater version and the roomier, slightly longer 2 + 2 (4-seater) version. Either version can be obtained with the GL or GLL styling/convenience package.

Further information on the various systems and components, as well as complete specifications, can be found in the appropriate individual Chapters.

1986 Nissan 300ZX Turbo

1986 Nissan 300ZX 2 + 2

General dimensions

2-seat model
- Overall length .... 170.7 in
- Overall width .... 66.5 in
- Overall height .... 51.0 in
- Wheelbase .... 91.3 in

2 + 2 model
- Overall length .... 178.5 in
- Overall width .... 67.9 in
- Overall height .... 51.6 in
- Wheelbase .... 99.2 in

Vehicle identification numbers

Modifications are a continuing and unpublicized process in vehicle manufacturing. Since spare parts manuals and lists are compiled on a numerical basis, the individual vehicle numbers are essential to correctly identify the component required.

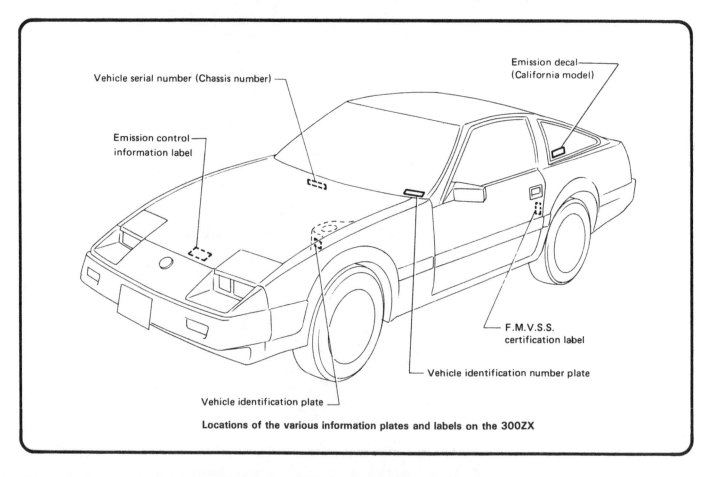

Locations of the various information plates and labels on the 300ZX

Vehicle identification number (VIN)

This very important identification number is located on a plate attached to the top left corner of the dashboard of the vehicle. The VIN also appears on the Vehicle Certificate of Title and Registration. It contains valuable information such as where and when the vehicle was manufactured, the model year and the body style. Also, the VIN label contains information about the way in which the vehicle is equipped.

This plate is especially useful for matching the color and type of paint during repair work.

Engine identification numbers

The engine identification number is located on a pad at the rear of the cylinder block.

Prefix and suffix designations

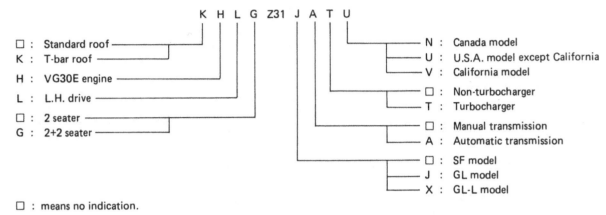

K H L G Z31 J A T U

☐ : Standard roof
K : T-bar roof
H : VG30E engine
L : L.H. drive
☐ : 2 seater
G : 2+2 seater

N : Canada model
U : U.S.A. model except California
V : California model
☐ : Non-turbocharger
T : Turbocharger
☐ : Manual transmission
A : Automatic transmission
☐ : SF model
J : GL model
X : GL-L model

☐ : means no indication.

300ZX vehicle identification code chart

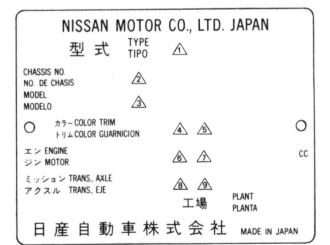

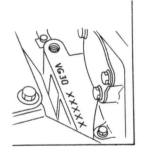

1 Type
2 Vehicle identification number (Chassis number)
3 Model
4 Body color code
5 Trim color code
6 Engine model
7 Engine displacement
8 Transmission model
9 Axle model

300ZX vehicle identification number (VIN) plate

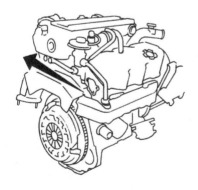

Location of the engine serial number

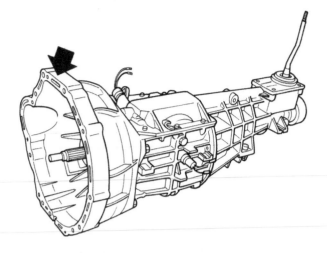

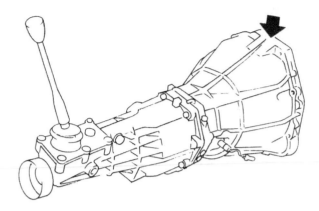

Non-turbocharged model manual transmission serial number location

Turbocharged model manual transmission serial number location

Manual transmission number

The 5-speed manual transmission ID number is located on a pad on the top of the bellhousing.

Automatic transmission numbers

The automatic transmission ID number is located on the right side of the housing above the oil pan.

Alternator numbers

The alternator ID number is located on top of the drive end frame.

Starter numbers

The starter ID number is stamped toward the rear of the outer case.

Battery numbers

The battery ID number is located on the cell cover segment on top of the battery.

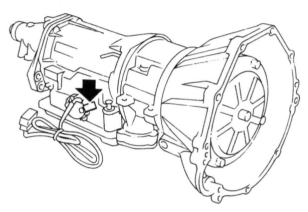

Automatic transmission serial number location

Vehicle Emissions Control Information label

The Emissions Control Information label is attached to the hood inside the engine compartment.

Buying parts

Replacement parts are available from many sources, which generally fall into one of two categories – authorized dealer parts departments and independent retail auto parts stores. Our advice concerning these parts is as follows:

Retail auto parts stores: Good auto parts stores will stock frequently needed components which wear out relatively fast, such as clutch components, exhaust systems, brake parts, tune-up parts, etc. These stores often supply new or reconditioned parts on an exchange basis, which can save a considerable amount of money. Discount auto parts stores are often very good places to buy materials and parts needed for general vehicle maintenance such as oil, grease, filters, spark plugs, belts, touch-up paint, bulbs, etc. They also usually sell tools and general accessories, have con-venient hours, charge lower prices and can often be found not far from home.

Authorized dealer parts department: This is the best source for parts which are unique to the vehicle and not generally available elsewhere (such as major engine parts, transmission parts, trim pieces, etc.).

Warranty information: If the vehicle is still covered under warranty, be sure that any replacement parts purchased – regardless of the source – do not invalidate the warranty!

To be sure of obtaining the correct parts, have engine and chassis numbers available and, if possible, take the old parts along for positive identification.

Maintenance techniques, tools and working facilities

Maintenance techniques

There are a number of techniques involved in maintenance and repair that will be referred to throughout this manual. Application of these techniques will enable the home mechanic to be more efficient, better organized and capable of performing the various tasks properly, which will ensure that the repair job is thorough and complete.

Fasteners

Fasteners are nuts, bolts, studs and screws used to hold two or more parts together. There are a few things to keep in mind when working with fasteners. Almost all of them use a locking device of some type, either a lockwasher, locknut, locking tab or thread adhesive. All threaded fasteners should be clean and straight, with undamaged threads and undamaged corners on the hex head where the wrench fits. Develop the habit of replacing all damaged nuts and bolts with new ones. Special locknuts with nylon or fiber inserts can only be used once. If they are removed, they lose their locking ability and must be replaced with new ones.

Rusted nuts and bolts should be treated with a penetrating fluid to ease removal and prevent breakage. Some mechanics use turpentine in a spout-type oil can, which works quite well. After applying the rust penetrant, let it work for a few minutes before trying to loosen the nut or bolt. Badly rusted fasteners may have to be chiseled or sawed off or removed with a special nut breaker, available at tool stores.

If a bolt or stud breaks off in an assembly, it can be drilled and removed with a special tool commonly available for this purpose. Most automotive machine shops can perform this task, as well as other repair procedures, such as the repair of threaded holes that have been stripped out.

Flat washers and lockwashers, when removed from an assembly, should always be replaced exactly as removed. Replace any damaged washers with new ones. Never use a lockwasher on any soft metal surface (such as aluminum), thin sheet metal or plastic.

Fastener sizes

For a number of reasons, automobile manufacturers are making wider and wider use of metric fasteners. Therefore, it is important to be able to tell the difference between standard (sometimes called U.S. or SAE) and metric hardware, since they cannot be interchanged.

All bolts, whether standard or metric, are sized according to diameter, thread pitch and length. For example, a standard 1/2 — 13 x 1 bolt is 1/2 inch in diameter, has 13 threads per inch and is 1 inch long. An M12 — 1.75 x 25 metric bolt is 12 mm in diameter, has a thread pitch of 1.75 mm (the distance between threads) and is 25 mm long. The two bolts are nearly identical, and easily confused, but they are not interchangeable.

In addition to the differences in diameter, thread pitch and length, metric and standard bolts can also be distinguished by examining the bolt heads. To begin with, the distance across the flats on a standard bolt head is measured in inches, while the same dimension on a metric bolt is sized in millimeters (the same is true for nuts). As a result, a standard wrench should not be used on a metric bolt and a metric wrench should not be used on a standard bolt. Also, most standard bolts have slashes radiating out from the center of the head to denote the grade or strength of the bolt, which is an indication of the amount of torque that can be applied to it. The greater the number of slashes, the greater the strength of the bolt. Grades 0 through 5 are commonly used on automobiles. Metric bolts have a property class (grade) number, rather than a slash, molded into their heads to indicate bolt strength. In this case, the higher the number, the stronger the bolt. Property class numbers 8.8, 9.8 and 10.9 are commonly used on automobiles.

Strength markings can also be used to distinguish standard hex nuts from metric hex nuts. Many standard nuts have dots stamped into one side, while metric nuts are marked with a number. The greater the number of dots, or the higher the number, the greater the strength of the nut.

Metric studs are also marked on their ends according to property class (grade). Larger studs are numbered (the same as metric bolts),

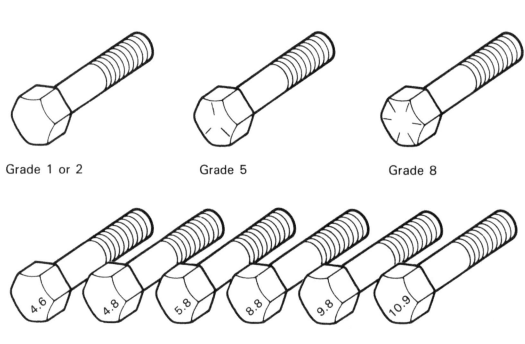

Grade 1 or 2   Grade 5   Grade 8

Bolt strength markings (top — standard/SAE/U.S.; bottom — metric)

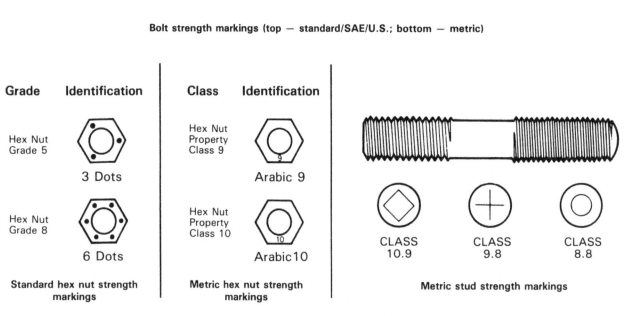

Grade	Identification
Hex Nut Grade 5	3 Dots
Hex Nut Grade 8	6 Dots

Standard hex nut strength markings

Class	Identification
Hex Nut Property Class 9	Arabic 9
Hex Nut Property Class 10	Arabic 10

Metric hex nut strength markings

CLASS 10.9   CLASS 9.8   CLASS 8.8

Metric stud strength markings

while smaller studs carry a geometric code to denote grade.

It should be noted that many fasteners, especially Grades 0 through 2, have no distinguishing marks on them. When such is the case, the only way to determine whether it is standard or metric is to measure the thread pitch or compare it to a known fastener of the same size.

Standard fasteners are often referred to as SAE, as opposed to metric. However, it should be noted that SAE technically refers to a non-metric *fine thread* fastener only. Coarse thread non-metric fasteners are referred to as U.S.S. sizes.

Since fasteners of the same size (both standard and metric) may have different strength ratings, be sure to reinstall any bolts, studs or nuts removed from your vehicle in their original locations. Also, when replacing a fastener with a new one, make sure that the new one has a strength rating equal to or greater than the original.

Tightening sequences and procedures

Most threaded fasteners should be tightened to a specific torque value (torque is the twisting force applied to a threaded component such as a nut or bolt). Overtightening the fastener can weaken it and cause it to break, while undertightening can cause it to eventually come loose. Bolts, screws and studs, depending on the material they are made of and their thread diameters, have specific torque values, many of which are noted in the Specifications at the beginning of each Chapter. Be sure to follow the torque recommendations closely. For fasteners not assigned a specific torque, a general torque value chart is presented here as a guide. As was previously mentioned, the size and grade of a fastener determine the amount of torque that can safely be applied to it. The figures listed here are approximate for Grade 2 and Grade 3 fasteners. Higher grades can tolerate higher torque values.

Metric thread sizes	Ft-lb	Nm/m
M-6	6 to 9	9 to 12
M-8	14 to 21	19 to 28
M-10	28 to 40	38 to 54
M-12	50 to 71	68 to 96
M-14	80 to 140	109 to 154

Pipe thread sizes		
1/8	5 to 8	7 to 10
1/4	12 to 18	17 to 24
3/8	22 to 33	30 to 44
1/2	25 to 35	34 to 47

U.S. thread sizes		
1/4 — 20	6 to 9	9 to 12
5/16 — 18	12 to 18	17 to 24
5/16 — 24	14 to 20	19 to 27
3/8 — 16	22 to 32	30 to 43
3/8 — 24	27 to 38	37 to 51
7/16 — 14	40 to 55	55 to 74
7/16 — 20	40 to 60	55 to 81
1/2 — 13	55 to 80	75 to 108

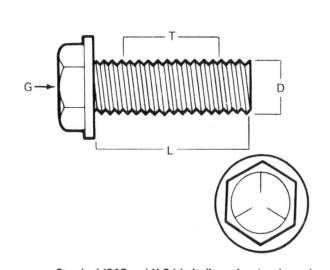

Standard (SAE and U.S.) bolt dimensions/grade marks

G Grade marks (bolt strength)
L Length (in inches)
T Thread pitch (number of threads per inch)
D Nominal diameter (in inches)

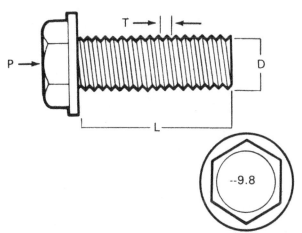

Metric bolt dimensions/grade marks

P Property class (bolt strength)
L Length (in millimeters)
T Thread pitch (distance between threads in millimeters)
D Diameter

Fasteners laid out in a pattern, such as cylinder head bolts, oil pan bolts, differential cover bolts, etc., must be loosened or tightened in sequence to avoid warping the component. This sequence will normally be shown in the appropriate Chapter. If a specific pattern is not given, the following procedures can be used to prevent warping.

Initially, the bolts or nuts should be assembled finger-tight only. Next, they should be tightened one full turn each, in a criss-cross or diagonal pattern. After each one has been tightened one full turn, return to the first one and tighten them all one-half turn, following the same pattern. Finally, tighten each of them one-quarter turn at a time until each fastener has been tightened to the proper torque. To loosen and remove the fasteners, the procedure would be reversed.

Component disassembly

Component disassembly should be done with care and purpose to help ensure that the parts go back together properly. Always keep track of the sequence in which parts are removed. Make note of special characteristics or marks on parts that can be installed more than one way, such as a grooved thrust washer on a shaft. It is a good idea to lay the disassembled parts out on a clean surface in the order that they were removed. It may also be helpful to make sketches or take instant photos of components before removal.

When removing fasteners from a component, keep track of their locations. Sometimes threading a bolt back in a part, or putting the washers and nut back on a stud, can prevent mix-ups later. If nuts and bolts cannot be returned to their original locations, they should be kept in a compartmented box or a series of small boxes. A cupcake or muffin tin is ideal for this purpose, since each cavity can hold the bolts and nuts from a particular area (i.e. oil pan bolts, valve cover bolts, engine mount bolts, etc.). A pan of this type is especially helpful when working on assemblies with very small parts, such as the carburetor, alternator, valve train or interior dash and trim pieces. The cavities can be marked with paint or tape to identify the contents.

Whenever wiring looms, harnesses or connectors are separated, it is a good idea to identify the two halves with numbered pieces of masking tape so they can be easily reconnected.

Gasket sealing surfaces

Throughout any vehicle, gaskets are used to seal the mating surfaces between two parts and keep lubricants, fluids, vacuum or pressure contained in an assembly.

Many times these gaskets are coated with a liquid or paste-type gasket sealing compound before assembly. Age, heat and pressure can sometimes cause the two parts to stick together so tightly that they are very difficult to separate. Often, the assembly can be loosened by striking it with a soft-face hammer near the mating surfaces. A regular hammer can be used if a block of wood is placed between the hammer and the part. Do not hammer on cast parts or parts that could be easily damaged. With any particularly stubborn part, always recheck to make sure that every fastener has been removed.

Avoid using a screwdriver or bar to pry apart an assembly, as they can easily mar the gasket sealing surfaces of the parts, which must remain smooth. If prying is absolutely necessary, use an old broom handle, but keep in mind that extra clean up will be necessary if the wood splinters.

After the parts are separated, the old gasket must be carefully scraped off and the gasket surfaces cleaned. Stubborn gasket material can be soaked with rust penetrant or treated with a special chemical to soften it so it can be easily scraped off. A scraper can be fashioned from a piece of copper tubing by flattening and sharpening one end. Copper is recommended because it is usually softer than the surfaces to be scraped, which reduces the chance of gouging the part. Some gaskets can be removed with a wire brush, but regardless of the method used, the mating surfaces must be left clean and smooth. If for some reason the gasket surface is gouged, then a gasket sealer thick enough to fill scratches will have to be used during reassembly of the components. For most applications, a non-drying (or semi-drying) gasket sealer should be used.

Hose removal tips

Warning: *If the vehicle is equipped with air conditioning, do not disconnect any of the A/C hoses without first having the system depressurized by a dealer service department or an air conditioning specialist.*

Hose removal precautions closely parallel gasket removal precautions. Avoid scratching or gouging the surface that the hose mates against or the connection may leak. This is especially true for radiator hoses. Because of various chemical reactions, the rubber in hoses can bond itself to the metal spigot that the hose fits over. To remove a hose, first loosen the hose clamps that secure it to the spigot. Then, with slip-joint pliers, grab the hose at the clamp and rotate it around the spigot. Work it back and forth until it is completely free, then pull it off. Silicone or other lubricants will ease removal if they can be applied between the hose and the outside of the spigot. Apply the same lubricant to the inside of the hose and the outside of the spigot to simplify installation.

As a last resort (and if the hose is to be replaced with a new one anyway), the rubber can be slit with a knife and the hose peeled from the spigot. If this must be done, be careful that the metal connection is not damaged.

If a hose clamp is broken or damaged, do not reuse it. Wire-type clamps usually weaken with age, so it is a good idea to replace them with screw-type clamps whenever a hose is removed.

Tools

A selection of good tools is a basic requirement for anyone who plans to maintain and repair his or her own vehicle. For the owner who has few tools, the initial investment might seem high, but when compared to the spiraling costs of professional auto maintenance and repair, it is a wise one.

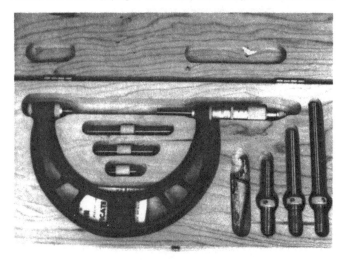

Micrometer set

Dial indicator set

Dial caliper

Hand-operated vacuum pump

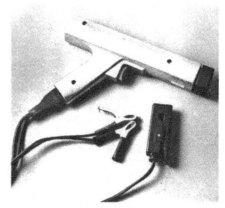

Timing light

Compression gauge with spark plug
hole adapter

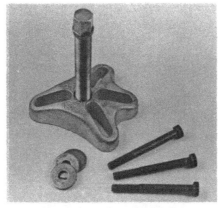

Damper/steering wheel puller

General purpose puller

Hydraulic lifter removal tool

Valve spring compressor

Valve spring compressor

Ridge reamer

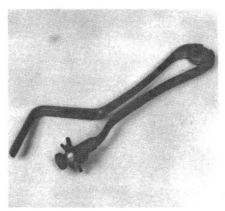

Piston ring groove cleaning tool

Ring removal/installation tool

Ring compressor

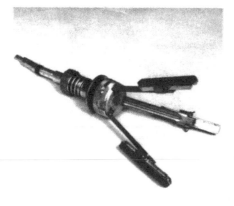

Cylinder hone

Brake hold-down spring tool

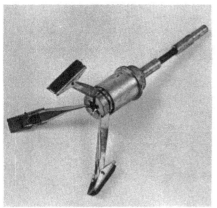

Brake cylinder hone

Clutch plate alignment tool

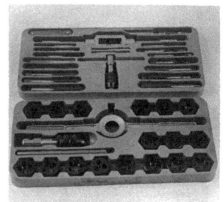

Tap and die set

To help the owner decide which tools are needed to perform the tasks detailed in this manual, the following tool lists are offered: *Maintenance and minor repair, Repair/overhaul* and *Special*.

The newcomer to practical mechanics should start off with the maintenance and minor repair tool kit, which is adequate for the simpler jobs performed on a vehicle. Then, as confidence and experience grow, the owner can tackle more difficult tasks, buying additional tools as they are needed. Eventually the basic kit will be expanded into the repair and overhaul tool set. Over a period of time, the experienced do-it-yourselfer will assemble a tool set complete enough for most repair and overhaul procedures and will add tools from the special category when it is felt that the expense is justified by the frequency of use.

Maintenance and minor repair tool kit

The tools in this list should be considered the minimum required for performance of routine maintenance, servicing and minor repair work. We recommend the purchase of combination wrenches (box-end and open-end combined in one wrench). While more expensive than open end wrenches, they offer the advantages of both types of wrench.

Combination wrench set (1/4-inch to 1 inch or 6 mm to 19 mm)
Adjustable wrench, 8 inch
Spark plug wrench with rubber insert
Spark plug gap adjusting tool
Feeler gauge set
Brake bleeder wrench
Standard screwdriver (5/16-inch x 6 inch)
Phillips screwdriver (No. 2 x 6 inch)
Combination pliers — 6 inch
Hacksaw and assortment of blades
Tire pressure gauge
Grease gun
Oil can
Fine emery cloth
Wire brush

Battery post and cable cleaning tool
Oil filter wrench
Funnel (medium size)
Safety goggles
Jackstands (2)
Drain pan

Note: *If basic tune-ups are going to be part of routine maintenance, it will be necessary to purchase a good quality stroboscopic timing light and combination tachometer/dwell meter. Although they are included in the list of special tools, it is mentioned here because they are absolutely necessary for tuning most vehicles properly.*

Repair and overhaul tool set

These tools are essential for anyone who plans to perform major repairs and are in addition to those in the maintenance and minor repair tool kit. Included is a comprehensive set of sockets which, though expensive, are invaluable because of their versatility, especially when various extensions and drives are available. We recommend the 1/2-inch drive over the 3/8-inch drive. Although the larger drive is bulky and more expensive, it has the capacity of accepting a very wide range of large sockets. Ideally, however, the mechanic should have a 3/8-inch drive set and a 1/2-inch drive set.

Socket set(s)
Reversible ratchet
Extension — 10 inch
Universal joint
Torque wrench (same size drive as sockets)
Ball peen hammer — 8 ounce
Soft-face hammer (plastic/rubber)
Standard screwdriver (1/4-inch x 6 inch)
Standard screwdriver (stubby — 5/16-inch)
Phillips screwdriver (No. 3 x 8 inch)
Phillips screwdriver (stubby — No. 2)

Pliers — vise grip
Pliers — lineman's
Pliers — needle nose
Pliers — snap-ring (internal and external)
Cold chisel — 1/2-inch
Scribe
Scraper (made from flattened copper tubing)
Centerpunch
Pin punches (1/16, 1/8, 3/16-inch)
Steel rule/straightedge — 12 inch
Allen wrench set (1/8 to 3/8-inch or 4 mm to 10 mm)
A selection of files
Wire brush (large)
Jackstands (second set)
Jack (scissor or hydraulic type)

Note: *Another tool which is often useful is an electric drill motor with a chuck capacity of 3/8-inch and a set of good quality drill bits.*

Special tools

The tools in this list include those which are not used regularly, are expensive to buy, or which need to be used in accordance with their manufacturer's instructions. Unless these tools will be used frequently, it is not very economical to purchase many of them. A consideration would be to split the cost and use between yourself and a friend or friends. In addition, most of these tools can be obtained from a tool rental shop on a temporary basis.

This list primarily contains only those tools and instruments widely available to the public, and not those special tools produced by the vehicle manufacturer for distribution to dealer service departments. Occasionally, references to the manufacturer's special tools are inluded in the text of this manual. Generally, an alternative method of doing the job without the special tool is offered. However, sometimes there is no alternative to their use. Where this is the case, and the tool cannot be purchased or borrowed, the work should be turned over to the dealer service department or an automotive repair shop.

Valve spring compressor
Piston ring groove cleaning tool
Piston ring compressor
Piston ring installation tool
Cylinder compression gauge
Cylinder ridge reamer
Cylinder surfacing hone
Cylinder bore gauge
Micrometers and/or dial calipers
Hydraulic lifter removal tool
Balljoint separator
Universal-type puller
Impact screwdriver
Dial indicator set
Stroboscopic timing light (inductive pick-up)
Hand operated vacuum/pressure pump
Tachometer/dwell meter
Universal electrical multimeter
Cable hoist
Brake spring removal and installation tools
Floor jack

Buying tools

For the do-it-yourselfer who is just starting to get involved in vehicle maintenance and repair, there are a number of options available when purchasing tools. If maintenance and minor repair is the extent of the work to be done, the purchase of individual tools is satisfactory. If,

on the other hand, extensive work is planned, it would be a good idea to purchase a modest tool set from one of the large retail chain stores. A set can usually be bought at a substantial savings over the individual tool prices, and they often come with a tool box. As additional tools are needed, add-on sets, individual tools and a larger tool box can be purchased to expand the tool selection. Building a tool set gradually allows the cost of the tools to be spread over a longer period of time and gives the mechanic the freedom to choose only those tools that will actually be used.

Tool stores will often be the only source of some of the special tools that are needed, but regardless of where tools are bought, try to avoid cheap ones, especially when buying screwdrivers and sockets, because they won't last very long. The expense involved in replacing cheap tools will eventually be greater than the initial cost of quality tools.

Care and maintenance of tools

Good tools are expensive, so it makes sense to treat them with respect. Keep them clean and in usable condition and store them properly when not in use. Always wipe off any dirt, grease or metal chips before putting them away. Never leave tools lying around in the work area. Upon completion of a job, always check closely under the hood for tools that may have been left there so they won't get lost during a test drive.

Some tools, such as screwdrivers, pliers, wrenches and sockets, can be hung on a panel mounted on the garage or workshop wall, while others should be kept in a tool box or tray. Measuring instruments, gauges, meters, etc. must be carefully stored where they cannot be damaged by weather or impact from other tools.

When tools are used with care and stored properly, they will last a very long time. Even with the best of care, though, tools will wear out if used frequently. When a tool is damaged or worn out, replace it. Subsequent jobs will be safer and more enjoyable if you do.

Working facilities

Not to be overlooked when discussing tools is the workshop. If anything more than routine maintenance is to be carried out, some sort of suitable work area is essential.

It is understood, and appreciated, that many home mechanics do not have a good workshop or garage available, and end up removing an engine or doing major repairs outside. It is recommended, however, that the overhaul or repair be completed under the cover of a roof.

A clean, flat workbench or table of comfortable working height is an absolute necessity. The workbench should be equipped with a vise that has a jaw opening of at least four inches.

As mentioned previously, some clean, dry storage space is also required for tools, as well as the lubricants, fluids, cleaning solvents, etc. which will soon become necessary.

Sometimes waste oil and fluids, drained from the engine or cooling system during normal maintenance or repairs, present a disposal problem. To avoid pouring them on the ground or into a sewage system, pour the used fluids into large containers, seal them with caps and take them to an authorized disposal site or recycling center. Plastic jugs, such as old antifreeze containers, are ideal for this purpose.

Always keep a supply of old newspapers and clean rags available. Old towels are excellent for mopping up spills. Many mechanics use rolls of paper towels for most work because they are readily available and disposable. To help keep the area under the vehicle clean, a large cardboard box can be cut open and flattened to protect the garage or shop floor.

Whenever working over a painted surface, such as when leaning over a fender to service something under the hood, always cover it with an old blanket or bedspread to protect the finish. Vinyl covered pads, made especially for this purpose, are available at auto parts stores.

Booster battery (jump) starting

Certain precautions must be observed when using a booster battery to jump start a vehicle.

a) Before connecting the booster battery, make sure that the ignition switch is in the Off position.
b) Turn off the lights, heater and other electrical loads.
c) The eyes should be shielded. Safety goggles are a good idea.
d) Make sure the booster battery is the same voltage as the dead one in the vehicle.
e) The two vehicles must not touch each other.
f) Make sure the transmission is in Neutral (manual transmission) or Park (automatic transmission).
g) If the booster battery is not a maintenance-free type, remove the vent caps and lay a cloth over the vent holes.

Connect the red jumper cable to the *positive* (+) terminals of each battery.

Connect one end of the black jumper cable to the *negative* (–) terminal of the booster battery. The other end of this cable should be connected to a good ground on the vehicle to be started, such as a bolt or bracket on the engine block. Use caution to ensure that the cable will not come into contact with the fan, drivebelts or other moving parts of the engine.

Start the engine using the booster battery, then, with the engine running at idle speed, disconnect the jumper cables in the reverse order of connection.

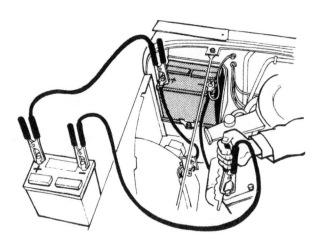

Booster cable connections (note that the negative cable is *not* attached to the negative terminal of the dead battery)

Jacking and towing

Jacking

The jack supplied with the vehicle should only be used for raising the vehicle when changing a tire or placing jackstands under the frame. **Caution:** *Never work under the vehicle or start the engine while this jack is being used as the only means of support.*

The vehicle should be on level ground with the wheels blocked and the transmission in Park (automatic) or Reverse (manual). If the wheel is being replaced, loosen the wheel nuts one-half turn and leave them in place until the wheel is raised off the ground. Refer to Chapter 10 for the tire changing procedure.

Place the jack under the side of the vehicle in the indicated position and raise it until the jack head groove fits into the rocker flange notch. **Caution:** *Carefully read the caution label attached to the jack body. Operate the jack with a slow, smooth motion until the wheel is raised off the ground.*

Lower the vehicle, remove the jack and tighten the nuts (if loosened or removed) in a criss-cross sequence by turning the wrench clockwise.

Towing

The vehicle can be towed with all four wheels on the ground, provided that speeds do not exceed 30 mph and the distance is not over 40 miles, otherwise transmission damage can result. **Caution:** *Front towing is not recommended with conventional sling type equipment on the turbocharged models because the towing rope or chain will come into contact with the spoiler and damage may result. With manual transmission equipped models, restrict towing speed to below 50 mph and towing distance to less than 50 miles. If the speed or distance to be towed is greater than the recommended speed and distance (both models), remove the driveshaft to prevent damage to the transmission.*

Towing equipment specifically designed for this purpose should be used and should be attached to the main structural members of the vehicle and not the bumper or brackets.

Safety is a major consideration when towing and all applicable state and local laws must be obeyed. A safety chain system must be used for all towing.

While towing, the parking brake should be released and the transmission should be in Neutral. The steering must be unlocked (ignition switch in the Off position). Remember that power steering and power brakes will not work with the engine off.

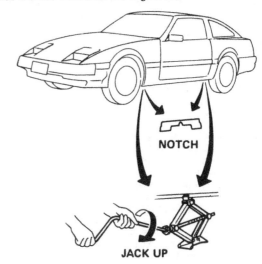

The jack head groove fits into the rocker panel flange notch

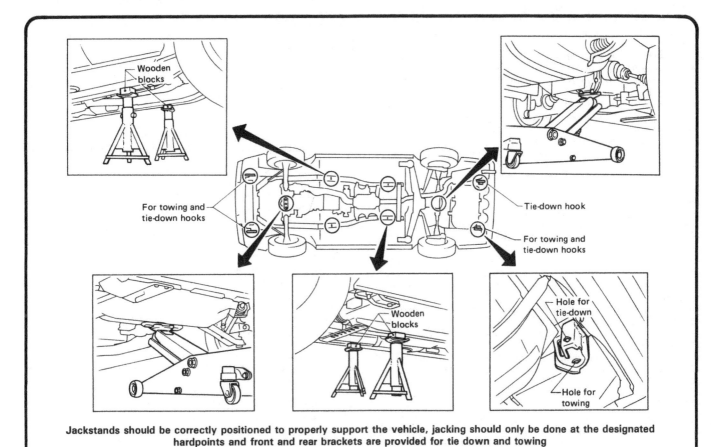

Jackstands should be correctly positioned to properly support the vehicle, jacking should only be done at the designated hardpoints and front and rear brackets are provided for tie down and towing

Safety first!

Regardless of how enthusiastic you may be about getting on with the job at hand, take the time to ensure that your safety is not jeopardized. A moment's lack of attention can result in an accident, as can failure to observe certain simple safety precautions. The possibility of an accident will always exist, and the following points should not be considered a comprehensive list of all dangers. Rather, they are intended to make you aware of the risks and to encourage a safety conscious approach to all work you carry out on your vehicle.

Essential DOs and DON'Ts

DON'T rely on a jack when working under the vehicle. Always use approved jackstands to support the weight of the vehicle and place them under the recommended lift or support points.

DON'T attempt to loosen extremely tight fasteners (i.e. wheel lug nuts) while the vehicle is on a jack — it may fall.

DON'T start the engine without first making sure that the transmission is in Neutral (or Park where applicable) and the parking brake is set.

DON'T remove the radiator cap from a hot cooling system — let it cool or cover it with a cloth and release the pressure gradually.

DON'T attempt to drain the engine oil until you are sure it has cooled to the point that it will not burn you.

DON'T touch any part of the engine or exhaust system until it has cooled sufficiently to avoid burns.

DON'T siphon toxic liquids such as gasoline, antifreeze and brake fluid by mouth, or allow them to remain on your skin.

DON'T inhale brake lining dust — it is potentially hazardous (see *Asbestos* below)

DON'T allow spilled oil or grease to remain on the floor — wipe it up before someone slips on it.

DON'T use loose fitting wrenches or other tools which may slip and cause injury.

DON'T push on wrenches when loosening or tightening nuts or bolts. Always try to pull the wrench toward you. If the situation calls for pushing the wrench away, push with an open hand to avoid scraped knuckles if the wrench should slip.

DON'T attempt to lift a heavy component alone — get someone to help you.

DON'T rush or take unsafe shortcuts to finish a job.

DON'T allow children or animals in or around the vehicle while you are working on it.

DO wear eye protection when using power tools such as a drill, sander, bench grinder, etc. and when working under a vehicle.

DO keep loose clothing and long hair well out of the way of moving parts.

DO make sure that any hoist used has a safe working load rating adequate for the job.

DO get someone to check on you periodically when working alone on a vehicle.

DO carry out work in a logical sequence and make sure that everything is correctly assembled and tightened.

DO keep chemicals and fluids tightly capped and out of the reach of children and pets.

DO remember that your vehicle's safety affects that of yourself and others. If in doubt on any point, get professional advice.

Asbestos

Certain friction, insulating, sealing, and other products — such as brake linings, brake bands, clutch linings, torque converters, gaskets, etc. — contain asbestos. *Extreme care must be taken to avoid inhalation of dust from such products since it is hazardous to health.* If in doubt, assume that they *do* contain asbestos.

Fire

Remember at all times that gasoline is highly flammable. Never smoke or have any kind of open flame around when working on a vehicle. But the risk does not end there. A spark caused by an electrical short circuit, by two metal surfaces contacting each other, or even by static electricity built up in your body under certain conditions, can ignite gasoline vapors, which in a confined space are highly explosive. Do not, under any circumstances, use gasoline for cleaning parts. Use an approved safety solvent.

Always disconnect the battery ground (–) cable *at the battery* before working on any part of the fuel system or electrical system. Never risk spilling fuel on a hot engine or exhaust component.

It is strongly recommended that a fire extinguisher suitable for use on fuel and electrical fires be kept handy in the garage or workshop at all times. Never try to extinguish a fuel or electrical fire with water.

Fumes

Certain fumes are highly toxic and can quickly cause unconsciousness and even death if inhaled to any extent. Gasoline vapor falls into this category, as do the vapors from some cleaning solvents. Any draining or pouring of such volatile fluids should be done in a well ventilated area.

When using cleaning fluids and solvents, read the instructions on the container carefully. Never use materials from unmarked containers.

Never run the engine in an enclosed space, such as a garage. Exhaust fumes contain carbon monoxide, which is extremely poisonous. If you need to run the engine, always do so in the open air, or at least have the rear of the vehicle outside the work area.

If you are fortunate enough to have the use of an inspection pit, never drain or pour gasoline and never run the engine while the vehicle is over the pit. The fumes, being heavier than air, will concentrate in the pit with possibly lethal results.

The battery

Never create a spark or allow a bare light bulb near the battery. The battery normally gives off a certain amount of hydrogen gas, which is highly explosive.

Always disconnect the battery ground (–) cable *at the battery* before working on the fuel or electrical systems.

If possible, loosen the filler caps or cover when charging the battery from an external source. Do not charge at an excessive rate or the battery may burst.

Take care when adding water and when carrying a battery. The electrolyte, even when diluted, is very corrosive and should not be allowed to contact clothing or skin.

Always wear eye protection when cleaning the battery to prevent the caustic deposits from entering your eyes.

Household current

When using an electric power tool, inspection light, etc., which operates on household current, always make sure that the tool is correctly connected to its plug and that, where necessary, it is properly grounded. Do not use such items in damp conditions and, again, do not create a spark or apply excessive heat in the vicinity of fuel or fuel vapor.

Secondary ignition system voltage

A severe electric shock can result from touching certain parts of the ignition system (such as the spark plug wires) when the engine is running or being cranked, particularly if components are damp or the insulation is defective. In the case of an electronic ignition system, the secondary system voltage is much higher and could prove fatal.

Automotive chemicals and lubricants

A number of automotive chemicals and lubricants are available for use during vehicle maintenance and repair. They include a wide variety of products ranging from cleaning solvents and degreasers to lubricants and protective sprays for rubber, plastic and vinyl.

Cleaners

Carburetor cleaner and choke cleaner is a strong solvent for gum, varnish and carbon. Most carburetor cleaners leave a dry-type lubricant film which will not harden or gum up. Because of this film it is not recommended for use on electrical components.

Brake system cleaner is used to remove grease and brake fluid from the brake system where clean surfaces are absolutely necessary. It leaves no residue and often eliminates brake squeal caused by contaminants.

Electrical cleaner removes oxidation, corrosion and carbon deposits from electrical contacts, restoring full current flow. It can also be used to clean spark plugs, carburetor jets, voltage regulators and other parts where an oil-free surface is desired.

Demoisturants remove water and moisture from electrical components such as alternators, voltage regulators, electrical connectors and fuse blocks. It is non-conductive, non-corrosive and non-flammable.

Degreasers are heavy-duty solvents used to remove grease from the outside of the engine and from chassis components. They can be sprayed or brushed on, and, depending on the type, are rinsed off either with water or solvent.

Lubricants

Motor oil is the lubricant formulated for use in engines. It normally contains a wide variety of additives to prevent corrosion and reduce foaming and wear. Motor oil comes in various weights (viscosity ratings) from 5 to 80. The recommended weight of the oil depends on the season, temperature and the demands on the engine. Light oil is used in cold climates and under light load conditions. Heavy oil is used in hot climates and where high loads are encountered. Multi-viscosity oils are designed to have characteristics of both light and heavy oils and are available in a number of weights from 5W-20 to 20W-50.

Gear oil is designed to be used in differentials, manual transaxles and other areas where high-temperature lubrication is required.

Chassis and wheel bearing grease is a heavy grease used where increased loads and friction are encountered, such as for wheel bearings, balljoints, tie rod ends and universal joints.

High temperature wheel bearing grease is designed to withstand the extreme temperatures encountered by wheel bearings in disc brake equipped vehicles. It usually contains molybdenun disulfide (moly), which is a dry-type lubricant.

White grease is a heavy grease for metal to metal applications where water is a problem. White grease stays soft under both low and high temperatures (usually from −100°F to + 190°F), and will not wash off or dilute in the presence of water.

Assembly lube is a special extreme pressure lubricant, usually containing moly, used to lubricate high-load parts such as main and rod bearings and cam lobes for initial start-up of a new engine. The assembly lube lubricates the parts without being squeezed out or washed away until the engine oiling system begins to function.

Silicone lubricants are used to protect rubber, plastic, vinyl and nylon parts.

Graphite lubricants are used where oils cannot be used due to contamination problems, such as in locks. The dry graphite will lubricate metal parts while remaining uncontaminated by dirt, water, oil or acids. It is electrically conductive and will not foul electrical contacts in locks such as the ignition switch.

Moly penetrants loosen and lubricate frozen, rusted and corroded fasteners and prevent future rusting or freezing.

Heat-sink grease is a special electrically non-conductive grease that is used for mounting HEI ignition modules where it is essential that heat be transferred away from the module.

Sealants

RTV sealant is one of the most widely used gasket compounds. Made from silicone, RTV is air curing, it seals, bonds, waterproofs, fills surface irregularities, remains flexible, doesn't shrink, is relatively easy to remove, and is used as a supplementary sealer with almost all low and medium temperature gaskets.

Anaerobic sealant is much like RTV in that it can be used either to seal gaskets or to form gaskets by itself. It remains flexible, is solvent resistant and fills surface imperfections. The difference between an anaerobic sealant and an RTV-type sealant is in the curing. RTV cures when exposed to air, while an anaerobic sealant cures only in the absence of air. This means that an anaerobic sealant cures only after the assembly of parts, sealing them together.

Thread and pipe sealant is used for sealing hydraulic and pneumatic fittings and vacuum lines. It is usually made from a teflon compound, and comes in a spray, a paint-on liquid and as a wrap-around tape.

Chemicals

Anti-seize compound prevents seizing, galling, cold welding, rust and corrosion in fasteners. High temperature anti-seize, usually made with copper and graphite lubricants, is used for exhaust system and manifold bolts.

Anaerobic locking compounds are used to keep fasteners from vibrating or working loose, and cure only after installation, in the absence of air. Medium strength locking compound is used for small nuts, bolts and screws that you expect to be removing later. High strength locking compound is for large nuts, bolts and studs which you don't intend to be removing on a regular basis.

Oil additives range from viscosity index improvers to chemical treatments that claim to reduce internal engine friction. It should be noted that most oil manufacturers caution against using additives with their oils.

Gas additives perform several functions, depending on their chemical makeup. They usually contain solvents that help dissolve gum and varnish that build up on carburetor and intake parts. They also serve to break down carbon deposits that form on the inside surfaces of the combustion chambers. Some additives contain upper cylinder lubricants for valves and piston rings, and others chemicals to remove condensation from the gas tank.

Other

Brake fluid is specially formulated hydraulic fluid that can withstand the heat and pressure encountered in brake systems. Care must be taken that this fluid does not come in contact with painted surfaces or plastics. An opened container should always be resealed to prevent contamination by water or dirt.

Weatherstrip adhesive is used to bond weatherstripping around doors, windows and trunk lids. It is sometimes used to attach trim pieces.

Undercoating is a petroleum-based tar-like substance that is designed to protect metal surfaces on the underside of the vehicle from corrosion. It also acts as a sound-deadening agent by insulating the bottom of the vehicle.

Waxes and polishes are used to help protect painted and plated surfaces from the weather. Different types of paint may require the use of different types of wax and polish. Some polishes utilize a chemical or abrasive cleaner to help remove the top layer of oxidized (dull) paint on older vehicles. In recent years many non-wax polishes that contain a wide variety of chemicals such as polymers and silicones have been introduced. These non-wax polishes are usually easier to apply and last longer than conventional waxes and polishes.

Conversion factors

Length (distance)
Inches (in)	X	25.4	= Millimetres (mm)	X 0.0394	= Inches (in)
Feet (ft)	X	0.305	= Metres (m)	X 3.281	= Feet (ft)
Miles	X	1.609	= Kilometres (km)	X 0.621	= Miles

Volume (capacity)
Cubic inches (cu in; in³)	X	16.387	= Cubic centimetres (cc; cm³)	X 0.061	= Cubic inches (cu in; in³)
Imperial pints (Imp pt)	X	0.568	= Litres (l)	X 1.76	= Imperial pints (Imp pt)
Imperial quarts (Imp qt)	X	1.137	= Litres (l)	X 0.88	= Imperial quarts (Imp qt)
Imperial quarts (Imp qt)	X	1.201	= US quarts (US qt)	X 0.833	= Imperial quarts (Imp qt)
US quarts (US qt)	X	0.946	= Litres (l)	X 1.057	= US quarts (US qt)
Imperial gallons (Imp gal)	X	4.546	= Litres (l)	X 0.22	= Imperial gallons (Imp gal)
Imperial gallons (Imp gal)	X	1.201	= US gallons (US gal)	X 0.833	= Imperial gallons (Imp gal)
US gallons (US gal)	X	3.785	= Litres (l)	X 0.264	= US gallons (US gal)

Mass (weight)
Ounces (oz)	X	28.35	= Grams (g)	X 0.035	= Ounces (oz)
Pounds (lb)	X	0.454	= Kilograms (kg)	X 2.205	= Pounds (lb)

Force
Ounces-force (ozf; oz)	X	0.278	= Newtons (N)	X 3.6	= Ounces-force (ozf; oz)
Pounds-force (lbf; lb)	X	4.448	= Newtons (N)	X 0.225	= Pounds-force (lbf; lb)
Newtons (N)	X	0.1	= Kilograms-force (kgf; kg)	X 9.81	= Newtons (N)

Pressure
Pounds-force per square inch (psi; lbf/in²; lb/in²)	X	0.070	= Kilograms-force per square centimetre (kgf/cm²; kg/cm²)	X 14.223	= Pounds-force per square inch (psi; lbf/in²; lb/in²)
Pounds-force per square inch (psi; lbf/in²; lb/in²)	X	0.068	= Atmospheres (atm)	X 14.696	= Pounds-force per square inch (psi; lbf/in²; lb/in²)
Pounds-force per square inch (psi; lbf/in²; lb/in²)	X	0.069	= Bars	X 14.5	= Pounds-force per square inch (psi; lbf/in²; lb/in²)
Pounds-force per square inch (psi; lbf/in²; lb/in²)	X	6.895	= Kilopascals (kPa)	X 0.145	= Pounds-force per square inch (psi; lbf/in²; lb/in²)
Kilopascals (kPa)	X	0.01	= Kilograms-force per square centimetre (kgf/cm²; kg/cm²)	X 98.1	= Kilopascals (kPa)
Millibar (mbar)	X	100	= Pascals (Pa)	X 0.01	= Millibar (mbar)
Millibar (mbar)	X	0.0145	= Pounds-force per square inch (psi; lbf/in²; lb/in²)	X 68.947	= Millibar (mbar)
Millibar (mbar)	X	0.75	= Millimetres of mercury (mmHg)	X 1.333	= Millibar (mbar)
Millibar (mbar)	X	0.401	= Inches of water (inH₂O)	X 2.491	= Millibar (mbar)
Millimetres of mercury (mmHg)	X	0.535	= Inches of water (inH₂O)	X 1.868	= Millimetres of mercury (mmHg)
Inches of water (inH₂O)	X	0.036	= Pounds-force per square inch (psi; lbf/in²; lb/in²)	X 27.68	= Inches of water (inH₂O)

Torque (moment of force)
Pounds-force inches (lbf in; lb in)	X	1.152	= Kilograms-force centimetre (kgf cm; kg cm)	X 0.868	= Pounds-force inches (lbf in; lb in)
Pounds-force inches (lbf in; lb in)	X	0.113	= Newton metres (Nm)	X 8.85	= Pounds-force inches (lbf in; lb in)
Pounds-force inches (lbf in; lb in)	X	0.083	= Pounds-force feet (lbf ft; lb ft)	X 12	= Pounds-force inches (lbf in; lb in)
Pounds-force feet (lbf ft; lb ft)	X	0.138	= Kilograms-force metres (kgf m; kg m)	X 7.233	= Pounds-force feet (lbf ft; lb ft)
Pounds-force feet (lbf ft; lb ft)	X	1.356	= Newton metres (Nm)	X 0.738	= Pounds-force feet (lbf ft; lb ft)
Newton metres (Nm)	X	0.102	= Kilograms-force metres (kgf m; kg m)	X 9.804	= Newton metres (Nm)

Power
Horsepower (hp)	X	745.7	= Watts (W)	X 0.0013	= Horsepower (hp)

Velocity (speed)
Miles per hour (miles/hr; mph)	X	1.609	= Kilometres per hour (km/hr; kph)	X 0.621	= Miles per hour (miles/hr; mph)

Fuel consumption*
Miles per gallon, Imperial (mpg)	X	0.354	= Kilometres per litre (km/l)	X 2.825	= Miles per gallon, Imperial (mpg)
Miles per gallon, US (mpg)	X	0.425	= Kilometres per litre (km/l)	X 2.352	= Miles per gallon, US (mpg)

Temperature

Degrees Fahrenheit = ($^{\circ}$C x 1.8) + 32

Degrees Celsius (Degrees Centigrade; $^{\circ}$C) = ($^{\circ}$F - 32) x 0.56

*It is common practice to convert from miles per gallon (mpg) to litres/100 kilometres (l/100km), where mpg (Imperial) x l/100 km = 282 and mpg (US) x l/100 km = 235

Troubleshooting

Contents

This section provides an easy reference guide to the more common problems which may occur during the operation of your vehicle. These problems and possible causes are grouped under various components or systems; i.e. Engine, Cooling system, etc., and also refer to the Chapter and/or Section which deals with the problem.

Remember that successful troubleshooting is not a mysterious *black art* practiced only by professional mechanics. It's simply the result of a bit of knowledge combined with an intelligent, systematic approach to the problem. Always work by a process of elimination, starting with the simplest solution and working through to the most complex — and never overlook the obvious. Anyone can forget to fill the gas tank or leave the lights on overnight, so don't assume that you are above such oversights.

Finally, always get clear in your mind why a problem has occurred and take steps to ensure that it doesn't happen again. If the electrical system fails because of a poor connection, check all other connections in the system to make sure that they don't fail as well. If a particular fuse continues to blow, find out why — don't just go on replacing fuses. Remember, failure of a small component can often be indicative of potential failure or incorrect functioning of a more important component or system.

Engine

1 Engine will not rotate when attempting to start

1   Battery terminal connections loose or corroded. Check the cable terminals at the battery. Tighten the cable or remove corrosion as necessary.
2   Battery discharged or faulty. If the cable connections are clean and tight on the battery posts, turn the key to the On position and switch on the headlights and/or windshield wipers. If they fail to function, the battery is discharged.
3   Automatic transmission not completely engaged in Park or clutch not completely depressed.
4   Broken, loose or disconnected wiring in the starting circuit. Inspect all wiring and connectors at the battery, starter solenoid and ignition switch.
5   Starter motor pinion jammed in flywheel ring gear. If manual transmission, place transmission in gear and rock the vehicle to manually turn the engine. Remove starter and inspect pinion and flywheel at earliest convenience.
6   Starter solenoid faulty (Chapter 5).
7   Starter motor faulty (Chapter 5).
8   Ignition switch faulty (Chapter 12).

2 Engine rotates but will not start

1   Fuel tank empty.
2   Battery discharged (engine rotates slowly). Check the operation of electrical components as described in previous Section.
3   Battery terminal connections loose or corroded. See previous Section.
4   Fuel injector or fuel pump faulty (Chapter 4).
5   Excessive moisture on, or damage to, ignition components (Chapter 5).
6   Worn, faulty or incorrectly gapped spark plugs (Chapter 1).
7   Broken, loose or disconnected wiring in the starting circuit (see previous Section).
8   Distributor loose, causing ignition timing to change. Turn the distributor as necessary to start engine, then set ignition timing as soon as possible (Chapter 1).
9   Broken, loose or disconnected wires at the ignition coil or faulty coil (Chapter 5).

3 Starter motor operates without rotating engine

1   Starter pinion sticking. Remove the starter (Chapter 5) and inspect.
2   Starter pinion or flywheel teeth worn or broken. Remove the cover at the rear of the engine and inspect.

4 Engine hard to start when cold

1   Battery discharged or low. Check as described in Section 1.
2   Fuel supply not reaching the injection system (see Section 2).
3   Fuel injection system in need of overhaul (Chapter 4).
4   Distributor rotor carbon tracked and/or mechanical advance mechanism rusted (Chapter 5).

5 Engine hard to start when hot

1   Air filter clogged (Chapter 1).
2   Fuel not reaching the fuel injection system (see Section 2).

6 Starter motor noisy or excessively rough in engagement

1   Pinion or flywheel gear teeth worn or broken. Remove the cover at the rear of the engine (if so equipped) and inspect.
2   Starter motor mounting bolts loose or missing.

7 Engine starts but stops immediately

1   Loose or faulty electrical connections at distributor, coil or alternator.
2   Insufficient fuel reaching the fuel injector(s). See fuel pressure relief caution in Chapter 4, open line and place a container under the disconnected fuel line. Observe the flow of fuel from the line. If little or none at all, check for blockage in the lines and/or replace the fuel pump (Chapter 4).
3   Vacuum leak at the gasket surfaces of the intake manifold and/or fuel injection unit. Make sure that all mounting bolts/nuts are tightened securely and that all vacuum hoses connected to the fuel injection unit and manifold are positioned properly and in good condition.

8 Oil puddle under engine

1   Oil pan gasket and/or oil plug seal leaking. Check and replace if necessary.
2   Oil pressure sending unit leaking. Replace unit or seal threads with teflon tape (Chapter 5).
3   Rocker cover gaskets leaking at front or rear of engine.
4   Engine oil seals leaking at front or rear of engine.

9 Engine lopes while idling or idles erratically

1   Vacuum leakage. Check mounting bolts/nuts at the fuel injection unit and intake manifold for tightness. Make sure that all vacuum hoses are connected and in good condition. Use a stethoscope or a length of fuel hose held against your ear to listen for vacuum leaks while the engine is running. A hissing sound will be heard. A soapy water solution will also detect leaks. Check the fuel injector and intake manifold gasket surfaces.
2   Leaking EGR valve or plugged PCV valve (see Chapters 1 and 6).
3   Air filter clogged (Chapter 1).
4   Fuel pump not delivering sufficient fuel to the fuel injector (see Section 7).
5   Fuel injection system out of adjustment (Chapter 4).
6   Leaking head gasket. If this is suspected, take the vehicle to a repair shop or dealer where the engine can be pressure checked.
7   Timing belt worn (Chapter 2).
8   Camshaft lobes worn (Chapter 2).

10 Engine misses at idle speed

1   Spark plugs worn or not gapped properly (Chapter 1).
2   Faulty spark plug wires (Chapter 1).

11 Engine misses throughout driving speed range

1 Fuel filter clogged and/or impurities in the fuel system (Chapter 1). Also check fuel output at the fuel injector (see Section 7).
2 Faulty or incorrectly gapped spark plugs (Chapter 1).
3 Incorrect ignition timing (Chapter 1).
4 Check for cracked distributor cap, disconnected distributor wires and damaged distributor components (Chapter 1).
5 Leaking spark plug wires (Chapter 1).
6 Faulty emissions system components (Chapter 6).
7 Low or uneven cylinder compression pressures. Remove spark plugs and test compression with gauge (Chapter 1).
8 Weak or faulty ignition system (Chapter 5).
9 Vacuum leaks at fuel injection unit, intake manifold or vacuum hoses (see Section 8).
10 Loose injector harness.

12 Engine stumbles on acceleration

1 Spark plugs fouled (Chapter 1). Clean or replace.
2 Fuel injection system needs adjustment or repair (Chapter 4).
3 Fuel filter clogged. Replace filter.
4 Incorrect ignition timing (Chapter 1).
5 Intake manifold air leak (Chapters 4 and 6).

13 Engine surges while holding accelerator steady

1 Intake air leak (Chapter 4).
2 Fuel pump faulty (Chapter 4).
3 Loose fuel injector harness connections.
4 Defective control unit.

14 Engine stalls

1 Idle speed incorrect (Chapter 1).
2 Fuel filter clogged and/or water and impurities in the fuel system (Chapter 1).
3 Distributor components damp or damaged (Chapter 5).
4 Faulty emissions system components (Chapter 6).
5 Faulty or incorrectly gapped spark plugs (Chapter 1). Also check spark plug wires (Chapter 1).
6 Vacuum leak at the carburetor/fuel injection unit, intake manifold or vacuum hoses. Check as described in Section 8.
7 Valve clearances incorrectly set (Chapter 2).

15 Engine lacks power

1 Incorrect ignition timing (Chapter 1).
2 Excessive play in distributor shaft. At the same time, check for worn rotor, faulty distributor cap, wires, etc. (Chapters 1 and 5).
3 Faulty or incorrectly gapped spark plugs (Chapter 1).
4 Fuel injection unit not adjusted properly or excessively worn (Chapter 4).
5 Faulty coil (Chapter 5).
6 Brakes binding (Chapter 1).
7 Automatic transmission fluid level incorrect (Chapter 1).
8 Clutch slipping (Chapter 8).
9 Fuel filter clogged and/or impurities in the fuel system (Chapter 1).
10 Emissions control system not functioning properly (Chapter 6).
11 Use of substandard fuel. Fill tank with proper octane fuel.
12 Low or uneven cylinder compression pressures. Test with compression tester, which will detect leaking valves and/or blown head gasket (Chapter 1).

16 Engine backfires

1 Emissions system not functioning properly (Chapter 6).
2 Ignition timing incorrect (Chapter 1).

3 Faulty secondary ignition system. Cracked spark plug insulator, faulty plug wires, distributor cap and/or rotor (Chapters 1 and 5).
4 Fuel injection unit in need of adjustment or worn excessively (Chapter 4).
5 Vacuum leak at fuel injection unit, intake manifold or vacuum hoses. Check as described in Section 8.
6 Valve clearances incorrect due to valves sticking or worn valve components (Chapter 2).

17 Pinging or knocking engine sounds during acceleration or uphill

1 Incorrect grade of fuel. Fill tank with fuel of the proper octane rating.
2 Ignition timing incorrect (Chapter 1).
3 Fuel injection unit in need of adjustment (Chapter 4).
4 Improper spark plugs. Check plug type against Emissions Control Information label located in engine compartment. Also check plugs and wires for damage (Chapter 1).
5 Worn or damaged distributor components (Chapter 5).
6 Faulty emissions system (Chapter 6).
7 Vacuum leak. Check as described in Section 8.

18 Engine runs with oil pressure light on

1 Low oil level. Check oil level and add oil if necessary (Chapter 1).
2 Idle rpm below specification (Chapter 1).
3 Short in wiring circuit. Repair or replace damaged wire.
4 Faulty oil pressure sender. Replace sender.
5 Worn engine bearings and/or oil pump.

19 Engine diesels (continues to run) after switching off

1 Idle speed too high (Chapter 1).
2 Ignition timing incorrectly adjusted (Chapter 1).
3 Thermo-controlled air cleaner heat valve not operating properly (Chapter 6).
4 Excessive engine operating temperature. Probable causes of this are malfunctioning thermostat, clogged radiator, faulty water pump (Chapter 3).
5 EGR system malfunction.

Engine electrical system

20 Battery will not hold a charge

1 Alternator drivebelt defective or not adjusted properly (Chapter 1).
2 Electrolyte level low or battery discharged (Chapter 1).
3 Battery terminals loose or corroded (Chapter 1).
4 Alternator not charging properly (Chapter 5).
5 Loose, broken or faulty wiring in the charging circuit (Chapter 5).
6 Short in vehicle wiring causing a continual drain on battery.
7 Battery defective internally.

21 Alternator light fails to go out

1 Fault in alternator or charging circuit (Chapter 5).
2 Alternator drivebelt defective or not properly adjusted (Chapter 1).
3 Alternator voltage regulator inoperative (Chapter 5).

22 Ignition light fails to come on when key is turned on

1 Warning light bulb defective (Chapter 12).
2 Alternator faulty (Chapter 5).
3 Fault in the printed circuit, dash wiring or bulb holder (Chapter 12).

Fuel system

23   Excessive fuel consumption

1   Dirty or clogged air filter element (Chapter 1).
2   Incorrectly set ignition timing (Chapter 1).
3   Emissions system not functioning properly (see Chapter 6).
4   Fuel injection system idle speed and/or mixture not adjusted properly (Chapter 1).
5   Fuel injection internal parts excessively worn or damaged (Chapter 4).
6   Low tire pressure or incorrect tire size (Chapter 1).

24   Fuel leakage and/or fuel odor

1   Leak in a fuel feed or vent line (Chapter 4).
2   Tank overfilled. Fill only to automatic shut off.
3   Emissions system filter clogged (Chapter 1).
4   Vapor leaks from system lines (Chapter 4).
5   Fuel injection internal parts excessively worn or out of adjustment (Chapter 4).

Cooling system

25   Overheating

1   Insufficient coolant in system (Chapter 1).
2   Water pump drivebelt defective or not adjusted properly (Chapter 1).
3   Radiator core blocked or radiator grille dirty and restricted (Chapter 3).
4   Thermostat faulty (Chapter 3).
5   Fan blades broken or cracked (Chapter 3).
6   Radiator cap not maintaining proper pressure. Have cap pressure tested by gas station or repair shop.
7   Ignition timing incorrect (Chapter 1).
8   Defective water pump (Chapter 3).

26   Overcooling

1   Thermostat faulty (Chapter 3).
2   Inaccurate temperature gauge (Chapter 12)

27   External coolant leakage

1   Deteriorated or damaged hoses or loose clamps. Replace hoses and/or tighten clamps at hose connections (Chapter 1).
2   Water pump seals defective. If this is the case, water will drip from the weep hole in the water pump body (Chapter 1).
3   Leakage from radiator core or header tank. This will require the radiator to be professionally repaired (see Chapter 3 for removal procedures).
4   Engine drain plugs or water jacket core plugs leaking (see Chapter 2).

28   Internal coolant leakage

Note: *Internal coolant leaks can usually be detected by examining the oil. Check the dipstick and inside of the rocker arm cover for water deposits and an oil consistency like that of a milkshake.*
1   Leaking cylinder head gasket. Have the cooling system pressure tested.
2   Cracked cylinder bore or cylinder head. Dismantle engine and inspect (Chapter 2).
3   Loose cylinder head bolts (Chapter 2).

29   Coolant loss

1   Too much coolant in system (Chapter 1).
2   Coolant boiling away due to overheating (see Section 16).
3   Internal or external leakage (see Sections 25 and 26).
4   Faulty radiator cap. Have the cap pressure tested.

30   Poor coolant circulation

1   Inoperative water pump. A quick test is to pinch the top radiator hose closed with your hand while the engine is idling, then let it loose. You should feel the surge of coolant if the pump is working properly (Chapter 1).
2   Restriction in cooling system. Drain, flush and refill the system (Chapter 1). If necessary, remove the radiator (Chapter 3) and have it reverse flushed.
3   Water pump drivebelt defective or not adjusted properly (Chapter 1).
4   Thermostat sticking (Chapter 3).

Clutch

31   Fails to release (pedal pressed to the floor — shift lever does not move freely in and out of Reverse)

1   Improper linkage free play adjustment (Chapter 8).
2   Clutch fork off ball stud. Look under the vehicle, on the left side of transmission.
3   Clutch plate warped or damaged (Chapter 8).

32   Clutch slips (engine speed increases with no increase in vehicle speed)

1   Linkage out of adjustment (Chapter 8).
2   Clutch plate oil soaked or lining worn. Remove clutch (Chapter 8) and inspect.
3   Clutch plate not seated. It may take 30 or 40 normal starts for a new one to seat.
4   Warped flywheel (Chapter 2).

33   Grabbing (chattering) as clutch is engaged

1   Oil on clutch plate lining. Remove (Chapter 8) and inspect. Correct any leakage source.
2   Worn or loose engine or transmission mounts. These units move slightly when clutch is released. Inspect mounts and bolts.
3   Worn splines on clutch plate hub. Remove clutch components (Chapter 8) and inspect.
4   Warped pressure plate or flywheel. Remove clutch components and inspect.
5   Hardened or warped clutch disc facing.

34   Squeal or rumble with clutch fully engaged (pedal released)

1   Improper adjustment — no free play (Chapter 1).
2   Release bearing binding on transmission bearing retainer. Remove clutch components (Chapter 8) and check bearing. Remove any burrs or nicks, clean and relubricate before reinstallation.
3   Weak linkage return spring. Replace the spring.
4   Cracked clutch disc.

35   Squeal or rumble with clutch fully disengaged (pedal depressed)

1   Worn, defective or broken release bearing (Chapter 8).
2   Worn or broken pressure plate springs or diaphragm fingers (Chapter 8).

36 Clutch pedal stays on floor when disengaged

1   Bind in linkage or release bearing. Inspect linkage or remove clutch components as necessary.
2   Linkage springs being over extended. Adjust linkage for proper free play. Make sure proper pedal stop (bumper) is installed.

Manual transmission

37 Noisy in Neutral with engine running

1   Input shaft bearing worn.
2   Damaged main drive gear bearing.
3   Worn countershaft bearings.
4   Worn or damaged countershaft end play shims.

38 Noisy in all gears

1   Any of the above causes, and/or:
2   Insufficient lubricant (see checking procedures in Chapter 1).
3   Worn or damaged output shaft or bearings.

39 Noisy in one particular gear

1   Worn, damaged or chipped gear teeth for that particular gear.
2   Worn or damaged synchronizer for that particular gear.

40 Slips out of gear

1   Transmission loose on clutch housing (Chapter 7).
2   Shift rods interfering with engine mounts or clutch lever (Chapter 7).
3   Shift rods not working freely (Chapter 7).
4   Damaged mainshaft pilot bearing.
5   Dirt between transmission case and engine or misalignment of transmission (Chapter 7).
6   Worn or improperly adjusted linkage (Chapter 7).
7   Worn synchro units.

41 Difficulty in engaging gears

1   Clutch not releasing completely (see clutch adjustment in Chapter 8).
2   Loose, damaged or out-of-adjustment shift linkage. Make a thorough inspection, replacing parts as necessary (Chapter 7).

42 Oil leakage

1   Excessive amount of lubricant in transmission (see Chapter 1 for correct checking procedures). Drain lubricant as required.
2   Side cover loose or gasket damaged.
3   Rear oil seal or speedometer oil seal in need of replacement (Chapter 7).

Automatic transmission

Note: *Due to the complexity of the automatic transmission, it is difficult for the home mechanic to properly diagnose and service this component. For problems other than the following, the vehicle should be taken to a dealer or reputable mechanic.*

43 Fluid leakage

1   Automatic transmission fluid is a deep red color. Fluid leaks should not be confused with engine oil, which can easily be blown by air flow to the transmission.
2   To pinpoint a leak, first remove all built-up dirt and grime from around the transmission. Degreasing agents and/or steam cleaning will achieve this. With the underside clean, drive the vehicle at low speeds so air flow will not blow the leak far from its source. Raise the vehicle and determine where the leak is coming from. Common areas of leakage are:
   a)   Pan: Tighten mounting bolts and/or replace pan gasket as necessary (see Chapters 1 and 7).
   b)   Filler pipe: Replace the rubber seal where pipe enters transmission case.
   c)   Transmission oil lines: Tighten connectors where lines enter transmission case and/or replace lines.
   d)   Speedometer connector: Replace the O-ring where speedometer cable enters transmission case (Chapter 7).

44 Transmission fluid brown or has a burned smell

1   Transmission fluid burned. Replace fluid. Do not overfill.

45 General shift mechanism problems

1   Chapter 7B deals with checking and adjusting the shift linkage on automatic transmissions. Common problems which may be attributed to poorly adjusted linkage are:
   a)   Engine starting in gears other than Park or Neutral.
   b)   Indicator on shifter pointing to a gear other than the one actually being used.
   c)   Vehicle moves when in Park.
2   Refer to Chapter 7B to adjust the linkage.

46 Transmission will not downshift with accelerator pedal pressed to the floor

1   Chapter 7B deals with adjusting the throttle valve (TV) cable to enable the transmission to downshift properly.

47 Engine will start in gears other than Park or Neutral

1   Chapter 7B deals with adjusting the neutral start switches used on automatic transmissions.

48 Transmission slips, shifts rough, is noisy or has no drive in forward or reverse gears

1   There are many probable causes for the above problems, but the home mechanic should be concerned with only one possibility — fluid level.
2   Before taking the vehicle to a repair shop, check the level and condition of the fluid as described in Chapter 1. Correct fluid level as necessary or change the fluid and filter if needed. If the problem persists, have a professional diagnose the probable cause.

Driveshaft

49 Leakage of fluid at front of driveshaft

1   Defective transmission rear oil seal. See Chapter 7B for replacement procedures. While this is done, check the splined yoke for burrs or a rough condition which may be damaging the seal. If found, these can be removed with emery clotch or a fine abrasive stone.

50 Knock or clunk when transmission is put under initial load (just after transmission is put into gear)

1   Loose or disconnected rear suspension components. Check all mounting bolts and bushings (Chapter 10).
2   Loose driveshaft bolts. Inspect all bolts and nuts and tighten to specification (Chapter 8).
3   Worn or damaged universal joint bearings. Test for wear (Chapter 8).
4   Worn mainshaft splines.

51 Metallic grating sound consistent with road speed

1   Pronounced wear in the universal joint bearings. Test for wear (Chapter 8).

52 Vibration

Note: *Before it can be assumed that the driveshaft is at fault, make sure the tires are perfectly balanced and perform the following test.*
1   Install a tachometer inside the car to monitor engine speed as the car is driven. Drive the car and note the engine speed at which the vibration is most pronounced. Now shift the transmission to a different gear and bring the engine speed to the same point.
2   If the vibration occurs at the same engine speed (rpm) regardless of which gear the transmission is in, the driveshaft is *Not* at fault since the driveshaft speed varies.
3   If the vibration decreases or is eliminated when the transmission is in a different gear at the same engine speed, refer to the following probable causes:
4   Bent or dented driveshaft. Inspect and replace as necessary (Chapter 8).
5   Undercoating or built-up dirt, etc. on the driveshaft. Clean the shaft thoroughly and test.
6   Worn universal joint bearings. Remove and inspect (Chapter 8).
7   Driveshaft and/or companion flange out of balance. Check for missing weights on the shaft. Remove the driveshaft and reinstall 180° from original position. Retest. Have driveshaft professionally balanced if problem persists.
8   Driveshaft improperly installed (Chapter 8).
9   Worn transmission rear bushing (Chapter 7).

Rear axle

53 Noise

1   Road noise. No corrective procedures available.
2   Tire noise. Inspect tires and check tire pressures (Chapter 1).
3   Front wheel bearings loose, worn or damaged (Chapter 10).
4   Insufficient differential oil (whining noise consistent with vehicle speed changes).

54 Vibration

1   See probable causes under *Driveshaft*. Proceed under the guidelines listed for the driveshaft. If the problem persists, check the rear wheel bearings by raising the rear of the car and spinning the wheels by hand. Listen for evidence of rough bearings. Remove and inspect (Chapter 8).

55 Oil leakage

1   Pinion oil seal damaged (Chapter 8).
2   Axleshaft oil seals damaged (Chapter 8).
3   Differential cover leaking. Tighten mounting bolts or replace the gasket as required (Chapter 8).

Brakes

Note: *Before assuming that a brake problem exists, make sure that the tires are in good condition and inflated properly (see Chapter 1), that the front end alignment is correct and that the vehicle is not loaded with weight in an unequal manner.*

56 Vehicle pulls to one side during braking

1   Defective, damaged or oil contaminated disc brake pads on one side. Inspect as described in Chapter 9.
2   Excessive wear of brake pad material or disc on one side. Inspect and correct as necessary.
3   Loose or disconnected front suspension components. Inspect and tighten all bolts to the specified torque (Chapter 10).
4   Defective caliper assembly. Remove caliper and inspect for stuck piston or other damage (Chapter 9).

57 Noise (high-pitched squeal without the brakes applied)

1   Disc brake pads worn out. The noise comes from the wear sensor rubbing against the disc (does not apply to all vehicles). Replace pads with new ones immediately (Chapter 9).

58 Excessive brake pedal travel

1   Partial brake system failure. Inspect entire system (Chapter 9) and correct as required.
2   Insufficient fluid in master cylinder. Check (Chapter 1), add fluid and bleed system if necessary (Chapter 9).
3   Rear brakes not adjusting properly. Make a series of starts and stops while the vehicle is in Reverse. If this does not correct the situation, remove drums and inspect self-adjusters (Chapter 9).

59 Brake pedal feels spongy when depressed

1   Air in hydraulic lines. Bleed the brake system (Chapter 9).
2   Faulty flexible hoses. Inspect all system hoses and lines. Replace parts as necessary.
3   Master cylinder mounting bolts/nuts loose.
4   Master cylinder defective (Chapter 9).

60 Excessive effort required to stop vehicle

1   Power brake booster not operating properly (Chapter 9).
2   Excessively worn linings or pads. Inspect and replace if necessary (Chapter 9).
3   One or more caliper pistons or wheel cylinders seized or sticking. Inspect and rebuild as required (Chapter 9).
4   Brake linings or pads contaminated with oil or grease. Inspect and replace as required (Chapter 9).
5   New pads or shoes installed and not yet seated. It will take a while for the new material to seat against the drum or rotor.

61 Pedal travels to the floor with little resistance

1   Little or no fluid in the master cylinder reservoir caused by leaking wheel cylinder(s), leaking caliper piston(s), loose, damaged or disconnected brake lines. Inspect entire system and correct as necessary.

62 Brake pedal pulsates during brake application

1   Wheel bearings not adjusted properly or in need of replacement (Chapter 1).
2   Caliper not sliding properly due to improper installation or obstruc-

tions. Remove and inspect (Chapter 9).
3 Rotor defective. Remove the rotor (Chapter 9) and check for excessive lateral runout and parallelism. Have the rotor resurfaced or replace it with a new one.

63 Parking brake does not hold

1 Mechanical parking brake linkage improperly adjusted. Adjust according to procedure in Chapter 9.

Suspension and steering systems

64 Vehicle pulls to one side

1 Tire pressures uneven (Chapter 1).
2 Defective tire (Chapter 1).
3 Excessive wear in suspension or steering components (Chapter 10).
4 Front end in need of alignment.
5 Front brakes dragging. Inspect brakes as described in Chapter 9.
6 Wheel bearings improperly adjusted (Chapter 10).

65 Shimmy, shake or vibration

1 Tire or wheel out-of-balance or out-of-round. Have professionally balanced.
2 Loose, worn or out-of-adjustment wheel bearings (Chapters 1 and 10).
3 Shock absorbers and/or suspension components worn or damaged (Chapter 10).

66 Excessive pitching and/or rolling around corners or during braking

1 Defective shock absorbers. Replace as a set (Chapter 10).
2 Broken or weak springs and/or suspension components. Inspect as described in Chapter 10.

67 Excessively stiff steering

1 Lack of fluid in power steering fluid reservoir (Chapter 1).
2 Incorrect tire pressures (Chapter 1).
3 Lack of lubrication at steering joints (Chapter 1).
4 Front end out of alignment.

5 Air in power steering system (Chapter 10)
6 Low tire pressure.

68 Excessive play in steering

1 Loose front wheel bearings (Chapter 1).
2 Excessive wear in suspension or steering components (Chapter 10).
3 Steering gearbox out of adjustment (Chapter 10).

69 Lack of power assistance

1 Steering pump drivebelt faulty or not adjusted properly (Chapter 1).
2 Fluid level low (Chapter 1).
3 Hoses or lines restricted. Inspect and replace parts as necessary.
4 Air in power steering system. Bleed system (Chapter 10).

70 Excessive tire wear (not specific to one area)

1 Incorrect tire pressures (Chapter 1).
2 Tires out of balance. Have professionally balanced.
3 Wheels damaged. Inspect and replace as necessary.
4 Suspension or steering components excessively worn (Chapter 10).

71 Excessive tire wear on outside edge

1 Inflation pressures incorrect (Chapter 1).
2 Excessive speed in turns.
3 Front end alignment incorrect (excessive toe-in). Have professionally aligned.
4 Suspension arm bent or twisted (Chapter 10).

72 Excessive tire wear on inside edge

1 Inflation pressures incorrect (Chapter 1).
2 Front end alignment incorrect (toe-out). Have professionally aligned.
3 Loose or damaged steering components (Chapter 10).

73 Tire tread worn in one place

1 Tires out of balance.
2 Damaged or buckled wheel. Inspect and replace if necessary.
3 Defective tire (Chapter 1).

Chapter 1   Tune up and routine maintenance

Contents

Specifications

Note: *Additional specifications and torque figures can be found in each individual Chapter*

Recommended fluids and lubricants

Note: *Listed here are manufacturer recommendations at the time this manual was written. Manufacturers occasionally upgrade their fluid and lubricant specifications, so check with your local auto parts store for current recommendations.*

Engine oil	Consult your owner's manual or local Nissan dealer for recommendations on the particular service grade and viscosity oil recommended for your area
Gear lubricant	Consult your owner's manual or local Nissan dealer for recommendations on the particular service grade and viscosity gear lubricant recommended for your area
Automatic transmission fluid	Dexron
Power steering fluid	Dexron
Brake and clutch fluid	Dot 3
Antifreeze	ethylene glycol base
Chassis lubrication	multi-purpose grease NLGI No. 2

Capacities

Engine oil	
without oil filter	3-7/8 qt
with oil filter	4-1/4 qt
Fuel tank	19 gal
Engine coolant	
reservior tank	7/8 qt
turbo model	11-5/8 qt
non-turbo model	11-1/8 qt
Transmission	
manual	4 pt
1984 thru 1986	4 pt
1987 on	
FS5W71C	4-1/4 pt
FS5R90A	5-1/8 pt
automatic	7-3/8 qt
Power steering	1 qt
Differential carrier	1-3/8 qt
Windshield and rear window washer tank	3-1/8 qt
Headlight washer tank	2-3/8 qt

General

Clutch

pedal height	7.68 to 8.07 in
free play	0.04 to 0.12 in
Brake pad wear limit	0.079 in

Rotor thickness limit (minimum)

front	
1984 thru 1986	0.787 in
1987 on	
CL28VB*	0.787 in
CL28VE*	0.940 in
rear	
1984 thru 1986	0.354 in
1987 on	0.710 in

Brake pedal free height

manual transmission	7.17 to 7.56 in
automatic transmission	7.24 to 7.64 in
Brake pedal depressed height	3.54 in under force of 110 lbs with engine running

*Rotor part no.

Engine

Ignition timing (1984 thru 1987 only)	20-degrees ± 2-degrees @ 700 rpm
Spark plug type and gap	NGK BCPR6ES-11 or equivalent @ 0.039 to 0.043 in
Engine idle speed	See Emission Control Label in engine compartment

Engine compression pressure

standard	
non-turbo	173 psi @ 300 rpm
turbo	165 psi @ 300 rpm (169 on 1988 and later models)
minimum	
non-turbo	128 psi @ 300 rpm
turbo	121 psi @ 300 rpm (125 on 1988 and later models)
Differential limit between cylinders	14 psi @ 300 rpm

Drive belt adjustment deflection

alternator	
non-turbo	0.24 to 0.31 in
turbo	0.24 to 0.35 in
air conditioning compressor	0.35 to 0.43 in
power steering pump	0.51 to 0.63 in

Torque specifications

	Ft-lbs
Spark plugs	14 to 22
Exhaust sensor	30 to 37
Oil pan drain plug	22 to 29
Manual transmission drain and filler plugs	
FS5W71C	18 to 25
FS5R90A	14 to 25
Automatic transmission	
pan bolts	3.6 to 5.1
filter bolts	2.2 to 2.9
Differential carrier drain and filler plugs	29 to 43
Wheel lug nuts	
aluminum	72 to 87
steel	58 to 72
Seat belt anchor bolts	17 to 23

The blackened terminal shown on the distributor cap indicates the Number One spark plug wire position

Cylinder location and distributor rotation

1 Introduction

Refer to illustrations 1.1a and 1.1b

This Chapter was designed to help the home mechanic maintain his or her vehicle for peak performance, economy, safety and long life.

On the following pages you will find a maintenance schedule and Sections which deal specifically with each item on the schedule. Included are visual checks, adjustments and item replacements.

Using the time/mileage maintenance schedule and the sequenced Sections will give you a planned program of maintenance for servicing your vehicle. Keep in mind that this is a comprehensive plan. Maintaining only a few items at the specified intervals will not give you the same results.

You will find as you service your vehicle that many of the procedures can be grouped together, due to the nature of the job at hand. For example:

If the vehicle is fully raised for a chassis lubrication check the exhaust system, suspension, steering and fuel system.

If the tires and wheels are removed — during a routine tire rotation, for example — check the suspension, brakes and wheel bearings.

If you must borrow or rent a torque wrench to perform a specific procedure, check the torque on other critical bolts, such as cylinder head bolts, brake caliper bolts, etc. the same day to save time and money.

The first step of any maintenance plan is to decide how you wish to proceed before the actual work begins. In other words, prepare a plan of action. Read through the appropriate Sections to get an idea of what must be done before you begin. Make a list of, then gather together, the necessary parts and tools. If you anticipate possible problem areas, don't hesitate to seek advice from your local parts man or dealer service department.

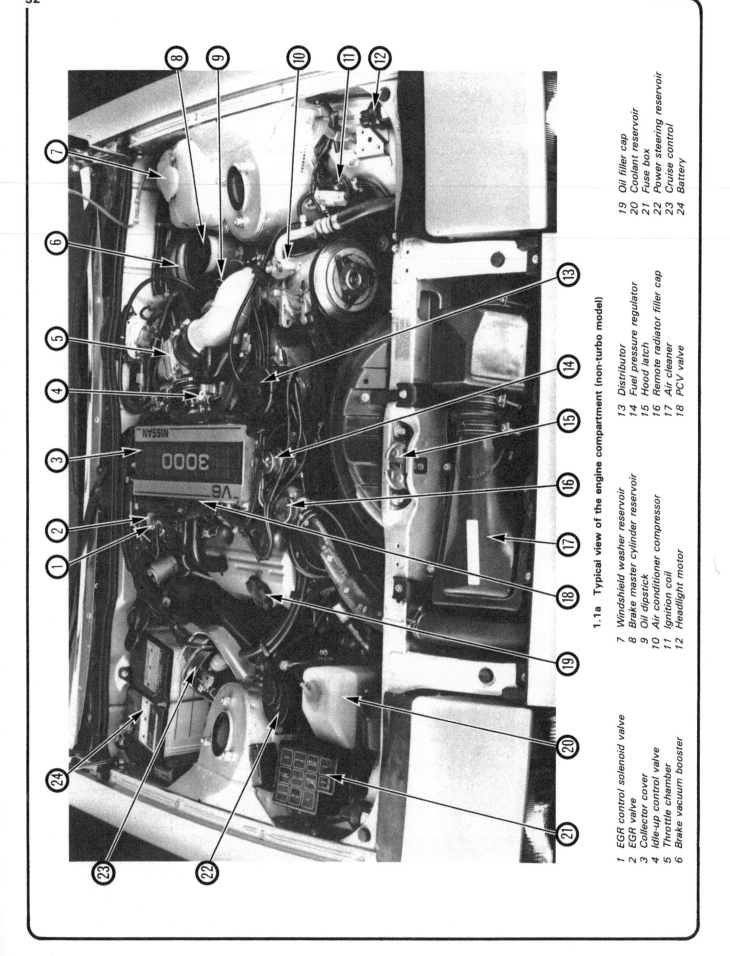

1.1a Typical view of the engine compartment (non-turbo model)

1 EGR control solenoid valve
2 EGR valve
3 Collector cover
4 Idle-up control valve
5 Throttle chamber
6 Brake vacuum booster

7 Windshield washer reservoir
8 Brake master cylinder reservoir
9 Oil dipstick
10 Air conditioner compressor
11 Ignition coil
12 Headlight motor

13 Distributor
14 Fuel pressure regulator
15 Hood latch
16 Remote radiator filler cap
17 Air cleaner
18 PCV valve

19 Oil filler cap
20 Coolant reservoir
21 Fuse box
22 Power steering reservoir
23 Cruise control
24 Battery

1.1b Typical view of the engine compartment (turbo model). The turbo equipped version of the 300ZX is very similar to the normally aspirated version, with these noted differences:

1 Auxiliary air control valve
2 Separate PCV hose
3 Injector cooling fan
4 Turbocharger
5 Emergency relief valve
6 Detonation sensor

2   Routine maintenance schedule

The following recommendations are provided with the assumption that the vehicle owner will be doing the maintenance or service work, as opposed to having a dealer service department do the work. These are factory maintenance recommendations. However, if you want to keep your vehicle in peak condition at all times — or if you just want to maintain good resale value — many of these procedures should be performed more frequently.

When the vehicle is new it should be serviced by a factory authorized dealer service department to protect the factory warranty. In many cases the initial maintenance check is done at no cost to the owner.

Every 250 miles or weekly, whichever comes first

Check the engine oil level (Section 4)
Check the engine coolant level (Section 4)
Check the windshield washer fluid level (Section 4)
Check the tires and tire pressures (Section 5)

Every 5000 miles or 6 months, whichever comes first

Check the power steering fluid level (Section 4)
Check the torque of the turbocharger and/or EFI mounting bolts (Chapter 4)
Change the oil and filter in turbo models (Section 7)
Check the cooling system (Section 21)
Check the brake master cylinder fluid level (Section 4)
Check and service the battery (Section 8)
Check and replace (if necessary) the windshield wiper blades (Section 6)

Every 7500 miles or 6 months, whichever comes first

Change the oil and filter in non-turbo models (Section 7)

Every 15,000 miles or 12 months, whichever comes first

Check the engine idle rpm (Section 14)
Inspect the brake lines and hoses (Section 25)
Inspect the brake pads and discs (Section 25)
Check parking brake (Chapter 9)
Check the manual transmission lubricant, automatic transmission fluid and differential gear oil (Section 4)
Change the rear axle oil (if vehicle is used to pull trailer) (Section 35)
Inspect the power steering lines and hoses (Section 30)
Inspect the steering gear and linkage (Section 26)
Inspect the rear axle and suspension components (Section 26)
Lubricate the locks, hinges and hood latch (Section 24)
Inspect the exhaust system (Section 27)
Inspect the seat belts, buckles, retractors, anchors and adjuster (Section 36)
Rotate the tires (Section 5)
Check the throttle linkage (Section 22)
Check the air filter (Section 13)

Every 30,000 miles or 24 months, whichever comes first

Drain, flush and refill the cooling system (Section 21)
Replace the fuel filter (Section 19)
Check the EGO sensor (Section 31)
Check the operation of the EGR valve and sensor (Section 29)
Inspect and replace, if necessary, the PCV valve (Section 20)
Replace the air filter and PCV filter (Section 13)
Inspect the vapor lines (Section 30)

Inspect and replace, if necessary, the distributor cap and rotor (Section 11)
Inspect the spark plug wires (Section 10)
Replace the spark plugs (Section 9)
Inspect the fuel system hoses and fuel tank filler cap (Section 18)
Check the drivebelt tension and condition (Section 12)
Inspect the front wheel bearings (Section 23)

Every 36,000 miles or 36 months

Check and adjust clutch pedal (Chapter 8)
Check the drivetrain (Section 28)
Check EEC filter (Chapter 6)

Every 60,000 miles or 48 months

Change the transmission and differential lubricant (Sections 33 and 34)
Compression the test engine (Section 16)
Replace the timing belt (Chapter 2)
Inspect the tie rod ends (Section 26)
Inspect the balljoints (Section 26)

Severe driving conditions

Severe driving conditions include repeated short distance trips, extensive idling, driving in dusty areas, extremely low or high ambient temperatures, towing, driving in areas using salt or other corrosives on the road or continuous driving on rough and/or muddy roads.

Under these conditions service should be performed more frequently. As a general rule, divide the service time in half. For example, instead of changing the oil and filter in the turbo equipped engine every 5000 miles, change them every 2500 miles.

Here is a guideline for maintenance of items that can deteriorate more rapidly under severe driving conditions:

Replace the air cleaner more frequently.
Change the engine oil and filter every 3000 miles or three months on non-turbo models.
Change the engine oil and filter every 2500 miles or three months on turbo models.
Inspect the brake pads and discs every 7500 miles.
Change the manual/automatic transmission and differential gear oil every 30,000 miles or 24 months.
Inspect the steering gear and linkage, axle, and suspension parts every 7500 miles or 6 months.
Inspect the locks, hinges and hood latch every 7500 miles or 6 months.
Inspect the exhaust system every 7500 miles or 6 months.

3   Tune-up — general information

The term *tune-up* is loosely used for any general operation that restores the engine to its proper running condition. A tune-up is not a specific operation, but rather a series of procedures, such as replacing the spark plugs, adjusting the idle speed, setting the ignition timing, etc.

If the routine maintenance schedule (Section 1) is followed closely, fluid levels are checked frequently and high wear items are inspected regularly from the beginning of the car's service life, as suggested throughout this manual, the engine will remain in good running condition and the need for additional tune-ups will be minimized.

More likely though, there will be times when the engine runs poorly because of irregular maintenance. Such a possibility is even more likely if a used vehicle is purchased. It's difficult to verify the kind of care it has received. In such cases, a complete engine tune-up outside of the regular routine maintenance intervals is highly recommended.

The following series of procedures are often necessary to bring a generally poor running engine back into a proper state of tune.

Minor tune-up

Clean, inspect and test the battery
Check all engine-related fluids
Check and adjust the drivebelts
Replace the spark plugs
Inspect the distributor cap and rotor
Inspect the spark plug and coil wires
Check and adjust the idle speed
Check and adjust the timing
Replace the fuel filter
Check the PCV valve
Tighten the EFI mounting bolts
Check the air and PCV filters
Clean and lubricate the throttle linkage
Check all underhood hoses

Major tune-up

All of the above, plus
Check the EGR system
Check the charging system
Check the fuel system
Test the battery
Check the engine compression
Check the cooling system
Replace the distributor cap and rotor
Replace the spark plug wires
Replace the air and PCV filters

---

4 Fluid level checks

Refer to illustrations 4.2, 4.4, 4.6, 4.9, 4.15a, 4.15b, 4.19, 4.26, 4.34, 4.40, 4.48a and 4.48b

1   There are a number of components on a vehicle which rely on the use of fluids to perform their job. During normal operation of the vehicle, these fluids are used up and must be replenished before damage occurs. See the Specifications at the front of this Chapter for the specific fluid to be used when addition is required. When checking fluid levels, it is important to have the vehicle on a level surface.

Engine oil

2   The engine oil level is checked with a dipstick located at the left side of the engine. The dipstick extends through a tube to the oil pan at the bottom of the engine (see illustration).
3   The oil level should be checked either before the vehicle has been driven or about 15 minutes after the engine has been shut off. If the oil is checked immediately after driving the vehicle, some of it will remain in the upper engine oil galleys, producing an inaccurate reading on the dipstick.
4   Pull the dipstick from the tube and wipe all the oil off the end with a clean rag. Insert the clean dipstick all the way back into the oil pan

and pull it out again. Look at the oil-darkened portion of the dipstick. The oil level should be between the L and H marks (see illustration).
5   It takes one quart of oil to raise the level from the L mark to the H mark on the dipstick. Do not allow the level to drop below the L mark — oil starvation induced engine damage may result. Conversely, do not overfill the engine by adding oil above the H mark — it may result in fouled spark plugs, oil leaks or seal failures.
6   To add oil, remove the twist-off cap located on the right rocker arm cover (see illustration). An oil can spout or funnel will reduce spills.
7   Checking the oil level can also be an important preventive maintenance step. You should always check the condition of the oil as well as the level. Take your thumb and index finger and wipe the oil off the dipstick. Look, and feel, for small dirt or metal particles clinging to the dipstick. This is an indication that the oil is dirty and should be changed (Section 6). If you notice that the oil level is dropping quickly, there's either a leak or some serious engine wear that must be corrected. If there are water droplets in the oil, or if it's milky looking, there is a leak between the cooling system and engine, such as a blown head gasket, cracked head, etc., which must be repaired immediately before further damage occurs.

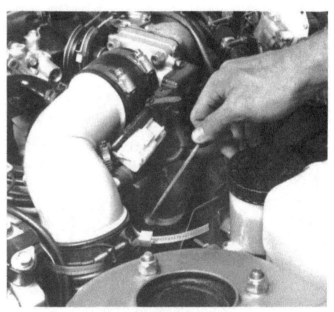

4.2   The oil dipstick is located on the left side of the engine

4.6   Twist-off the oil cap located on the right rocker arm cover to add engine oil

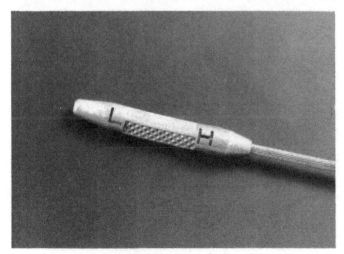

4.4   Keep the oil level between the L and H marks indicated on the dipstick

Engine coolant

8   All vehicles covered by this manual are equipped with a pressurized coolant recovery system which simplifies coolant level checks. A white plastic coolant reservoir is connected by a hose to the radiator cap. As the engine heats up during operation, coolant vapor escapes from the radiator via the pressure cap, through the connecting tube and into the reservoir. As the engine cools, the coolant is automatically drawn back into the radiator to keep the level correct.

9   The coolant level should be checked when the engine is hot. Observe the level of fluid in the reservoir, which should be between the *MAX* and *MIN* marks on the side of the reservoir (see illustration).

10 **Warning:** *Under no circumstances should the radiator cap or the coolant recovery reservoir cap be removed when the system is hot, because escaping steam and scalding liquid can cause serious personal injury.* To remove the radiator cap, wait until the system has cooled completely, then wrap a thick cloth around the cap and turn it to the first stop. If any steam escapes, wait until the system has cooled further before removing the cap. The coolant recovery cap may be carefully removed after it is apparent that no further boiling is occurring in the recovery tank.

11 If only a small amount of coolant is required to bring the system up to the proper level, plain water can be used. However, to maintain the proper antifreeze/water mixture in the system, both should be mixed together to replenish a low level. Coolant offering protection to –20 °F should be mixed with water in the proportion specified on the container. **Warning:** *Do not allow antifreeze to come in contact with your skin*

or painted surfaces of the vehicle. Flush contacted areas immediately with plenty of water.

12 Add small amounts of coolant at a time to the reservoir, waiting for it to be absorbed by the cooling system before adding more.

13 Note the condition of the coolant while checking the level. It should be relatively clean. If it is rust colored, the system should be drained, flushed and refilled (Section 20).

14 If the cooling system requires repeated additions to maintain the proper level, have the radiator cap pressure checked for proper operation. Also look for leaks in the system such as cracked hoses, loose hose connections, leaking gaskets, or a bad bearing in the water pump (Chapter 3).

Windshield washer fluid

15 Fluid for the windshield washer system is located in the engine compartment in a plastic reservoir attached to the left fender well (see illustration). A separate tank is located near the radiator for vehicles equipped with headlight washers (see illustration). The level in these reservoirs should be maintained near the top, except during periods when freezing temperatures are expected, at which time the fluid level should be no higher than 3/4-full to allow for expansion should the fluid freeze. The use of an additive such as Optikleen will help lower the freezing point of the fluid and will result in better cleaning of the windshield surface. Do not use antifreeze — it will cause damage to the vehicle's paint.

16 In cold weather warm the windshield with the defroster before using the washer to help prevent icing.

Battery electrolyte

Certain precautions must be taken when working on or near the battery:

    a) Never expose a battery to open flame or sparks which could ignite the hydrogen gas given off by the battery

    b) Wear protective clothing and eye protection to reduce the possibility of the corrosive sulfuric acid solution inside the battery harming you. If the fluid is splashed or spilled, flush the contacted area immediately with plenty of water

    c) Remove all metal jewelry which could contact the positive terminal and another grounded metal source, causing a short circuit

    d) Always keep batteries and battery acid out of the reach of children.

17 All vehicles with which this manual is concerned are equipped with a maintenance-free battery which is permanently sealed (except for vent holes) and has no filler caps. Water does not have to be added to these batteries at any time.

Brake fluid

Caution: *Be careful not to spill brake fluid on any painted surfaces of*

4.9   Check the engine coolant level while the engine is warm. The fluid level should be between the MAX and MIN marks

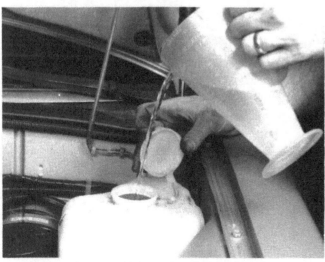

4.15a   The windshield washer fluid level should be maintained near the top of the reservoir

4.15b   Vehicles equipped with headlight washers have a reservoir (arrow) located in the front of the engine compartment

the car. It can damage the paint finish. Wash spills off immediately with plenty of water.

18 The brake fluid reservoir is attached to the brake master cylinder. Remove any dirt or loose particles from the cover before opening the master cylinder. If checking the brake fluid on a vehicle with water or snow on it, be sure to completely clean the hood and cowling before opening the reservoir. Serious contamination of the brake fluid can result if water is allowed to enter the brake fluid reservoir.

19 To check the level of the brake fluid, note how high the fluid is in relation to the *MAX* and *MIN* marks on the side of the reservoir (see illustration).

20 Additional fluid must be added to the reservoir anytime the brake fluid warning light comes on or the brake fluid level falls below the *MAX* mark on the reservoir.

21 Carefully pour the specified brake fluid into the master cylinder. Be sure to use the recommended fluid. Mixing different types of brake fluid can cause damage to the system. Consult your owner's manual or the Specifications in this Chapter for the correct fluid.

22 At this time the fluid and master cylinder should be inspected for contamination. Normally, the braking system will need no more than periodic draining and refilling, but if rust deposits, dirt particles or water droplets contaminate the fluid, the system should be drained and refilled with fresh fluid and bled.

23 Reinstall the master cylinder cap. Make sure it is properly seated to prevent fluid leakage.

24 The brake fluid in the master cylinder will drop slightly as the brake pads at each wheel wear down during normal operation. If the master cylinder requires repeated replenishing to keep it at the proper level, it's an indication of leakage in the brake system. Correct it immediately. Check all brake lines and calipers and their connections immediately (Chapter 9).

25 If, upon checking the master cylinder fluid level, you discover the reservoir is below the *MIN* mark or empty, the hydraulic system should be bled (Chapter 9).

Manual transmission oil

26 Manual transmissions do not have a dipstick. The oil level is checked with the transmission cold. Locate the plug in the side of the transmission (see illustration) and use a rag to clean the plug and the area around it, then remove the plug.

27 If oil immediately starts leaking out, thread the plug back into the transmission. The level is correct. If there is no leakage, completely remove the plug and place your little finger inside the hole. The oil level should be right at the bottom edge of the plug hole.

28 If it isn't, the transmission needs more oil. Use a syringe to squeeze the recommended lubricant into the plug hole until the fluid level is again just at the bottom edge of the plug hole. See your owner's manual or the front of this Chapter for the correct lubricant.

29 Thread the plug back into the transmission case and tighten it to

the specified torque.

30 Drive the vehicle a short distance, then check for leaks around the plug.

Automatic transmission fluid

31 The level of the automatic transmission fluid should be carefully maintained. Low fluid level can lead to slipping or loss of drive, while overfilling can cause foaming and loss of fluid.

32 With the parking brake set, start the engine, then move the shift lever through all the gear ranges, ending in Park. The fluid level must be checked with the vehicle on level ground and with the engine running at idle.

33 **Note:** *Incorrect fluid level readings will result if the vehicle has just been driven at high speeds for an extended period, in city traffic during hot weather or if it's been pulling a trailer. If any of these conditions apply, wait until the fluid cools (about 30 minutes).*

34 Remove the transmission dipstick from the filler tube located on the right side of the engine compartment at the rear of the right cylinder bank (see illustration).

35 Carefully touch the end of the dipstick to determine if the fluid is cool (about room temperature), warm or hot (uncomfortable to the touch).

4.19 Maintain the brake reservoir fluid level between the MAX and MIN marks

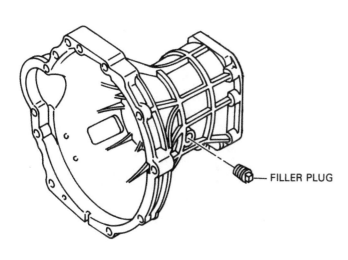

FILLER PLUG

4.26 The manual transmission filler plug is located on the side of the case

4.34 The automatic transmission dipstick is located at the right rear of the engine

36 Wipe the fluid from the dipstick with a clean rag and push the dipstick back into the filler tube until the cap seats.
37 Pull the dipstick out again and note the fluid level.
38 If the fluid feels cool, the level should be near the *L* mark.
39 If the fluid feels warm or hot, the level should be near the *H* mark.
40 Add just enough of the recommended fluid to fill the transmission to the proper level (see illustration).
41 It takes about one pint to raise the level from the *L* mark to the *H* mark, so add the fluid a little at a time and keep checking the level until it is correct.
42 The condition as well as the level of the automatic transmission fluid should be checked. If the fluid is a dark reddish-brown color or if it has a burned smell, it should be changed. If you are in doubt about the condition of the fluid, purchase some new fluid and compare the two for color and smell.

Power steering fluid

43 Unlike manual steering, the power steering system relies on fluid which periodically requires replenishing.
44 The reservoir for the power steering pump is located on the engine compartment right fender well.
45 Before beginning a fluid level check, the front wheels should be pointed straight ahead and the engine should be off.
46 Use a clean rag to wipe off the reservoir cap and the area around the cap. This will help prevent any foreign matter from entering the reservoir during the check.
47 Make sure the engine is at normal operating temperature.
48 Remove the dipstick (see illustration), wipe it off with a clean rag, reinsert it, then withdraw it and read the fluid level. The level should

be between the marks (see illustration).
49 If additional fluid is required, pour the specified type directly into the reservoir, using a funnel to prevent spills.
50 If the reservoir requires frequent fluid additions, carefully check all power steering hoses, hose connections, the power steering pump and the steering assembly for leaks.

Hydraulic clutch fluid

Caution: *Be careful not to spill brake fluid on any painted surfaces of the car, as it can damage the paint finish. If spilled, wash immediately with water.*
51 The fluid master cylinder for the hydraulic clutch is located in the engine compartment on the left side of the firewall, next to the brake master cylinder.
52 The clutch master cylinder reservoir fluid level should be checked periodically. If the fluid is significantly below the *MAX* mark, additional fluid should be added to the reservoir.
53 Before unscrewing the cap, clean all dirt and grease from around the cap thoroughly. If any foreign matter enters the master cylinder, blockage in the clutch system line can occur.
54 Twist the cap counterclockwise to remove it from the reservoir.
55 Inspect the master cylinder for contamination. If rust deposits, dirt particles or water droplets are visible, the system should be dismantled, drained and refilled with fresh fluid (Chapter 8).
56 Carefully pour the specified brake fluid into the clutch master cylinder.
57 Reinstall the master cylinder cap.
58 If the system requires repeated fluid replenishment, check the system for leakage. Inspect all clutch lines and connections.
59 If the master cylinder fluid level falls below the *MIN* mark or appears empty, the clutch system should be bled (Chapter 8).

---

5   Wheels and tires — inspection, maintenance and servicing

Refer to illustrations 5.3a, 5.3b, 5.17, 5.20, 5.24, 5.28, 5.33 and 5.37

Inspection

1   Periodic inspection of the tires may not only prevent you from being stranded with a flat tire, but can also provide you with clues to steering and suspension system problems before major damage occurs.
2   Proper tire inflation adds miles to the lifespan of the tires, allows the vehicle to achieve maximum miles per gallon and contributes to overall ride quality.
3   When inspecting the tires, check tread wear first. An irregular tread pattern — such as cupping, flat spots, more wear on one side than the other — is an indication of front end alignment and/or balance problems. If you note any of these symptoms, take the vehicle to an automotive repair shop or a dealer to correct the problem (see illustration). Tread wear indicators are molded into the tire (see illustration) to indicate the limit of useful tread life.

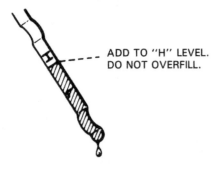

ADD TO "H" LEVEL.
DO NOT OVERFILL.

4.40   Maintain the automatic transmission fluid level near the H mark

4.48a   Clean the top of the power steering reservoir before removing the cap

4.48b   The power steering fluid level should be within the two marks on the dipstick

4   Check the tread area for cuts and punctures. Many times a nail or tack will embed itself into the tire tread, yet the tire will hold air pressure for a short time. In most cases, a repair shop or gas station can repair the punctured tire.

5   It is also important to check both inner and outer sidewalls of the tires. Look for deteriorated rubber, cuts and punctures. Inspect the inboard sides of the tires for signs of brake fluid leakage, indicating that a thorough brake inspection is needed immediately.

6   Incorrect tire pressure cannot be determined merely by looking at the tires, especially radial tires. A tire pressure gauge must be used. If you do not have a reliable gauge, purchase one and keep it in the glovebox. Pressure gauges built into the nozzles of air hoses at gas stations are often inaccurate.

7   Always check tire inflation when the tires are cold. In this case, cold means that the vehicle has not been driven more than one mile after sitting for three hours or more. It is normal for the pressure to increase four to eight pounds when the tires warm up.

8   Unscrew the valve cap and press the gauge firmly onto the valve. Observe the reading on the gauge and compare it to the recommended tire pressure listed on the tire placard. The tire placard is usually attached to the driver's door.

9   Check all tires and add air as necessary to bring them up to the specified pressure. Do not forget the spare tire. Be sure to reinstall the valve caps to keep dirt and moisture out of the valve stem mechanism.

Wheel and tire replacement

10   Refer to *Jacking and towing* at the front of this manual for the correct procedure for raising the vehicle.

11   Park the car on a level surface. Set the parking brake and put the transmission in gear. Manual transmissions should be in Reverse, automatic transmissions should be in Park.

12   Loosen, but do not remove, the wheel lug nuts by turning them counterclockwise.

13   Raise the car just enough so the tire clears the ground.

14   Remove the lug nuts.

15   Remove the wheel and tire.

16   If a flat tire is being replaced, ensure that there is adequate ground clearance for the new inflated tire, then mount the wheel and tire on the wheel studs.

17   Apply a light coat of spray lubricant or light oil to the wheel stud threads and install the lug nuts. **Caution:** *Aluminum wheels require washers between the wheel and the lug nut. On steel wheels, be sure the cone-shaped end of the lug nut faces the wheel (see illustration).*

18   Lightly tighten the wheel lug nuts alternately and evenly until there is no wheel play.

19   Adjust tire pressure to the specified value indicated on the tire placard.

20   Lower the vehicle until the wheel touches the ground, then tighten the wheel lug nuts to the specified torque in a criss-cross pattern (see illustration).

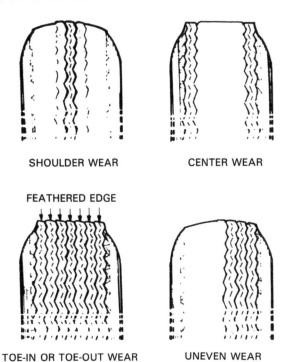

SHOULDER WEAR   CENTER WEAR

FEATHERED EDGE

TOE-IN OR TOE-OUT WEAR   UNEVEN WEAR

5.3a   An irregular tire wear pattern is indicative of various problems in the front end

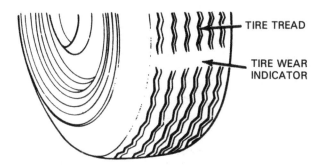

TIRE TREAD

TIRE WEAR INDICATOR

5.3b   Tire wear indicators are designed into the tread of the tires as a warning that they need replacing

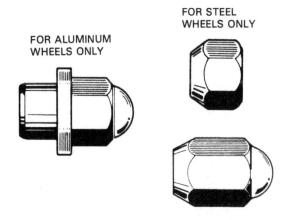

FOR STEEL WHEELS ONLY

FOR ALUMINUM WHEELS ONLY

5.17   Different wheel lugs are used for aluminum and steel wheels and should not be interchanged

5.20   Use a torque wrench to make sure the wheel lug nuts are properly tightened

Spare tire inflation with air canister

Caution: *Prior to inflation, carefully read the directions and caution notes on both the air canister and spare tire.*

21   Raise the vehicle and install the spare tire (Section 5).
22   Position the spare tire so the tire valve is at the 6 o'clock position.
23   **Note:** *If the outside temperature is below 14°F, the air canister should be warmed prior to operation. This can be done by placing the canister next to the windshield defroster for 5 to 10 minutes.*
24   Remove the valve cap from the spare tire and position the air canister on the valve. Holding the canister tightly in place, depress the release button (see illustration). You should be able to hear the gas entering the spare tire. **Warning:** *The metal parts of the canister become extremely cold during inflation and can cause frostbite, so adequate protection should be used to prevent contact.*
25   Hold the canister in position with the button depressed for an additional minute after the sound stops to make sure that as much gas as possible has entered the tire.
26   After inflating the tire, remove the canister from the tire valve and save it for later disposal. **Note:** *Buy another canister promptly so you'll be prepared for further emergencies.*
27   Replace the tire valve cap, lower the car to the ground, and fully tighten the wheel lug nuts in a criss-cross tightening pattern. **Note:** *In cold weather the tire may not appear fully inflated. In this case, drive slowly for the first mile or two until the tire temperature rises and the pressure increases.*

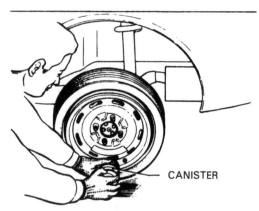

5.24   With the canister held tightly in place, push the release button and listen for the gas entering the tire. Protect your hands against the extremely cold can

5.33   The spare tire collapses to take up less storage space

Spare tire inflation with air compressor

28   Mount the spare tire onto the car, remove the valve cap from the spare and connect the air compressor hose to the valve (see illustration).
29   Insert the air compressor power cord into the cigarette lighter socket. It will take about six minutes for the compressor to inflate the tire to the recommended pressure.
30   Once the tire is fully inflated, remove the power cord plug from the cigarette lighter socket. Check the tire for correct pressure with a pressure gauge and adjust as necessary.
31   Disconnect the air compressor hose from the tire valve and replace the valve cap. Lower the car to the ground and tighten the wheel lug nuts to the specified torque.

Spare tire deflation

32   After removing the spare tire from the axle, the tire can be deflated by depressing the valve stem button or by removing the valve stem. **Warning:** *Do not inhale the gas escaping from the tire.*
33   After the spare tire is deflated, flatten the tire to its original shape (see illustration), replace the valve stem and cap, then put the spare back in its storage area in the rear compartment.

Tire rotation

34   Tires should be rotated at the specified intervals or any time that uneven wear becomes apparent. **Note:** *Since the car must be raised and the tires removed anyway, check the brakes (Section 24) and pack the wheel bearings (Chapter 10).*
35   Refer to the accompanying illustration for the preferred tire rotation pattern.
36   Refer to the information in *Jacking and towing* at the front of this manual for the proper procedures to follow when raising the vehicle and changing a tire. If the brakes are to be checked, do not apply the parking brake as stated. Make sure the tires are blocked to prevent the vehicle from rolling.
37   Preferably, the entire vehicle should be raised at the same time. This can be done on a hoist or by jacking up each corner and then lowering the vehicle onto jackstands placed under the frame. Always use four jackstands and make sure the vehicle is firmly supported.

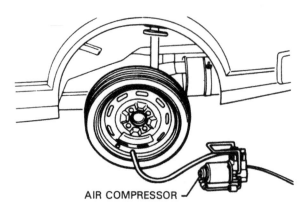

5.28   Inflate the spare tire with the optional air compressor by attaching the compressor hose to the wheel valve and connecting the compressor cord to the cigarette lighter

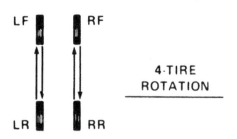

5.35   Correct rotation sequence for radial tires

38 After rotation check and adjust the tire pressures as necessary and be sure to check the lug nut tightness.

39 For further information on the wheels and tires, refer to Chapter 10.

6 Windshield wiper blades — inspection and replacement

Refer to illustrations 6.4 and 6.5

1   The windshield wiper and blade assembly should be inspected periodically for damage, loose components and cracked or worn blade elements.

2   Road film can build up on the wiper blades and affect their efficiency, so they should be washed regularly with a mild detergent solution.

3   The action of the wiper mechanism may loosen the nuts, bolts and fasteners, so check and tighten them as necessary at the same time the wiper blades are checked.

4   To remove a wiper blade assembly, remove the two screws attaching it to the arm (see illustration).

5   Grasping the wiper blade assembly in one hand, squeeze the retaining tabs of the metal tip with needle nose pliers, then pull the rubber element out of the blade assembly (see illustration).

6   With the blade assembly in one hand, position the new rubber wiper blade, then slide it back into place on the blade assembly until it seats. Be sure that the tangs of the metal tip are locked into place.

7   Reverse the disassembly procedure for installation.

7 Engine oil and filter change

Refer to illustrations 7.3 and 7.9

1   Frequent oil changes are the best preventive maintenance available to the home mechanic. As engine oil ages, it becomes diluted and contaminated. If allowed to remain in the engine, old oil can cause premature engine wear.

2   Some sources recommend oil filter changes every other oil change, but the minimal cost and relative ease of installation of an oil filter, plus the potential damage that can be caused by an ineffective filter, makes it more effective to change the filter along with the oil.

3   The tools necessary for a normal oil and filter change are a wrench to fit the oil pan drain plug, an oil filter wrench to remove the old filter, a six-quart capacity container into which to drain the old oil, a funnel or spout to pour in the new oil and clean shop rags or newspaper to mop up any spills (see illustration).

4   Access to the underside of the vehicle is greatly improved if the vehicle can be lifted on a hoist, driven onto ramps or supported by jackstands. **Warning:** *Do not work under a vehicle which is supported only with a bumper, hydraulic or scissor-type jack.*

5   If this is your first oil change on the vehicle, it is a good idea to crawl underneath and familiarize yourself with the locations of the oil drain plug and the oil filter. The engine and exhaust components will be hot, so figure out how to work around them before beginning.

6   Allow the engine to warm to normal operating temperature. Use this warm-up time to gather everything necessary for the job. The correct type of oil to buy for your application can be found in your owner's manual. Once the engine oil is warm, it will drain better and more built-up sludge will be removed.

7   Raise and support the vehicle. Notched jacking points are provided front and rear on both sides of the car for jacking. Make sure the car is firmly supported. Jackstands should be placed toward the front of the square-section frame rails running the length of the vehicle.

8   Move all necessary tools, rags and newspapers under the vehicle. Position the drain pan under the drain plug. Keep in mind that the oil will initially flow from the pan with some force, so place the pan accordingly.

6.4   Remove the two attaching screws holding the wiper blade to the arm

6.5   Squeeze the wiper blade retaining tabs inwards, then pull the metal tip of the rubber element from the blade

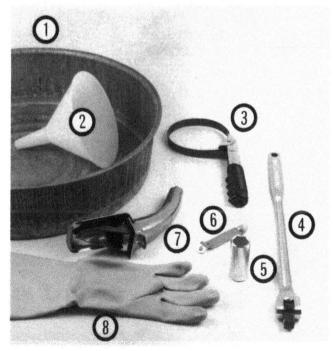

7.3   Typical tools required to perform an engine oil and filter change

9   Being careful not to touch any of the hot exhaust components, use a wrench to remove the drain plug (see illustration). Depending on oil temperature, you may want to wear gloves while unscrewing the plug the final few turns.
10   Drain the old oil into the pan. It may be necessary to move the pan farther under the engine as the oil flow slows to a trickle.
11   After all the oil has drained, wipe off the drain plug with a clean rag. Small metal particles clinging to the plug will quickly contaminate the new oil if not removed.
12   Clean the area around the drain plug opening and reinstall the plug. Tighten the plug to the specified torque.
13   Move the drain pan into position under the oil filter.
14   Working from above, reach down and loosen the oil filter canister, which is located on the lower right rear side of the engine block, with an oil filter wrench.
15   Sometimes the oil filter is screwed on so tightly that it cannot be loosened. As a last resort, you can punch a metal bar or long screwdriver directly through the canister and use it as a T-bar to turn the filter. If you do it this way, be prepared for oil to spurt out of the canister as it is punctured.
16   Completely unscrew the old filter. It will still have some oil in it,

so be careful. Dump the contents of the oil filter into the drain pan.
17   Compare the old filter with the new one to make sure they are the same size and type.
18   Use a clean rag to remove all oil, dirt and sludge from the area where the oil filter mounts to the engine. Check the old filter to make sure the rubber gasket is not stuck to the engine mounting surface. If it is, peel it off. The new filter comes with a new gasket and will not seal properly if the old gasket is left on the block.
19   Open one of the cans of oil apply a light coat of oil to the rubber gasket.
20   Attach the new filter to the engine block in accordance with the directions printed on the packing box or the canister itself. It's best to use only your hand to tighten the new filter canister. Most filter manufacturers recommend not using a filter wrench because over-tightening the canister will damage the seal.
21   Remove everything from under the vehicle before lowering the vehicle.
22   Unscrew the oil filler cap located on the right rocker arm cover of the engine. The cap will be labeled *Engine oil* or *Oil*.

23   Push the spout into the top of the oil can and pour the fresh oil through the filler opening. A funnel may also be used.
24   Pour three quarts of fresh oil into the engine. Wait a few minutes to allow the oil to drain into the pan, then check the level on the oil dipstick (see Section 3 if necessary). If the oil level is at or near the *L* mark, start the engine and allow the new oil to circulate.
25   Run the engine for only about a minute and then shut it off. Immediately look under the vehicle and check for leaks at the oil pan drain plug and around the oil filter.
26   With the new oil circulated and the filter now completely full, recheck the level on the dipstick and add enough oil to bring the level to the *F* mark on the dipstick.
27   During the first few trips after an oil change, make it a point to check frequently for leaks and proper oil level.
28   The old oil drained from the engine cannot be reused in its present state and should be disposed of. Oil reclamation centers, auto repair shops and gas stations will normally accept the oil, which can be refined and used again. After the oil has cooled it can be drained into a suitable container (capped plastic jugs, topped bottles, milk cartons, etc.) for transport to one of these disposal sites.

---

8   Battery — servicing and replacement

Refer to illustrations 8.1, 8.5 and 8.12

Servicing

1   Tools and materials required for battery maintenance include eye and hand protection, baking soda, petroleum jelly, a battery cable puller and a cable/terminal post cleaning tool (see illustration).

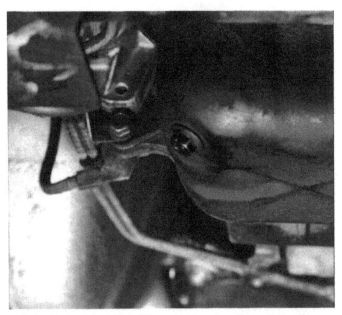

7.9   Drain the engine oil by removing the oil pan drain plug

8.1   Eye and hand protection, baking soda, petroleum jelly and tools required for battery maintenance

8.5   Remove the battery cable from the post and clean the battery post and cables

2 A sealed maintenance free battery is standard equipment on all vehicles with which this manual is concerned. Although this type of battery has many advantages over the older, capped cell type and never requires the addition of water, it should nevertheless be routinely maintained according to the procedures which follow. **Warning:** *Hydrogen gas in small quantities is present in the area of the two small side vents on sealed batteries, so keep lighted tobacco and open flames or sparks away from them.*

3 The external condition of the battery should be monitored periodically for damage such as cracked case or cover.

4 Check the tightness of the battery cable clamps to ensure a good electrical connection and check the entire length of each cable for cracks and frayed conductors.

5 If corrosion is evident, remove the cables from the terminals, clean them with a battery brush (see illustration) and reinstall the cables. Corrosion can be kept to a minimum by applying a layer of petroleum jelly or grease to the terminals and cable clamps after they are assembled.

6 If your vehicle is equipped with a rubber protector over the positive terminal, make sure that it's not torn or missing. It should completely cover the terminal.

7 Make sure that the battery tray is in good condition and that the hold-down clamp bolts are tight. If the battery is removed from the tray, make sure that no parts remain on the tray when the battery is reinstalled. Do not overtighten the hold-down clamp bolts.

8 Corrosion on the hold-down components, battery case and surrounding areas may be removed with a solution of water and baking soda. Protective gloves will protect your hands while working on corroded surfaces. Thoroughly wash all cleaned areas with plain water.

9 Any metal parts of the vehicle damaged by corrosion should be covered with a zinc-based primer then painted.

10 Further information on the battery, charging system and jump-starting can be found in Chapter 5 and at the front of this manual.

Replacement

11 Disconnect the negative cable from the battery by loosening the battery clamp hold-down bolt.

12 Install the cable puller onto the cable connector and turn the adjusting screw to raise the connector off of the battery post (see illustration).

13 Remove the positive cable from the battery in the same manner. **Caution:** *Don't twist the battery clamps. Twisting them can break the battery post loose inside the battery.*

14 Lift the battery off the battery tray and replace it with a new one of an equivalent or better reserve cold cranking capacity.

15 When installing the battery cable connectors to the battery carefully tap the connectors onto the post with a small hammer before tightening the hold-down clamp. **Caution:** *Always replace the battery cables in the reverse order of removal — positive cable first, then negative.*

9 Spark plugs — replacement

Refer to illustrations 9.2, 9.4 and 9.7

1 The best policy to follow when replacing spark plugs is to buy the new spark plugs beforehand, adjust them to the proper gap (see the Vehicle Emission Control Information label under the hood) and replace each plug one at a time. Be sure you obtain the correct type plugs for your specific engine.

2 The spark plugs are located in deep ports and replacing them requires patience. An accessory plug wrench is included in the kit (see illustration). **Note:** *To get at the left rear plugs, remove the harness support before inserting the plug wrench.*

3 Allow the engine to cool thoroughly before attempting to remove the plugs. During this time, each of the new spark plugs should be inspected for defects and the gap checked.

4 The plug gap is checked by inserting the proper thickness wire plug gauge, or a feeler gauge, between the electrodes at the tip of the plug. There should be a slight drag on the gauge as it is dragged between the two electrodes. If the gap is incorrect, use the notched adjuster on the gauge body to bend the curved side electrode slightly until the proper gap is achieved (see illustration). If the side electrode is not exactly over the center one, use the notched adjuster to align the two.

8.12 Remove both cables and unscrew the two hold-down clamp nuts (arrows) which will allow battery removal

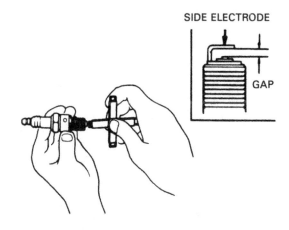

9.4 Use the notched adjuster on the feeler gauge body to bend the curved side of the electrode to adjust the gap or to center it

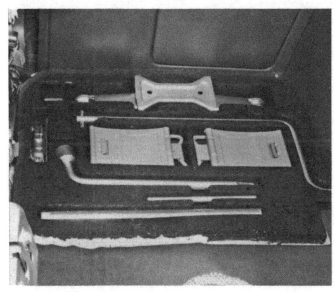

9.2 A special plug wrench is supplied in the tool kit

5　With the engine cold, remove the spark plug wire from one spark plug. A spark plug wire puller is available from your local parts store for removing the wires. Otherwise grab the boot at the end (base) of the wire, not by the wire itself. Sometimes it is necessary to use a twisting motion while the boot and plug wire are pulled free.

6　Before the plug is removed use compressed air to blow any dirt or foreign material away from the spark plug area. A common bicycle pump will also work.

7　Place the spark plug wrench or a socket with swivel and an extension over the plug and remove it by turning it in a counterclockwise direction (see illustration). **Note:** *A quick visual inspection of the old spark plugs will reveal a great deal about the running condition of the engine. Compare each spark plug to the accompanying chart for diagnosis.*

8　Carefully insert one of the new plugs into the spark plug hole and tighten it by hand. **Caution:** *Be careful not to cross-thread the spark plug in the hole. If resistance is felt as you thread the spark plug in by hand, back it out and start again. Do not, under any circumstances, force the spark plug into the hole with a wrench or socket or you may strip the threads.*

9　Once you have installed the plug as far as it will go by hand, tighten it to the specified torque.

10　Before pushing the spark plug wire onto the end of the plug, inspect it for cracks and corrosion.

11　Attach the plug wire to the new spark plug, again using a twisting motion on the boot until it is firmly seated on the spark plug. Make sure the wire is routed away from the exhaust manifold.

12　Follow the same procedure for each spark plug. If you replace them one at a time, you won't mix up the wires.

13　After the rear spark plug on the left side of the engine has been replaced, install the harness support in its original position.

10　Spark plug wires — inspection and replacement

Refer to illustration 10.8

1　The spark plug wires should be checked at the recommended intervals and whenever new spark plugs are installed in the engine.

2　The wires should be inspected one at a time to prevent mixing up the firing order.

3　Disconnect the plug wire from the spark plug. A removal tool can be used for this purpose or you can grab the rubber boot, twist slightly and pull the wire free. Do not pull on the wire itself — only on the rubber boot.

4　Inspect inside the boot for corrosion, which will look like a white or green crusty powder. Push the wire and boot back onto the end of the spark plug. It should be a tight fit on the plug. If it isn't, remove the wire and use pliers to carefully crimp the metal connector inside the boot until it fits securely on the end of the spark plug.

5　Using a clean rag, wipe the entire length of the wire to remove any built-up dirt and grease. Once the wire is clean, check for burns, cracks and other damage. Do not bend the wire sharply. The conductor inside might break.

6　Disconnect the wire from the distributor. Again, pull only on the rubber boot. Check for corrosion and a tight fit. Push the wire back onto the distributor terminal.

7　Check both ends of the remaining spark plug wires one at a time, making sure they are securely fastened at the distributor and the spark plug when the check is complete.

8　If new spark plug wires are required, purchase a new set for your specific engine model. Pre-cut wire sets are available with the rubber boots already installed. Remove and replace the wires one at a time to avoid mix-ups in the firing order (see illustration).

9　Check the distributor cap and rotor for wear (Section 11). It is common practice to install a new cap and rotor whenever new spark plug wires are installed. When installing a new cap, remove the wires from the old cap one at a time and attach them to the new cap in the exact same location — do not simultaneously remove all the wires from the old cap or firing order mix-ups may occur.

11　Distributor — check

Refer to illustrations 11.4 and 11.7
Caution: *When servicing the distributor cap or rotor, only spray cleaner on components while removed from the vehicle. The crank angle sensor located under the rotor is a delicate component and should not be exposed to any chemical, so use use cleaning agents cautiously to avoid blowing any dirt or dust on it.*

1　Although the breakerless distributor used on this vehicle requires much less maintenance than conventional distributors, periodic inspections should be performed when the plug wires are inspected.

2　Remove the distributor cap by first disconnecting the ignition coil wire from the coil, then removing the two retaining screws holding the cap to the distributor body.

3　Place the cap, with the spark plug and coil wires still attached, out of the way. If necessary, use a length of wire or tape to secure

9.7　On this engine the spark plugs are located on the intake side of the head

10.8　Remove and replace the spark plug wires one at a time to avoid mix-ups in the firing order

the cap so it won't flop back in your way.

4   Remove the set screw in the base of the rotor (see illustration) and lift the rotor from the shaft.

5   Inspect the rotor for cracks or damage. Carefully check the condition of the metal contact at the top of the rotor for excessive burning or pitting. If its condition is doubtful, replace it with a new one.

6   Install the rotor and tighten the set screw in the base.

7   Before installing the distributor cap, inspect the cap for cracks or other damage. Closely examine the contacts on the inside of the cap for excessive corrosion or damage. Slight scoring is normal. Again, if in doubt as to the quality of the cap, replace it with a new one (see illustration).

8   Replace the cap and secure it with the two hold down screws.

9   Connect the ignition coil wire to the coil.

12   Drivebelts — check and adjustment

Refer to illustrations 12.2a, 12.2b, 12.5, 12.6 and 12.11

1   The drivebelts, or V-belts as they are sometimes called, are located at the front of the engine and play an important role in the overall operation of the vehicle and its components. Due to their function and material make-up, drivebelts are prone to failure after extended use. They should be inspected and adjusted periodically to prevent failure in use.

2   Drivebelts are used to turn the alternator, power steering pump, water pump and air-conditioning compressor. There are three drivebelts on this vehicle (see illustrations).

3   With the engine off, open the hood and locate the belts at the front of the engine. Using your fingers (and a flashlight, if necessary), move along the belts checking for fraying, cracks and separation of the belt plies. Also watch for glazing, which gives the belt a shiny appearance. Both sides of the belt should be inspected, which means you will have to twist the belt to check the underside.

4   The tension of each belt is checked by pushing on the belt at a distance halfway between the pulleys. Push firmly with your thumb and see how much the belt deflects. Compare your measurements with the specified belt deflection for each drivebelt. A rule of thumb: If the distance from pulley center-to-pulley center is between 7 and 11 inches, the belt should deflect 1/4-inch. If the belt is longer and travels between pulleys spaced 12 to 16 inches apart, the belt should deflect 1/2-inch.

11.4   A set screw in the base of the rotor must be loosened for rotor removal

11.7   Check the distributor for cracks or other damage and check contacts for excessive corrosion

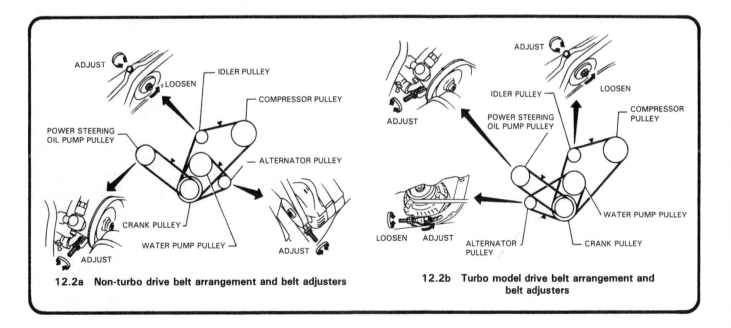

12.2a   Non-turbo drive belt arrangement and belt adjusters

12.2b   Turbo model drive belt arrangement and belt adjusters

Air conditioning compressor

5   Loosen the center bolt on the idler pulley (see illustration).
6   Turn the adjusting nut on the compressor bracket (clockwise to tighten) until tension of the air conditioning compressor drivebelt is correct (see illustration).
7   Tighten the center bolt on the idler pulley.

Alternator

8   The alternator is on the left side of the engine on non-turbo models and on the right side on turbo models.
9   Remove the six bolts retaining the engine splash shield from underneath the front of the car.
10 Loosen the top pivot bolt and the bottom locking bolt on the alternator.
11 Turn the adjusting nut clockwise to tighten the belt (see illustration).
12 Once the specified tension is achieved, tighten the locking and pivot bolts.
13 Install the engine splash shield.

Power steering pump

14 Remove the six bolts retaining the engine splash shield from underneath the front of the car.
15 Working from underneath the car, loosen the pump hold-down bolt.
16 From inside the engine compartment locate the power steering pump adjuster and turn the adjusting bolt until the specified tension is achieved.
17 From underneath the vehicle, tighten the pump hold down bolt.
18 Install the engine splash shield.

---

13   Air filter — replacement

Refer to illustrations 13.2 and 13.3

1   The air filter is located inside the air cleaner housing at the front of the engine compartment.
2   To remove the filter, unscrew the four retaining screws at the top of the cleaner housing and lift the cover (see illustration).

12.5   To adjust the air conditioner belt tension, first loosen the center bolt on the idler pulley

12.6   Turn the adjusting bolt clockwise to tighten the air conditioner belt

12.11   Loosen the bottom locking bolt and turn the adjusting nut to obtain specified alternator belt tension

13.2   Gain access to the air filter by removing the four retaining screws at the top of the air cleaner housing

3   Carefully lift the air filter out of the housing (see illustration).
4   The paper type air filter cannot be washed. If it is slightly dirty, you can blow it out with compressed air. If that doesn't restore it, replace it with a new unit.
5   Reverse the removal procedure for installation.

14   Engine idle speed — check and adjustment

Refer to illustrations 14.6 and 14.9

1   The engine idle speed is adjustable on VG3OE (non-turbo) models. It should be checked at the specified maintenance interval with the engine at normal operating temperature.
2   Set the parking brake and block both front wheels with wheel chocks.
3   Turn off all electrical accessories. There must be no electrical load on the engine during the idle speed adjustment.
4   Attach a tachometer in accordance with the manufacturer's instructions.
5   Run the engine at 2000 rpm for two minutes under no load.
6   Disconnect the idle-up solenoid harness connector (see illustration) and race the engine two or three times under no load, then run the engine at idle speed.
7   Place automatic transmission equipped vehicles in the *D* position.
8   See the VECI label under the hood for the correct idle speed for your vehicle.
9   If the idle speed is not correct, turn the idle speed adjusting screw with a screwdriver (see illustration) until the desired rpm is attained.
10   Once the correct idle speed is attained, stop the engine and reconnect the idle-up solenoid harness connector. Put the shift lever back in *P* on automatic transmission equipped models.
11   Disconnect the tachometer.
12   Remove the wheel chocks.

15   Ignition timing

Refer to illustrations 15.2, 15.6, 15.8 and 15.9

1   The ignition timing setting for your car is printed on the VECI decal located on the underside of the hood.
2   Locate the timing marks — a series of notches on the edge of the crankshaft pulley. With the engine turning in its normal clockwise operating direction (as you look at it from the front), the first mark to approach the stationary pointer attached to the oil pump housing is the 30° mark (see illustration). Each subsequent mark is in 5° increments. The last mark to approach the pointer is the 0° mark.
3   Locate the timing notch on the pulley specified for your vehicle and mark it with a dab of paint or chalk so it will be visible under the timing light.
4   Connect a tachometer in accordance with the manufacturer's in-

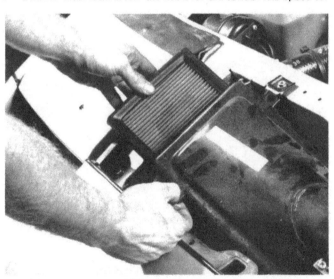

13.3   Lift up on the air cleaner housing cover and carefully lift the air filter from the housing

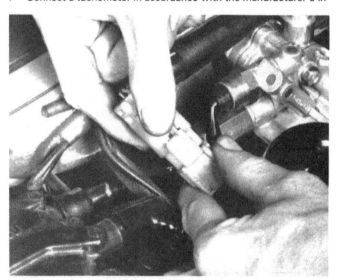

14.6   Disconnect the idle-up solenoid by pressing down on the locking tab and pulling the connector apart

14.9   Adjust the idle speed by turning the idle speed adjusting screw

15.2   As the crankshaft pulley turns clockwise (looking from the front of the engine), the first mark to approach the pointer is the 30° mark, and each subsequent mark indicates another 5°, so the last mark to approach the pointer will be the 0° mark

structions and check the idle speed (Section 14).

5   Allow the engine to reach normal operating temperature. Be sure the air conditioner and other electrical accessories are off.

6   Turn the ignition off. Connect the pick-up lead of the timing light to the number one spark plug wire (see illustration). Use either a jumper lead between the spark plug wire and plug or an inductive-type pick-up. Do not pierce the spark plug wire or attempt to insert a wire between the boot and the spark plug wire.

7   Make sure that the wires of the timing light are clear of all moving engine components, then start the engine.

8   Point the strobe timing light at the timing marks (see illustration), again being careful to avoid entanglement with moving parts. The notch that you painted should appear stationary and should be in alignment with the stationary pointer. If the notch and pointer are not in alignment, or if the painted notch appears blurry or jumpy, you must adjust the timing.

9   If the timing marks do not coincide, turn off the engine and loosen the adjusting nut at the base of the distributor (see illustration) just enough to rotate the distributor.

10   Start the engine and slowly twist the distributor either left or right until the painted notch and the pointer coincide.

11   Shut off the engine and tighten the distributor adjusting nut, being careful not to move the distributor.

12   Start the engine and recheck the timing to make sure the notch and pointer are still in alignment.

13   Disconnect the timing light.

14   Race the engine two or three times and then allow it to run at idle. Recheck the idle speed with the tachometer. If it has changed from its correct setting, readjust it.

---

16   Compression check

Refer to illustration 16.6

1   A compression check can tell you a lot about the condition of the engine. For example, it can tell you if compression is weak because of worn piston rings, defective valves and seats or a blown head gasket. Make sure the engine oil is the correct viscosity and that the battery is fully charged before doing a compression check.

2   Run the engine until it reaches normal operating temperature.

3   Turn the engine off and have an assistant depress the accelerator pedal to the floor so that the throttle valve is fully open.

4   Remove all spark plug wires from the plugs after marking them clearly so that they can be installed in their original positions. Clean the area around the spark plugs before you remove them to keep dirt from falling into the cylinders while performing the compression test. Remove the spark plugs.

5   Disconnect the coil wire from the coil.

6   Install the compression gauge into the number one cylinder spark plug hole (see illustration).

15.6   Clip the timing light connector to the number one spark plug wire

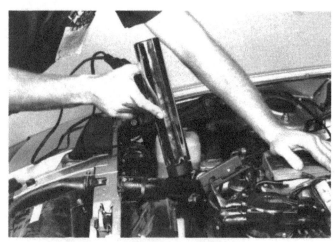

15.8   When you point the timing light at the crankshaft pulley and the stationary pointer with the engine running, the notch should appear stationary and should be aligned with the pointer

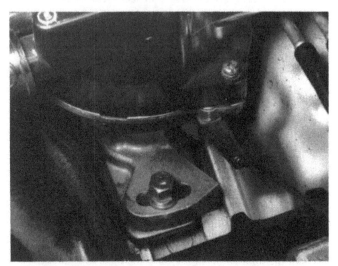

15.9   If the timing notch and pointer don't align, loosen the distributor hold-down bolt and turn the distributor until the notch and pointer line up

16.6   Install the compression gauge in the number one cylinder, then observe the number of compression strokes required to reach the highest setting and record the reading

Common spark plug conditions

NORMAL

Symptoms: Brown to grayish-tan color and slight electrode wear. Correct heat range for engine and operating conditions.
Recommendation: When new spark plugs are installed, replace with plugs of the same heat range.

WORN

Symptoms: Rounded electrodes with a small amount of deposits on the firing end. Normal color. Causes hard starting in damp or cold weather and poor fuel economy.
Recommendation: Plugs have been left in the engine too long. Replace with new plugs of the same heat range. Follow the recommended maintenance schedule.

CARBON DEPOSITS

Symptoms: Dry sooty deposits indicate a rich mixture or weak ignition. Causes misfiring, hard starting and hesitation.
Recommendation: Make sure the plug has the correct heat range. Check for a clogged air filter or problem in the fuel system or engine management system. Also check for ignition system problems.

ASH DEPOSITS

Symptoms: Light brown deposits encrusted on the side or center electrodes or both. Derived from oil and/or fuel additives. Excessive amounts may mask the spark, causing misfiring and hesitation during acceleration.
Recommendation: If excessive deposits accumulate over a short time or low mileage, install new valve guide seals to prevent seepage of oil into the combustion chambers. Also try changing gasoline brands.

OIL DEPOSITS

Symptoms: Oily coating caused by poor oil control. Oil is leaking past worn valve guides or piston rings into the combustion chamber. Causes hard starting, misfiring and hesitation.
Recommendation: Correct the mechanical condition with necessary repairs and install new plugs.

GAP BRIDGING

Symptoms: Combustion deposits lodge between the electrodes. Heavy deposits accumulate and bridge the electrode gap. The plug ceases to fire, resulting in a dead cylinder.
Recommendation: Locate the faulty plug and remove the deposits from between the electrodes.

TOO HOT

Symptoms: Blistered, white insulator, eroded electrode and absence of deposits. Results in shortened plug life.
Recommendation: Check for the correct plug heat range, over-advanced ignition timing, lean fuel mixture, intake manifold vacuum leaks, sticking valves and insufficient engine cooling.

PREIGNITION

Symptoms: Melted electrodes. Insulators are white, but may be dirty due to misfiring or flying debris in the combustion chamber. Can lead to engine damage.
Recommendation: Check for the correct plug heat range, over-advanced ignition timing, lean fuel mixture, insufficient engine cooling and lack of lubrication.

HIGH SPEED GLAZING

Symptoms: Insulator has yellowish, glazed appearance. Indicates that combustion chamber temperatures have risen suddenly during hard acceleration. Normal deposits melt to form a conductive coating. Causes misfiring at high speeds.
Recommendation: Install new plugs. Consider using a colder plug if driving habits warrant.

DETONATION

Symptoms: Insulators may be cracked or chipped. Improper gap setting techniques can also result in a fractured insulator tip. Can lead to piston damage.
Recommendation: Make sure the fuel anti-knock values meet engine requirements. Use care when setting the gaps on new plugs. Avoid lugging the engine.

MECHANICAL DAMAGE

Symptoms: May be caused by a foreign object in the combustion chamber or the piston striking an incorrect reach (too long) plug. Causes a dead cylinder and could result in piston damage.
Recommendation: Repair the mechanical damage. Remove the foreign object from the engine and/or install the correct reach plug.

7   Crank the engine over about five or six times with a remote starter switch or have an assistant operate the ignition switch.

8   Observe the number of compression strokes required to reach the highest reading. Also note the pattern in which the cylinder builds up to its highest reading. Compression should rise quickly in a healthy engine. Low compression on the first stroke, followed by gradually increasing pressure on successive strokes, indicates worn piston rings. An initially low compression reading which does not build up during successive compression strokes indicates leaking valves or a defective head gasket. Record the highest gauge reading obtained. Repeat this procedure for the remaining cylinders and compare the readings with the standards given in the Specifications.

9   Further diagnosis of the cylinders can be accomplished through a wet compression test.

10   To perform a wet compression test on a low reading cylinder, pour a couple of teaspoons of engine oil — a squirt can works well - into the cylinder through the spark plug hole and repeat the compression test.

11   If compression increases after oil is added, the piston rings are worn. If the compression does not increase significantly, leakage is occurring at the valves or head gasket.

12   If two adjacent cylinders have equally low compression, there is a good possibility that the head gasket between them is blown. The existence of coolant in the combustion chambers or the crankcase

would also be a strong indication of a blown head gasket.

13   If compression is higher than normal, the combustion chambers may be coated with carbon deposits. If they are, they must be removed and decarbonized.

14   If compression is way down, or varies greatly between cylinders, it would be a good idea to have a leak-down test performed by an automotive shop. A leak-down test can pinpoint exactly where the leakage is occurring.

15   Remove the compression gauge, replace the spark plugs and attach the spark plug wires. Plug the ignition coil wire back into the coil.

---

17   Fuel pressure — relief

Refer to illustrations 17.2 and 17.3
Warning: *There are certain precautions to take when inspecting or servicing the fuel system components. Work in a well-ventilated area and do not allow open flames (cigarettes, appliance pilot lights, etc.) to get near the work area. Mop up spills immediately and do not store fuel soaked rags where they could ignite.*

1   Start the engine and let it run at idle speed.

2   Remove the luggage floor mat and locate the fuel pump harness (blue connector) (see illustration).

3   Push down on the locking tabs on the connector and pull the connector apart (see illustration).

4   After the engine stalls, crank it over two or three times to make sure that all pressure is released.

5   Turn the ignition switch off.

6   Connect the fuel pump harness connector after servicing the fuel system.

---

18   Fuel system check

Refer to illustration 18.4
Warning: *There are certain precautions to take when inspecting or servicing the fuel system components. Work in a well-ventilated area and do not allow open flames (cigarettes, appliance pilot lights, etc.) near the work area. Mop up spills immediately and do not store fuel soaked rags where they could ignite.*

1   Relieve the fuel pressure (Section 17) before servicing any component of the fuel system.

2   The fuel system is most easily checked with the vehicle raised on a hoist or supported on jackstands so that the components underneath the vehicle are visible and readily accessible.

3   Any time the smell of gasoline is apparent while driving or after the vehicle has been in the sun, the system should be inspected.

4   Remove the gas filler cap and check for corrosion. The rubber seal imprint should be unbroken (see illustration). Replace the cap if any damage is apparent.

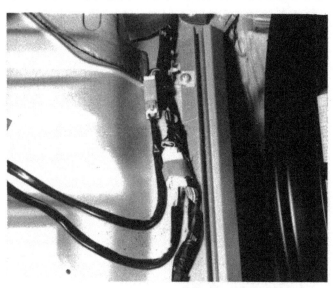

17.2   The fuel pump harness is located under the luggage floor mat

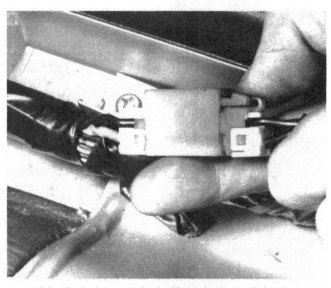

17.3   Push down on the locking tabs and pull the fuel pump connector apart

18.4   Remove the gas filler cap and check the O-ring seal for damage

5   With the vehicle raised, inspect the gas tank and filler neck for punctures, cracks and other damage. The connection between the filler neck and the tank is especially critical. Sometimes a rubber filler neck will leak because of loose clamps or rubber deterioration. **Warning:** *Do not, under any circumstances, try to repair a fuel tank yourself (except rubber components). A welding torch or any open flame can easily cause the fuel vapors to explode if the proper precautions are not taken.*
6   Carefully check all rubber hoses and metal lines leading away from the fuel tank. Check for loose connections, deteriorated hoses, crimped lines and other damage. Follow the lines up to the front of the vehicle, carefully inspecting them all the way. Repair or replace any damaged sections.
7   Ensure that hose clamp screws do not come in contact with adjacent parts when tightening.

19   Fuel filter — replacement

Refer to illustration 19.2
Warning: *There are certain precautions to take when inspecting or servicing the fuel system components. Work in a well-ventilated area and do not allow open flames (cigarettes, appliance pilot lights, etc.) to*

19.2   Loosen both hose clamps and the filter retaining clamp screw to remove the fuel filter

20.1   Squeeze the hose clamp, then pull it off the PCV valve

get near the work area. Mop up spills immediately and do not store fuel soaked rags where they could ignite.
1   Relieve the fuel pressure (Section 17).
2   After fuel pressure has been released, loosen both fuel hose clamps and loosen the filter retaining clamp screw (see illustration).
3   Note the flow direction of the filter before removing it. The top of the filter should read *out* and the bottom should say *in*.
4   Wrap a rag around the fuel filter to absorb spilled fuel.
5   Hold the filter with one hand and pull the hose off the top of the filter with the other. Remove the hose clamp.
6   With the filter in one hand, pull the hose off the bottom. Remove the old hose clamp.
7   Remove the old filter from its retaining clamp bracket and insert the new filter with the word *out* facing up.
8   Slide a screw clamp onto each hose.
9   Insert the bottom hose onto the *in* side of the filter and tighten the hose clamp.
10   Insert the top hose, leading to the fuel injection unit, onto the *out* side of the filter and tighten the hose clamp.
11   Tighten the fuel filter retaining clamp screw.

20   PCV valve — checking and replacement

Refer to illustration 20.1
1   Check the PCV valve by removing the ventilation hose (see illustration). The engine must be running. If the valve is working properly, a hissing noise will be heard as air passes through the valve.
2   Another way to check the PCV valve is to place your finger over the mouth of the valve. If the valve is working properly, a strong vacuum should be felt immediately.
3   To replace the PCV valve, remove the hose and screw the valve out of the intake manifold.
4   Screw the new valve into the intake manifold and replace the hose.
5   Tighten the hose clamp.

21   Cooling system — check and servicing

Refer to illustrations 21.3, 21.4a, 21.4b, 21.13 and 21.14
Check
1   Many engine and automatic transmission failures can be attributed to a faulty cooling system.
2   The cooling system should be checked with the engine cold.
3   Remove the remote coolant cap (see illustration) and thoroughly clean it inside and out with a scrub brush and clean water.

21.3   Thouroughly clean the remote radiator filler cap

4   Carefully check the upper and lower radiator hoses and the smaller diameter heater hoses. Inspect each hose along its entire length, replacing any hose which is cracked, swollen or shows signs of deterioration. Cracks may become more apparent if the hose is squeezed (see illustrations).

5   Make sure that all hose connections are tight. A leak in the cooling system will usually leave a telltale white or rust colored deposit on the area around the leak.

6   Use compressed air or a soft brush to remove bugs, leaves, etc. from the front of the radiator or air-conditioning condenser. Do not damage the delicate cooling fins or cut yourself on them.

7   Have the cap and system pressure tested. If you do not have a pressure tester, most gas stations and repair shops will do this for a minimal charge.

Servicing

8   Periodically the cooling system should be drained, flushed and re-filled with water/antifreeze mixture. This will prevent rust and corrosion which can impair the performance of the cooling system and ultimately cause engine damage. **Warning:** *Antifreeze is a poisonous solution. Do not to spill it on either yourself or your vehicle. Rinse off spills immediately with plenty of water. It is especially important to mop up coolant spills if you or your neighbors have pets because animals are attracted to the taste of coolant.* Before dumping any antifreeze down the kitchen sink or a storm drain it is advisable to consult your local authorities about the local laws regarding the proper disposal of coolant. In some areas, reclamation centers have been set up to collect both oil and coolant instead of allowing these liquids to be introduced into the sewage and water facilities.

9   On models without automatic air conditioning, slide the temperature control lever to the *Hot* position.

10 On automatic air conditioning equipped models, turn the ignition switch *On* and set the temperature to maximum. Turn off the ignition switch.

11 With the engine cold, remove the radiator remote fill pressure cap.

12 Place a large container under the radiator and engine block to catch the coolant mixture as it drains.

13 Drain the radiator by removing the drain plug from the bottom of the radiator (see illustration). Do not splash the solution on your skin or in your eyes.

14 Locate and open the drain valves located on the sides of the engine block (see illustration).

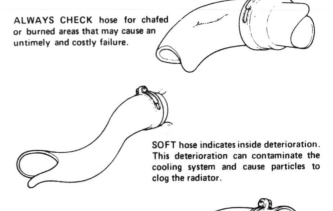

ALWAYS CHECK hose for chafed or burned areas that may cause an untimely and costly failure.

SOFT hose indicates inside deterioration. This deterioration can contaminate the cooling system and cause particles to clog the radiator.

HARDENED hose can fail at any time. Tightening hose clamps will not seal the connection or stop leaks.

SWOLLEN hose or oil soaked ends indicate danger and possible failure from oil or grease contamination. Squeeze the hose to locate cracks and breaks that cause leaks.

21.4a   Cracks may become more apparent if the hose is squeezed

21.4b   Radiator hose inspection procedure

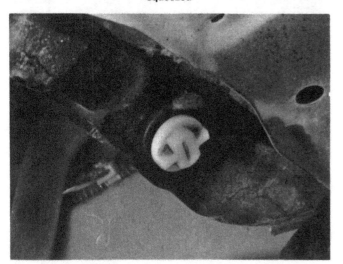

21.13   Drain the radiator by removing the drain plug

21.14   Open the engine block drain valve

15 Screw in the drain plug on the radiator and close the drain valves on the engine block.
16 Fill the radiator with water and warm the engine.
17 Stop the engine and let it cool.
18 Repeat Steps 13 through 17 two or three times.
19 Drain the cooling system a final time then fill the radiator and engine with a new antifreeze/water mixture up to the filler opening. Follow the instructions on the antifreeze container for the proper ratio of antifreeze to water.
20 Fill the reservoir tank to the *MAX* level, then install and tighten the radiator cap.
21 Run the engine at approximately 2000 rpm for about one minute.
22 Check for coolant leaks.
23 Stop the engine and let it cool. Add more mixture until the coolant level is up to the filler opening.
24 Fill the reservoir tank with coolant up to the *MAX* level.

22 Throttle linkage — check and adjustment

Refer to illustrations 22.3 and 22.4

1 With the engine off, depress the throttle pedal several times to make sure that the cable linkage is operating smoothly without jamming or dragging. Make sure that the pedal returns to its original position when released. If it doesn't, the cable is either binding or too tight.

22.3 The dog on the throttle valve actuator should meet the stop on the EFI throttle chamber

22.4 Adjust the slack out of the accelerator cable by loosening the locking nut and turning the adjusting nut until the desired tension is attained

2 To check the throttle cable for correct adjustment, have an assistant hold the throttle pedal to the floor.
3 The half-round throttle valve actuator should open until the dog meets the stop on the EFI throttle chamber (see illustration).
4 If the throttle valve does not open all the way it must be adjusted. Back off the cable adjustment locking nut and remove slack from the throttle cable by turning the adjusting nut until the dog reaches the stop (see illustration).

23 Wheel bearings — check

1 Generally, the wheel bearings will require infrequent service. However, they should be checked whenever the wheels are raised for any reason.
2 With the vehicle securely supported on jackstands, spin each wheel and check for noise, rolling resistance or free play.
3 Grasp the top of the tire with one hand and the bottom with the other. Rock the tire on the spindle. If the wheel moves (has side play), the bearings should be removed, inspected and, if necessary, replaced (Chapter 10).
4 Remove each wheel and tire. Inspect the area around the bearings for grease leaking past the wheel bearing seals. Seals must be replaced (Chapter 10) if any leakage is apparent.

24 Chassis and body — lubrication and inspection

Refer to illustrations 24.8, 24.11 and 24.12
Note: *The 300ZX steering connections are "sealed-for-life" and require no lubrication. But they should still be periodically checked for leaks, cracks and other obvious signs of wear.*

1 For easier access to the underside of the vehicle, raise it and place jackstands under the frame. Make sure the vehicle is securely supported by the jackstands.
2 From underneath the car, inspect the rear axle and suspension components for looseness, wear or damage.
3 Rock each front wheel sideways and listen for rattles indicating loose components. Adjust the wheel bearing preload, if necessary (Chapter 10).
4 Tighten all steering and suspension nuts and bolts.
5 Check the struts (shock absorbers) for oil leakage or damage. If any damage is apparent, replace the struts.
6 Check the balljoints for grease leakage or other damage.
7 Check the rear wheels for looseness by shaking them from side to side. Adjust the wheel bearing preload if necessary.
8 While you are under the vehicle, clean and lubricate the parking brake cable, along with the cable guides and levers. Smear multipurpose grease onto the cable and its guides with your fingers (see illustration).

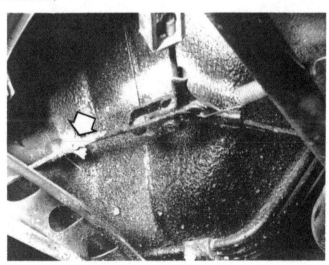

24.8 Multi-purpose grease is used to lubricate the parking brake cable, guides and levers

9   Lubricate the transmission shift linkage rods and swivels with a few drops of engine oil.

10   Lower the vehicle to lubricate the body components.

11   Open the hood and lubricate the hood latch mechanism with a few drops of oil (see illustration). The hood has an inside release, so have an assistant pull the release knob from inside the vehicle as you lubricate the cable at the latch.

12   Lubricate all the hinges — door, hood, hatch — with a few drops of engine oil to keep them in proper working order (see illustration).

13   The key lock cylinders can be lubricated with spray-on graphite, which is available at auto parts stores.

---

25   Brake check

Refer to illustrations 25.5 and 25.8
Note: *For detailed illustrations of the brake system, refer to Chapter 9.*

1   The brakes should be inspected every time the wheels are removed or whenever a defect is suspected. There are several symptoms of a potential brake system defect: The vehicle pulls to one side when the brake pedal is depressed; the brakes make odd noises — squealing, grating, grinding, etc. — when they are applied; brake pedal travel is excessive; the pedal pulsates; fluid is leaking, usually on the inside of the tire or wheel. **Warning:** *Wheel assemblies should be cleaned with a vacuum. Dust and dirt present on brake and clutch assemblies may contain asbestos fibers that are hazardous to your health. Cleaning off components with compressed air blows these fibers into the air around you, increasing the likelihood that you will inhale them.*

2   Check the brake hydraulic fluid reservoir atop the master cylinder. The fluid level should be near the *MAX* mark.

3   Raise the vehicle and support it securely on jackstands. Remove the wheels.

4   The disc brake pads inside the brake calipers should be inspected. There is an outer pad and an inner pad in each caliper.

5   Check the pad thickness by looking through the inspection holes in the caliper body (see illustration). If the lining material is worn thinner than the specified brake pad wear limit, the pads should be replaced. Keep in mind that the friction material is riveted or bonded to a metal backing shoe. The metal backing is not included in this measurement.

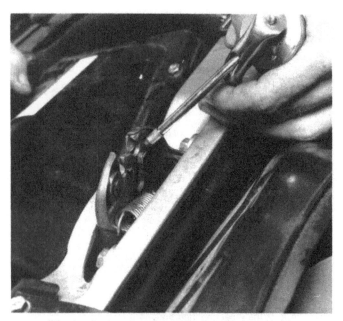

24.11   Apply a few drops of oil to the hood latch mechanism

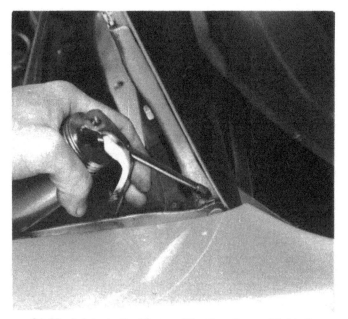

24.12   Lubricate the hinges with a few drops of light oil

25.5   Inspect the brake pad thickness through the inspection hole in the caliper

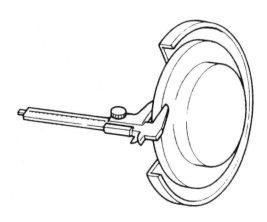

25.8   Measure the rotor with a caliper to make sure it's not thinner than the specified minimum thickness

Note: *All four pads — two per wheel — should be replaced at the same time.*

6   Since it may be difficult to measure the exact thickness of the re-maining friction material while the pads are installed in the calipers, remove the pads (Chapter 9) for further inspection or replacement if you are concerned about their condition.

7   Before installing the wheels, check the brake hose connections leading to the caliper for leakage and damage (cracking, splitting, etc.) to the brake hose. Replace the hose or fittings as necessary (Chapter 9).

8   Check the condition of the rotor. It should be measured at several different points along its circumference with a dial caliper (see illustration) to determine if it's beyond the thickness wear limit. Also look for scoring, gouging and bluish (burned) spots. If these conditions exist, the hub/rotor assembly should be removed for servicing (Chapter 9).

9   Check the hydraulic lines inside the engine compartment for any leakage. Also check the vacuum line running to the power brake booster. It should not be hardened or cracked and the connectors should be tight.

Parking brake

10   The easiest way to check the operation of the parking brake is to park the vehicle on a steep hill with the parking brake set and the transmission in Neutral. If the parking brake cannot prevent the vehicle from rolling, it is in need of adjustment (Chapter 9).

Brake pedal height

11   Check the brake pedal for smooth operation. Its range of motion should be within the limits of specified free height and depressed height (Chapter 9).

26   Suspension and steering check

1   Whenever the front of the vehicle is raised for service, visually check the suspension and steering components for wear.

2   Indications of a fault in these systems are excessive play in the steering wheel before the front wheels react, excessive sway around corners, body movement over rough roads or binding at some point as the steering wheel is turned.

3   Before the vehicle is raised for inspection, test the shock absorbers by pushing down to rock the vehicle at each corner. If you push down and the vehicle does not come back to a level position within one or two bounces, the shocks are worn and should be replaced. While you're compressing the suspension components, listen for squeaks and strange noises. If this quick check indicates that further inspection is warranted, the questionable suspension components must be removed (Chapter 10).

4   Raise the front of the vehicle and support it securely on jackstands. **Warning:** *The following procedures must be performed underneath the car, so make sure the jackstands are well placed so that there is no danger of the car falling.*

5   Crawl under the vehicle and retighten all nuts and bolts. Check for broken or disconnected parts and deteriorated rubber bushings on all suspension and steering components. Look for grease or fluid leaking from around the steering rack and pinion unit. Check the power steering hoses and connections for leaks.

6   Check the balljoint seals. Inspect them for wear, cracks and damage. Replace damaged seals (Chapter 10).

7   Check the wheel bearings for free play (Section 23). If play is evident, adjust the wheel bearing preload (Chapter 10).

8   Have an assistant turn the steering wheel from side-to-side and check the steering components for free play, chafing and binding. If the steering does not react briskly to steering wheel movement, try to determine where the slack is located.

9   Check for play at the outer ends of the tie rods, or excessive play within the rack-and-pinion steering mechanism itself.

10   A similar inspection of all suspension components should be con-ducted at the rear (Chapter 10). Again, tighten all bolts and check for damaged or disconnected parts and deteriorated rubber bushings.

27   Exhaust system — check

Refer to illustration 27.2

1   The engine and exhaust system must be cold. Check the complete exhaust system from the exhaust manifold to the muffler and tailpipe. It's easier to work on the exhaust system if the car is raised.

2   Check the exhaust pipes and all connections for leakage and cor-rosion that might cause leakage. Make sure that all brackets and rubber hangers are in good condition and tight (see illustration).

3   Inspect the underside of the body for holes, corrosion, open seams, etc. which may allow exhaust gases to enter the passenger compart-ment. Seal all body openings with silicone or body putty.

4   Rattles and other noises can often be traced to the exhaust system, especially the mounts and hangers. Try to move the pipes, muffler and catalytic converter. If the components are touching body or suspension parts, secure the exhaust system with new mounts.

5   While you're inspecting the exhaust system, check the running condition of the engine by inspecting inside the end of the tailpipe. The exhaust deposits here are an indication of engine state-of-tune. If the pipe is black and sooty or coated with white deposits, the engine is in need of a tune-up.

28   Drivetrain inspection

Refer to illustrations 28.7a and 28.7

1   Raise the rear of the vehicle and support it securely on jackstands. The transmission should be in Neutral.

2   From under the car inspect the condition of the driveshaft. Look for any dents or cracks in the tubing. If any damage is found, the driveshaft must be replaced (Chapter 8).

3   Check for oil leakage at the forward and rear ends of the driveshaft. Leakage at the forward end of the driveshaft, where it connects to the transmission, indicates a defective rear transmission seal. Leakage at the rear, where the driveshaft enters the differential carrier, indicates a defective pinion seal. If either seal is leaking, it must be replaced (Chapters 7 and 8 respectively).

4   Have an assistant turn the rear wheel so the driveshaft will rotate. As it does, check the universal joints. They should not be binding, loose or noisy. If they are, replace them (Chapter 8).

5   The universal joints can also be checked while stationary. Grip both sides of each joint and attempt to twist them. The slightest movement in either joint is a sign of wear. Also try lifting up the shaft. If you can push it up at all, this is also an indication that the joints are worn.

6   Check the driveshaft mounting bolts at both ends to make sure they are tight (Chapter 8).

27.2   Check pipes and connections for signs of leakage and/or corrosion indicating a potential failure

7   Check for free play in the universal joints of the rear drive axles in the same manner described above. Also check for grease or oil leakage from the rear drive axles and inspect the rubber boots at both ends of each axle (see illustration). Oil leakage at the differential junction indicates a defective rear axle grease seal (see illustration). Leakage from either boot means that it's torn and must be replaced (Chapter 8).
8   Check the differential carrier housing fluid level. Open the filler plug (the upper one) located on the back of the carrier. The fluid level should be even with the bottom of the filler hole. If it isn't, fill it to the bottom edge of the hole.

29   EGR valve — check

Refer to illustrations 29.1 and 29.5

1   Open the hood and locate the EGR valve at the right rear of the engine (see illustration).
2   Reach under the EGR valve and push the valve diaphragm up with a finger to make sure that the valve isn't stuck. It should spring back when released.
3   Start the engine and warm it to operating temperature.
4   With the engine still running, disconnect the vacuum hose attached to the EGR valve.
5   Connect a hand held vacuum pump to the nipple where the vacuum hose was removed (see illustration).
6   Pump up the vacuum pump. The diaphragm inside the valve should

open and the engine should cough and stumble badly, or stall. If it does, the valve works.
7   If the engine doesn't stall or run poorly when vacuum is applied to the EGR valve, the valve is defective. Remove it and look for blockage or carbon buildup. If there is no apparent cause for valve malfunction, replace it (Chapter 6).

30   Underhood hoses — check

Refer to illustrations 30.2 and 30.3
Warning: *Replacement of air-conditioning hoses must be left to a dealer or air-conditioning specialist who has the proper equipment to depressurize the system safely. Never remove air conditioning components or hoses until the system has been depressurized.*

1   The high temperatures present under the hood can cause deterioration of the numerous rubber and plastic hoses used for engine, accessory and emission systems operation.
2   Periodic inspection should be made for cracks, loose clamps, material hardening and leaks (see illustration).
3   Some, but not all, vacuum hoses use clamps to secure the hoses to fittings. Check all hose clamps for tension. Loose clamps allow hoses to leak. Make sure unclamped hoses have not expanded and/or hardened where they slip over fittings (see illustration) or they will leak.
4   It is quite common for vacuum hoses, especially those in the emissions system, to be color coded or identified by colored stripes molded

28.7a   Pull back on the boot and inspect for cracks and signs of leakage

28.7b   Inspect around the carrier housing for leaks where the inboard ends of the axles protrude from the carrier

29.1   The EGR valve is located at the right rear of the engine

29.5   Apply vacuum to the EGR valve with a vacuum pump and, if the valve is working properly, the engine should run rough or die

into the hose. Various systems require hoses with different wall thicknesses, collapse resistance and temperature resistance. When replacing hoses be sure to use the same hose material on the new hose.

5   Often the only effective way to check a hose is to remove it completely from the vehicle. Where more than one hose is removed, be sure to label the hoses and their attaching points to insure proper attachment.

6   When checking vacuum hoses, be sure to include any plastic T-fittings in the check. Check the fittings for cracks, and the hose where it fits over the fitting for enlargement, which could cause leakage.

7   A small piece of vacuum hose (1/4-inch inside diameter) can be used as a stethoscope to detect vacuum leaks. Hold one end of the hose to your ear and probe around the vacuum hoses and fittings, listening for the hissing sound characteristic of a vacuum leak. **Warning:** *When probing with the vacuum hose stethoscope, be careful not to allow your body or the hose to come into contact with moving engine components such as drivebelts, the cooling fan, etc.*

Fuel hose

Warning: *There are certain precautions which must be taken when inspecting or servicing fuel system components. Work in a well ventilated area and do not allow open flames (cigarettes, appliance pilot lights, etc.) or bare light bulbs near the work area. Mop up any spills immediately and do not store fuel soaked rags where they could ignite.*

8   The fuel lines are usually under a small amount of pressure, so be prepared to catch spilling fuel anytime fuel lines are disconnected. **Warning:** *Fuel injected engines generally have high pressure in the fuel lines even after shutting off the engine, for a considerable time. That's why you must relieve fuel pressure (Section 17) before disconnecting any fuel lines on fuel injected engines.*

9   Check all rubber fuel lines for deterioration and chafing. Check especially for cracking in areas where the hoses bend and near clamping points such as where hoses are attached to the fuel pump, fuel filter and carburetor or fuel injection unit.

10   High quality fuel lines should be used for fuel line replacement. Under no circumstances should unreinforced vacuum line, clear plastic tubing or water hose be used for fuel line replacement.

11   Fuel lines are commonly equipped with spring-type clamps. These clamps often lose their tension over a period of time or they are sprung during removal. That's why it's a good idea to replace spring-type clamps with screw clamps whenever a hose is replaced.

Metal lines

12   Sections of metal line are often used to route fuel between the fuel pump and carburetor or fuel injection unit. Check such lines for crimps and cracks in the vicinity of any bent sections. Watch for cracks or loose fittings where the brake lines are connected to the master cylinder and the proportioning valve. Any sign of leakage calls for an immediate and thorough inspection of the entire brake system (Chapter 9).

13   If a section of metal line must be replaced, seamless steel tubing must be used. Copper and aluminum tubing do not have the strength necessary to withstand engine vibration.

31   Exhaust gas sensor — servicing

Refer to illustrations 31.2, 31.3, 31.6, 31.10 and 31.13

1   The exhaust gas sensor is a screw-in unit located in the exhaust manifold. The sensor should be inspected at the specified intervals but need not be replaced unless found to be faulty.

2   The electronic control unit (ECU) is equipped with a self diagnostic system for checking the sensor. To reach the ECU, turn the plastic retaining screw on the passenger side kick panel 90° (see illustration), then remove the panel.

3   Remove the two screws attaching the ECU to the body and remove

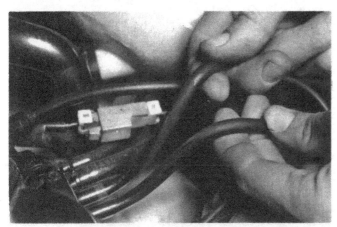

30.2   Check the vacuum hoses for cracks and hardening

30.3   Pull the hose from the component and check for a tight fit — a loose hose will not hold a proper vacuum

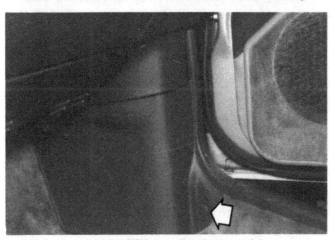

31.2   To get at the ECU, turn the plastic retaining screw 90° and pull out the kick panel

31.3   Pull back the insulation and remove the two retaining screws attaching the ECU to the body

it (see illustration).

4   Start the engine and warm it to normal operating temperature.

5   Run the engine at about 2000 rpm for about two minutes.

6   Verify that the diagnosis mode selector on the ECU is turned fully counterclockwise (see illustration).

7   With the engine running at 2000 rpm, check the green inspection light on the bottom of the ECU. It should blink on and off more than five times during a 10-second period.

8   If the control unit inspection light is operating as described, there is no problem. If it isn't, the system should be checked by a Nissan dealer.

9   Shut off the engine and replace the control unit by reversing the removal procedure.

10 If the exhaust gas sensor needs to be replaced, it may be difficult to remove when the engine is cold, so the operation is best done with the engine at operating temperature.

    a)   Remove the wiring connector from the sensor.

    b)   Apply a penetrating oil to the threads of the exhaust gas sensor and allow it to soak in for 5 minutes.

    c)   Carefully unscrew the sensor and remove it (see illustration). Be careful that you do not damage the threads in the exhaust manifold while removing the sensor.

    d)   Coat the threads of the new sensor with anti-seize compound and install it.

    e)   Carefully tighten the sensor to the specified torque.

    f)   Reconnect the wiring connector to the sensor.

Sensor warning lamp

11 The exhaust gas sensor warning lamp will come on at 30,000 miles, whether the sensor has been replaced during that service interval or not, to let you know that the sensor should be inspected and, if necessary, replaced. **Note:** *After the third inspection at 90,000 miles disconnect the warning lamp wiring harness so that the warning lamp will not come on again.*

12 To gain access to the sensor warning light connector, remove the three screws retaining the driver side under-dash cover and remove the cover.

13 Locate the connector taped to the left of the hood latch release handle and disconnect it (see illustration).

14 Replace the under dash cover.

---

32   Drive train insulators — check

Refer to illustration 32.3

1   When checking the underside of the chassis, inspect the condition of the engine, transmission and rear differential carrier insulators.

2   Inspect the rubber portion of the insulators for deterioration and check for loose mounting bolts.

3   Insert a large screwdriver or pry bar in between each insulator and drive train component. Carefully pry it apart to see if the rubber has separated from the steel base (see illustration).

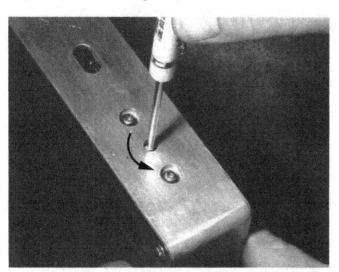

31.6   Make sure that the diagnosis mode selector is turned fully counterclockwise

31.10   Carefully unscrew the exhaust sensor from the exhaust manifold

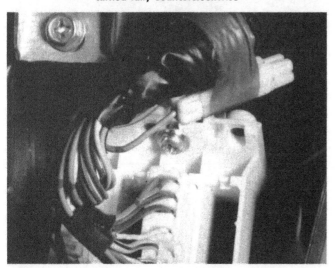

31.13   Disconnect the exhaust gas sensor connector, which is taped to a harness next to the hood latch release handle

32.3   Use a pry bar to determine whether the rubber has separated from its component

33 Automatic transmission fluid and filter change

Refer to illustrations 33.8, 33.9 and 33.11

1   At the specified time intervals, the transmission fluid should be changed and the filter replaced with a new one. Since there is no drain plug, the transmission oil pan must be removed from the bottom of the transmission to drain the fluid.

2   Before draining, purchase the specified transmission fluid (see *Recommended lubricants* near the front of this Chapter) a new filter and all necessary gaskets. **Caution:** *Because of the susceptibility of automatic transmissions to contamination, the old filter and gaskets should not be reused under any circumstances.*

3   Other tools necessary for this job include jackstands to support the vehicle in a raised position, a wrench to remove the oil pan bolts, a standard screwdriver, an 8-pint capacity drain pan, newspapers and clean rags.

4   The fluid should be drained while it's still warm (within 30 minutes after a drive) to help remove any built-up sediment. Wear protective gloves — fluid temperatures can exceed 350 °F in a hot transmission.

5   After the car has been driven to warm up the fluid, raise it and place it on jackstands.

6   Move the equipment you will need under the vehicle. Avoid contact with hot components.

7   Place the drain pan under the transmission oil pan and remove the oil pan bolts along the rear edge and both sides of the pan. Loosen, but do not remove, the front pan bolts.

8   Carefully pry the pan down at the rear, allowing the hot fluid to drain into the drain pan (see illustration). If necessary use a screwdriver to break the gasket seal at the rear of the pan, but do not damage the

pan or the mating surface of the transmission case in the process.

9   Support the pan and remove the remaining bolts at the front of the pan. Lower the pan and pour the remaining fluid into the drain pan. Check the fluid and the bottom of the pan for metal particles that may indicate transmission failure (see illustration).

10 Thoroughly clean the transmission oil pan with solvent. Dry with compressed air if available. It is important that all remaining gasket material be removed from the oil pan mounting flange. Use a gasket scraper or putty knife for this.

11 Remove the bolts holding the filter to the valve body (see illustration) and clean or replace the filter/strainer.

12 Replace the filter/strainer and tighten the bolts to the specified torque.

13 Apply a bead of gasket sealant around the oil pan mounting surface. Apply the sealant between the bolt holes and the inner edge of the pan flange. Press the new gasket into place on the pan, making sure all the bolt holes line up.

14 Lift the pan up to the bottom of the transmission and install the mounting bolts. Working around the pan, snug the bolts in a diagonal fashion, until they're finger tight. Then tighten them to the specified torque.

15 Remove the jackstands and lower the vehicle.

16 Open the hood and remove the dipstick from its guide tube. Insert a funnel into the dipstick tube to avoid spills.

17 It is best to add a little fluid at a time, continually checking the level with the dipstick. Allow the fluid time to drain into the pan. Add fluid until the level just registers on the dipstick. In most cases, a good starting point is 4 to 5 pints.

18 With the selector lever in the Park position, apply the brake and, if possible, start the engine without depressing the accelerator. Do not race the engine at high speed — run it at a low idle for at least two minutes.

19 Depress the brake pedal and shift the transmission through each gear. Place the selector back in Park and, with the engine still at idle, check the level on the dipstick. Look under the vehicle for leaks around the transmission oil pan mating surface.

20 Add more fluid through the dipstick tube until the level on the dipstick is 1/4-inch below the lower mark on the dipstick. Add no more fluid at this time or you may overfill the transmission.

21 Push the dipstick back into the tube and drive the vehicle until it reaches normal operating temperature. Park on a level surface and check the fluid level with the engine idling and the selector in Park. The level should now be at the upper mark on the dipstick. If it isn't, add more fluid to bring it up to this point. Again, do not overfill.

34 Manual transmission — oil change

Refer to illustration 34.3

1   Drive the vehicle for 15 minutes in stop-and-go traffic to warm the oil in the case.

33.8   Lower the back of the pan to start draining the fluid

33.9   A small amount of fine metal shavings indicates normal wear, but the size of the metal chips in this pan warrants further inspection of the transmission

33.11   Remove the bolts holding the filter/strainer to the valve body

2　Raise the vehicle and place it securely on four jackstands. Move a drain pan with at least a 5 pint capacity, rags, newspapers and a 1/2-inch drive ratchet wrench with extension under the car.

3　Insert the 1/2-inch drive socket extension into the drain plug located in the underside of the transmission case (see illustration) and remove the plug. If it's not too hot, you can usually unscrew the drain plug with your fingers after the first couple of turns, but if the plug is still too hot to touch, use the wrench and simply let the plug drop into the pan. You can retrieve it once the oil has cooled. Allow plenty of time for the oil to drain.

4　If the transmission is equipped with a magnetic drain plug, inspect the bits of metal clinging to it. If they are any bigger than dust particles, some of the gear components in the transmission may be excessively worn. If so, they should be inspected as soon as practical. If the transmission is not equipped with a magnetic drain plug, allow the oil in the pan to cool and then feel with your hands along the bottom of the drain pan for any metal bits.

5　After all the lubricant has drained, clean off the drain plug and reinstall it, tightening it to the specified torque.

6　Remove the oil inspection/fill plug located on the left side of the transmission and fill the transmission with a hand pump or syringe with

the specified amount and grade of lubricant until the level is just at the bottom edge of the inspection plug hole.

7　Replace the filler plug and tighten to the specified torque.

8　Remove the jackstands and lower the vehicle.

9　After driving the vehicle, check the drain and filler plugs for any signs of leakage.

35　Differential oil change

Refer to illustration 35.4

1　Drive the vehicle for 10 to 15 minutes at highway speeds to warm up the axle lubricant to operating temperature.

2　Raise the vehicle and support on jackstands and place a drain pan of at least 5 pint capacity under the differential carrier.

3　Wire brush the filler and drain plug areas of the axle carrier cover to prevent possible entry of rust, dirt, etc. into the axle assembly.

4　Remove the differential lubricant drain plug. The drain plug is the lower plug located on the rear of the differential carrier rear housing (see illustration).

5　Allow the lubricant sufficient time to drain completely. Replace the drain plug and tighten to the specified torque.

6　Feel with your hands along the bottom of the drain pan for any metal bits that may have come out with the oil. If you discover any bits coarser than a fine metal sludge, it's an indication of possibly excessive wear of differential components. The internals of the differential will have to be carefully inspected in the near future.

7　Remove the differential inspection/filler plug located above the drain plug. Using a hand pump or syringe, fill the transmission with the specified amount and grade of lubricant until the level is right at the bottom edge of the inspection plug hole.

8　Install the inspection/filler plug and tighten it to the specified torque.

9　Remove the jackstands and lower the vehicle.

36　Seat belts, buckles, retractors, anchors and adjuster — check

Refer to illustrations 36.3 and 36.4

1　Carefully inspect the seat belt webbing. If it is cut, frayed, or damaged, replace the belt assembly.

2　The front seat belt retractors are designed to lock and restrict belt movement when the belt is pulled quickly from the retractors.

3　If the operation of either seat belt is questionable, grasp the shoulder belt and pull sharply forward (see illustration). The retractor should lock instantly, thereby preventing further belt movement. If the retractor does not lock during this check replace it with a new one.

34.3　Use a 1/2-inch drive ratchet with an extension to remove the drain plug (arrow), located in the underside of the transmission case

35.4　The differential drain plug is the lower one

36.3　Check the shoulder harness for correct operation by yanking on it — a quick yank should lock the harness

4   Push the belt tongue into the the belt buckle until an audible click is heard. Grab the two parts of the belt and yank on them (see illustration). The belt should remain locked.

5   Press the release button and note whether the buckle disconnects decisively.

6   Check retractors for smooth operation by pulling each belt out and watching it retract — the retractor should reel it in smoothly and quickly.

7   Check all the anchor bolts for tightness. Tighten them to the specified torque as necessary.

37   Warning lights — general information

Refer to illustration 37.1

1   Your Nissan is equipped with several dash mounted warning lights (see illustration) and a voice warning system to inform you of the operating condition of your vehicle. Many times the warning is something simple, such as a door ajar or the headlights on. The system also warns of vital engine functions such as engine oil pressure loss, low coolant level, brake system failure and low electrical charging rate. When one of these lights comes on, it is a sign that the problem must be dealt with immediately.

Oil pressure warning light

2   The oil pressure warning light indicates that the engine oil pressure is low. Should the light flicker on and off or stay on during normal driving

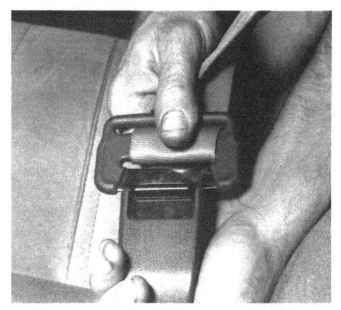

36.4   Buckle the belt and jerk on the buckle assembly to verify its locking ability

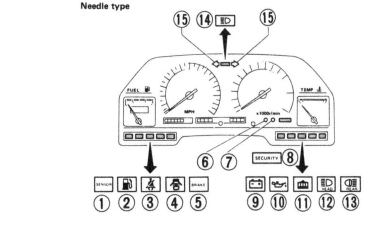

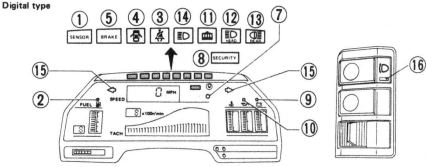

37.1   Instrument warning lamp description

1 Exhaust gas sensor warning light	7 Cruise control pilot light (Green)	12 Headlight warning light
2 Fuel level warning light	8 Theft warning indicator light	13 Tail light warning light
3 Seat belt warning light	9 Charge warning light	14 High beam indicator light
4 Door warning light	10 Oil pressure warning light	15 Turn signal/Hazard indicator lights
5 Brake warning light	11 Coolant level warning light	16 Retractable headlight warning light
6 Cruise control pilot light (Blue)		

speeds you should pull off the road immediately and stop the engine until the cause is found and corrected (Chapter 2). **Caution:** *Continued operation of the engine when the oil pressure warning light is on may damage the engine.*

Retractable headlight warning light

3   This light comes on while the retractable headlights are opening and closing. Should the warning light remain on after the headlights have opened or closed, something is wrong with the retracting mechanism and it should be checked (Chapter 12).

Charge (alternator) warning light

4   The charge warning light indicates an alternator or electrical wiring harness malfunction.
5   This light should glow only when the ignition switch is in the *ON* position (engine off), indicating the bulb itself and the electrical wiring are satisfactory. It should go out when the engine is started.
6   Should the light come on during normal operation, the alternator and electrical system should be checked as soon as possible (Chapter 5). **Note:** *If the alternator and electrical system are functioning normally, but the electrical load is too heavy for the system, the warning light may glow slightly. When this occurs, there is no need to check the alternator and electrical system as long as the light goes out once electrical accessories are turned off.*

Fuel level warning light

7   This light serves as a fuel reserve warning system. When it comes on, 2-1/8 US gallons of fuel remain in the tank. **Note:** *Generally speaking, it's good practice to keep the gas tanks of fuel injected automobiles full, especially in moist climates, because it helps to ward off moisture caused by condensation.*

Exhaust gas sensor warning light

8   This light, which comes on every 30,000 miles, means that the exhaust gas sensor should be inspected. The exhaust gas sensor warning light also comes on when the ignition switch is turned to the *ON* position before the engine starts.

Brake warning light

9   The brake warning light serves both the foot brake system and the parking brake.
10   The light glows when the ignition switch is turned to *ON* before the engine is started.
11   Should the light fail to come on when the ignition switch is turned to *ON*, check the electrical system for a burned-out bulb or look for an open circuit in the wiring for the brake light system (Chapter 12).
12   The light will continue to glow when the parking brake is set with the engine running.
13   If the warning light comes on while you are driving, it's an indication that the brake fluid level is lower than the prescribed level. Pull over and check the brake fluid level immediately. If the reservoir is low, fill it to the specified level.
14   **Warning:** *If these checks cannot be made immediately, pull off the road and stop carefully. Remember that your stopping distance may be longer and the pedal may go down farther than normal and be more difficult to operate. Test the brakes by carefully starting and stopping the vehicle on the shoulder of the road. If you judge it to be safe, drive carefully to the nearest service station for repairs. Otherwise, have your car towed. Driving the vehicle with the brake warning light on may be dangerous.*

Seat belt warning light and chime

15   The driver's seat is equipped with a seat belt warning light and chime system. When the ignition switch is turned to the *ON* position, the seat belt warning light will come on and remain on for about six seconds. The chime will also sound for about six seconds any time the ignition switch is turned to *ON* but it will be deactivated sooner if the driver's seat belt is securely fastened.

Lights 'ON' warning chime/voice

16   When the key is removed from the ignition and the driver's door is opened, a chime and/or voice warning will alert you if the headlight switch is still turned on. Both warnings cancel as soon as the headlights are turned off.

Key warning chime

17   A warning chime tells the driver that the key is in the ignition. The chime will sound any time the driver's door is open when the ignition key is in the *ACC*, *OFF* or *LOCK* position.

Door warning light

18   The door warning light comes on if one of the doors or the rear hatch is not closed securely when the engine is running. It also glows when the ignition switch is turned *ON* and the engine is not running.

Coolant level warning light

19   This light comes on when the coolant level in the reservoir tank drops below the *MIN* mark on the reservoir.

Headlight warning light

20   This light comes on when the headlight switch is in the *ON* position with the engine off, or if a headlight bulb is burned out.

Tail light warning light

21   This light comes on with the light switch in the *ON* position with the engine off or when a tail light bulb is burned out.

High beam indicator light

22   With the headlights on, the high beam indicator glows whenever the high beams are in use and goes out when the driver switches to low beams. If it fails to come on when the lights are switched to the high beam, check the headlights for a burned out high beam filament. If the headlights are both good, check the high beam indicator bulb itself (Chapter 12).

Turn signal/hazard indicator lights

23   The green indicator light on the dash panel should flash when the exterior turn signal lights flash. If it doesn't, check the flasher module, the wiring and the turn signal lights (Chapter 12).

Theft warning indicator light

24   The theft warning light is indicated by the word *Security* on the dash panel. Once the key is removed from the ignition and all the doors are locked the indicator will come on for 30 seconds.
25   Once the light goes off the theft alarm system remains armed until one of the doors is opened with a key. If the theft warning indicator light fails to come on for any reason, the theft warning system should be checked (Chapter 12).

Cruise control pilot light

26   When the cruise control unit is in operation, the green pilot light on the combination meter panel will glow.
27   Once the desired speed is set by pushing the *Coast* set switch, a blue pilot light will light up on an analog (needle type) dashboard and the set speed will light up in the graphic display on a digital dashboard. If either light fails to come on when the system is in use, inspect the dashboard bulbs and the cruise control wiring (Chapter 12).

Voice warning

28   A voice warning is sounded to warn the driver of an open left or right door, a low fuel level or a parking brake left on when the ignition switch is in the *ON* position. Each warning will continue until the condition is corrected, except the low fuel level warning, which repeats two or three times, then stops.
29   The voice warning will sound once the vehicle is moving over 5 mph if one of the doors is open or if the parking brake is on.
30   A voice warning indicating the lights are still on will sound if the driver's door is opened and the ignition switch is not in the *ON* position.

Chapter 2 Part A V6 engine

Contents

Specifications

Cylinder arrangement	V6
Displacement	2,960 cc (180.62 cu in)
Bore and stroke	87 x 83 mm (3.43 x 3.27 in)
Firing order	1-2-3-4-5-6
Compression ratio	
non-turbo	9.0
turbo	7.8
Cylinder numbers (see illustration 3.3a on page 68)	
right bank	1–3–5
left bank	2–4–6
Valve timing	
intake opens	20-degrees BTDC
intake closes	52-degrees ABDC
intake duration	252-degrees
exhaust opens	62-degrees BBDC
exhaust closes	10-degrees ATDC
exhaust duration	252-degrees

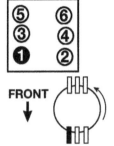

The blackened terminal shown on the distributor cap indicates the Number One spark plug wire position

Cylinder location and distributor rotation

Torque specifications	Kg-m	Ft-lbs
Collector cover	0.6 to 0.8	4.3 to 5.8
Collector	1.8 to 2.2	13 to 16
Throttle chamber	1.8 to 2.2	13 to 16
EGR control valve	1.8 to 2.3	13 to 17
Intake relief valve	3.0 to 4.0	22 to 29
Injector holder	0.25 to 0.33	1.8 to 2.4
Cylinder head temperature sensor	1.2 to 1.6	9 to 12
Thermal transmitter	1.5 to 2.0	11 to 14
Exhaust manifold	1.8 to 2.2	13 to 16
Exhaust manifold stay	2.2 to 2.8	16 to 20
Exhaust outlet	2.5 to 3.0	18 to 22
EGR tube	3.5 to 4.5	25 to 33
Exhaust connecting tube	2.2 to 2.8	16 to 20
Exhaust gas sensor		
1984 thru 1986/1987 and later non-turbo	4.1 to 5.1	30 to 37
1987 and later turbo	1.8 to 2.4	13 to 17
Crankshaft pulley	12.5 to 13.5	90 to 98
Water inlet	1.6 to 2.1	12 to 15
Detonation sensor	2.5 to 3.5	18 to 25
PCV valve	3.0 to 4.0	22 to 29
Distributor bolt	0.5 to 0.63	3.6 to 4.6
Alternator adjusting bar bolt	1.4 to 1.7	10 to 12
Air regulator	0.5 to 0.63	3.6 to 4.6
Rocker cover	0.1 to 0.3	0.7 to 2.2
Tensioner nut	4.4 to 5.9	32 to 43
Belt cover	0.3 to 0.5	2.2 to 3.6
Rocker shaft	1.8 to 2.2	13 to 16
Camshaft pulley	8.0 to 9.0	58 to 65
Camshaft plate	8.0 to 9.0	58 to 65
Water pump	1.6 to 2.1	12 to 15
Drain plug	3.0 to 4.0	22 to 29
Flywheel	10 to 11	72 to 80
Water drain connector	3.5 to 4.5	25 to 33
Water drain plug	2.2 to 2.8	16 to 20
Spark plug	2.0 to 3.0	14 to 22

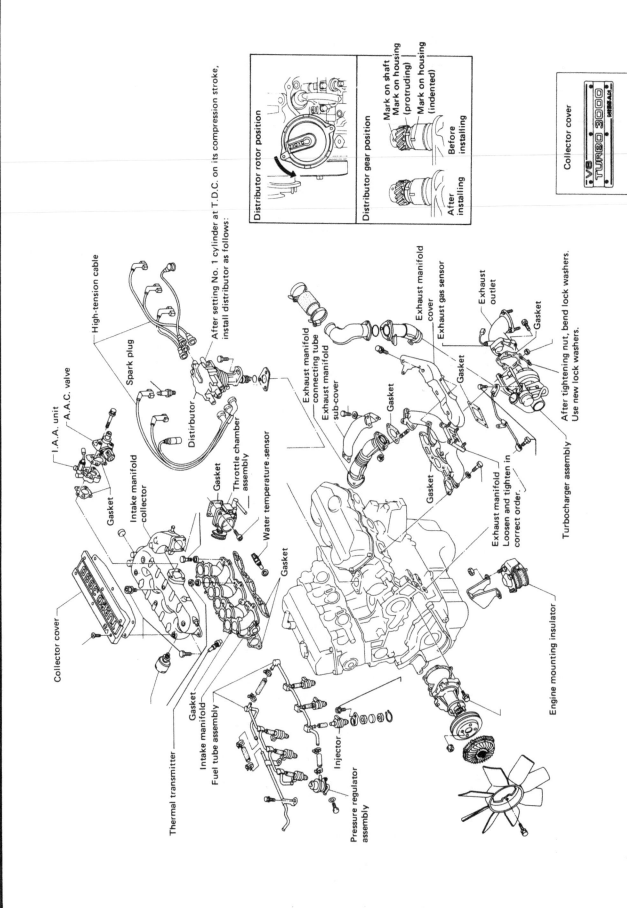

Distributor rotor position

After setting No. 1 cylinder at T.D.C. on its compression stroke, install distributor as follows:

Distributor gear position

Mark on shaft
Mark on housing (protruding)
Mark on housing (indented)

Before installing

After installing

Collector cover

High-tension cable

Spark plug

Distirbutor

I.A.A. unit
A.A.C. valve

Gasket

Intake manifold collector

Gasket

Throttle chamber assembly

Water temperature sensor

Gasket

Exhaust manifold connecting tube
Exhaust manifold sub-cover

Exhaust manifold cover
Exhaust gas sensor

Exhaust outlet

Gasket

Gasket

Gasket

After tightening nut, bend lock washers. Use new lock washers.

Gasket

Exhaust manifold Loosen and tighten in correct order.

Turbocharger assembly

Collector cover

Thermal transmitter

Gasket
Intake manifold
Fuel tube assembly

Injector

Pressure regulator assembly

Engine mounting insulator

1.1a Exploded view of the VG30ET (turbo) engine

Torque specifications

	Kg-m	Ft-lbs
Turbo related parts (1984 thru 1987)		
oil feed tube	1.5 to 2.0	11 to 14
oil return tube	1.0 to 1.2	7 to 9
water inlet and outlet tubes	3.2 to 4.2	23 to 30
turbocharger unit	4.5 to 5.5	33 to 40
Turbo related parts (1988 on)		
oil feed tube	1.6 to 2.4	12 to 17
oil return tube	1.3 to 1.6	9 to 12
water inlet and outlet tubes	2.0 to 3.2	14 to 23
exhaust outlet	2.2 to 3.0	16 to 22
turbocharger unit	2.2 to 3.0	16 to 22
Cylinder head – tighten in five stages		
1st stage	3.0	22
2nd stage	6.0	43
3rd stage	Loosen all bolts	
4th stage	3.0	22
5th stage	5.5 to 6.5	40 to 47
Intake manifold bolt – tighten in two stages		
1st stage	0.3 to 0.5	2.2 to 3.6
2nd stage	1.6 to 2.0	12 to 14
Intake manifold nut – tighten in two stages		
1st stage	0.3 to 0.5	2.2 to 3.6
2nd stage	2.35 to 2.76	17 to 20

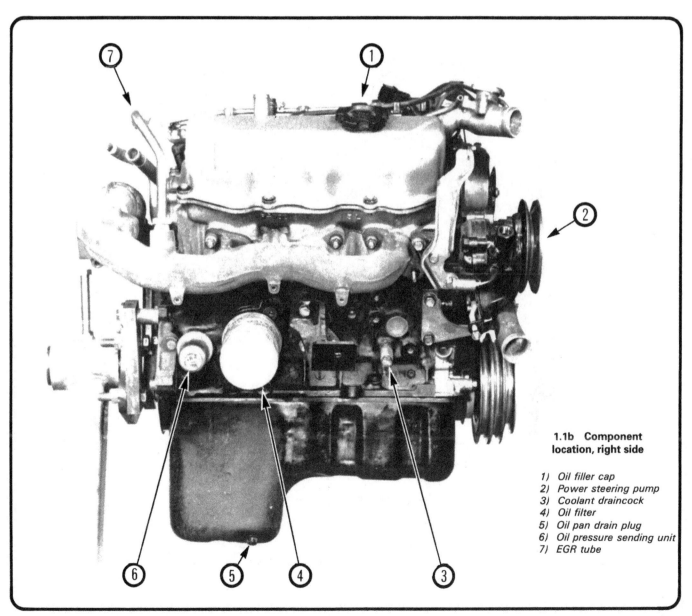

1.1b Component location, right side

1) Oil filler cap
2) Power steering pump
3) Coolant draincock
4) Oil filter
5) Oil pan drain plug
6) Oil pressure sending unit
7) EGR tube

1 General information

Refer to illustrations 1.1a, 1.1b, 1.1c and 1.1d

Your Nissan 300ZX is powered by a fuel injected SOHC V6 engine with a bore and stroke of 87 X 83 mm (3.43 X 3.27 inches). The 2960cc (181 cu in) engine is available in two versions: a 160 horsepower normally aspirated model (VG30E) or an optional 200 hp turbocharged version (VG30ET). Aside from a knock sensor circuit to ward off detonation and a lower compression ratio for the turbo version, the engines are virtually identical.

The 60-degree V6 has a cast iron block and aluminum crossflow heads with a camshaft in each head. The block has thin walled sections for light weight. A ''cradle frame'' main bearing casting — the main bearing caps are cast as a unit, with a bridge, or truss, connecting them — supports the cast ductile iron crankshaft.

Both camshafts are driven off the front of the crankshaft by a cog

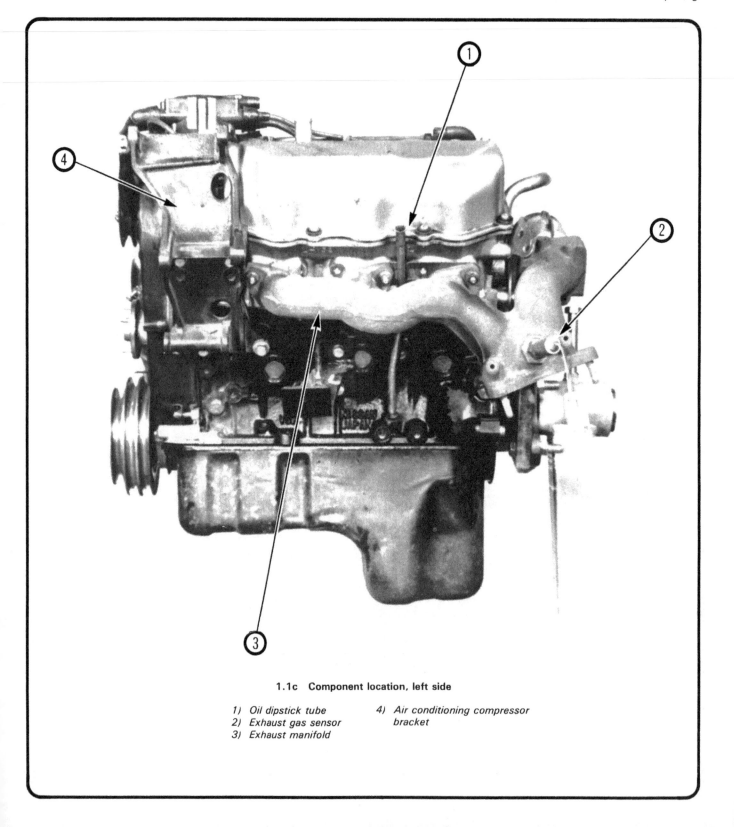

1.1c Component location, left side

1) Oil dipstick tube	4) Air conditioning compressor
2) Exhaust gas sensor	bracket
3) Exhaust manifold	

belt. A spring loaded tensioner, adjusted by an eccentric type locknut, maintains belt tension. Each camshaft actuates two valves per cylinder through hydraulic tappets and shaft-mounted forged aluminum rockers.

Each cast aluminum three-ring piston has two compression rings and a three-piece oil control ring. The piston pins are pressed into forged steel connecting rods. The normally aspirated engine's flat topped pistons deliver a 9.0:1 compression ratio; the turbo's dished pistons produce 7.8:1.

The distributor, which is mounted on the front of the left cylinder head, is driven by a helical gear on the front of the left camshaft. The water pump, which is bolted to the front of the block, is driven by the crankshaft by a drivebelt and pulley. The gear type oil pump is mounted on the front of the crankshaft.

From the oil pump, lubrication travels through the oil filter to the main oil gallery, from which it is routed either directly to the main bearings, crankshaft, connecting rod bearings and piston and cylinder walls or to the cylinder head. Turbocharged engines have an additional circuit for lubricating the turbo bearings. Turbo models equipped with automatic transmissions route the oil through an oil cooler situated between the oil filter and the main oil gallery.

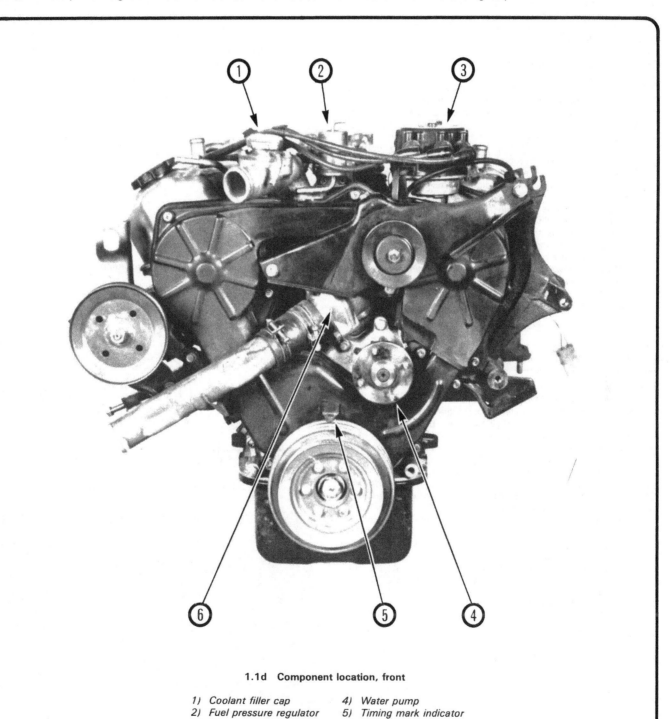

1.1d Component location, front

1) Coolant filler cap
2) Fuel pressure regulator
3) Distributor cap
4) Water pump
5) Timing mark indicator
6) Thermostat housing

2 Repair operations possible with engine in vehicle

The following engine repair operations can be performed with the engine installed in the vehicle:

Removal of the intake collector
Removal of the rocker arm cover
Removal of the rocker arm components
Removal of the lifters
Removal of the intake manifold
Removal of the exhaust manifolds
Removal of the vibration damper
Removal of the engine front oil seals
Removal of the timing belt
Removal of the cylinder heads
Removal of the rear main oil seal
Removal of the engine mounts

Whenever engine work is required, there are some basic steps which the home mechanic should perform before any work is begun. The following preliminary steps will help prevent delays during the operation:

    a) Read through the appropriate Sections in this manual to get an understanding of the procedures involved, the tools that are necessary and the replacement parts which will be needed.

    b) Contact your local dealer or auto parts store to check on the availability and cost of replacement parts. You will often have to decide whether to overhaul, or simply remove and replace, a faulty part. It is generally a good policy to know where to get new or rebuilt components before you start.

    c) If it is necessary to disassemble the A/C system in order to perform one of the procedures outlined in this Chapter, it is essential that a qualified specialist depressurize the system. **Warning:** *Never disconnect an air conditioning system hose while it is still pressurized. Performed improperly, depressurization can cause serious personal injury and damage to the air conditioning system.*

3 Top Dead Center (TDC) positioning

Refer to illustrations 3.3a and 3.3b

1   Top Dead Center (TDC) refers to the uppermost position of a piston in its travel up and down the cylinder on the firing stroke. Positioning the piston at TDC is an essential part of many procedures, such as ignition timing, valve timing and the removal and installation of the distributor and oil pump.

2   In order to bring a piston to its TDC position, the crankshaft must be rotated. Below, you will find several ways to do this. **Warning:** *Before you use any of these methods, put the transmission in the Neutral position and unplug the ignition coil wire from the distributor cap and ground it on the engine to prevent the ignition system from firing.*

    a) The most accurate method is to turn the crankshaft by hand with a wrench on the crankshaft pulley bolt at the front of the engine. When you're facing the engine, crankshaft rotation is in a clockwise direction.

    b) A remote self-starter switch hooked up to the starter motor is a quicker method. Once the piston is close to TDC, use a wrench to set it precisely.

    c) If you have an assistant who can turn the ignition key in short bursts, you can get the piston close to TDC without a remote starter switch. Again, use a wrench to position the piston exactly at TDC.

3   Various methods can be used to determine that the piston is at the TDC position.

    a) To bring the number 1 or 6 piston (see illustration) to TDC, the simplest checking method is to align the timing marks on the crankshaft pulley with the stationary timing indicator on the front of the timing belt cover (see illustration). When the "0" notch in the pulley aligns with the pointer on the timing plate, both the number 1 and 6 are at TDC. To determine which one is on its compression stroke, refer to Step B below.

    b) To verify that a piston is on its compression stroke, number each spark plug wire, then remove all the spark plugs from the engine. Locate the number 1 cylinder spark plug wire and

trace it back to the distributor. Write a number 1 on the distributor body directly below the terminal where the number one spark plug wire attaches to the distributor cap. Repeat this procedure for each of the remaining five cylinders — using numbers 2, 3, 4, etc. — and remove the distributor cap. Each time you establish TDC for a piston, look at the numbers you have written on the distributor body. For example, if you are trying to verify that the number 1 or 6 piston is at TDC when the 0 notch on the crank pulley is aligned with the pointer on the timing belt cover, the distributor rotor should be pointing directly at the number 1 you made on the distributor body. If it isn't, turn the crankshaft one more complete revolution (360°) in a clockwise direction. If the rotor is now pointing at the 1 on the distributor body, then the number 1 piston is at TDC on the compression stroke. This procedure applies to the other cylinders as well.

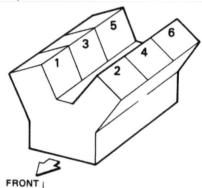

3.3a   Cylinder numbering

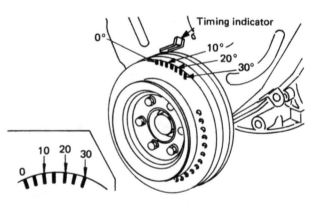

3.3b   Timing marks on the crankshaft pulley

4.3   Remove the six retaining screws on the intake manifold collector cover

4 Intake collector — removal and installation

Refer to illustrations 4.3, 4.4, 4.5, 4.7a, 4.7b, 4.9, 4.16, 4.19 and 4.20

1   Release the fuel pressure (Chapter 1).
2   Disconnect the negative cable at the battery. Place the cable out of the way so it cannot accidentally contact the negative terminal.

3   Remove the six attaching screws holding the intake collector cover to the collector (see illustration).
4   Disconnect the electrical connectors on the left side of the engine (see illustration).
5   Disconnect the electrical connectors on the right side of the engine (see illustration).
6   Disconnect the idle-up solenoid valve connector.

4.4 Label each connector clearly to facilitate installation, then disconnect the components listed for the left side of the engine

1) FICD/idle up
2) Air conditioner
3) Full throttle indicator
4) TVS switch
5) Power brake vacuum line
6) Fuel regulator

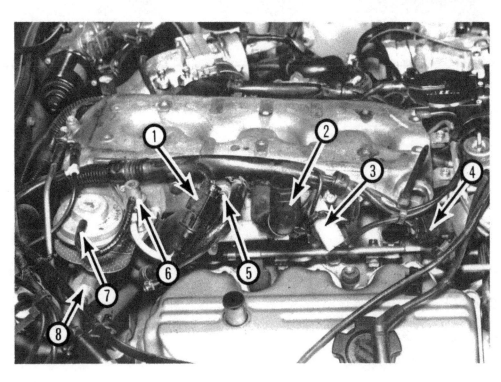

4.5 Label each connector clearly to facilitate installation, then disconnect the components listed for the right side of the engine

1) EGR solenoid
2) AIR regulator
3) Distributor TDC sensor
4) Cylinder head temperature sender
5) Air regulator
6) EGR solenoid
7) EGR valve
8) EGR tube

7   Remove the two heat shield retaining bolts (see illustration), then disconnect the exhaust gas sensor electrical connector (see illustration).

8   Remove the five wiring harness hold-downs — three on the left and two on the right — from the intake collector.

9   To remove the injector electrical connectors, insert a screwdriver behind the metal retaining clip, pry it away from the connector and pull the connector off the injector. Repeat this step for each injector (see illustration).

10 Disconnect the fuel pressure regulator electrical connector.

11 Disconnect the distributor TDC sensor electrical connector.

12 Disconnect the air regulator retaining bolts and remove the air regulator from the collector.

13 Disconnect the EGR tube from the EGR valve attached to the collector.

14 Disconnect the remaining hoses attached to the left side of the intake collector.

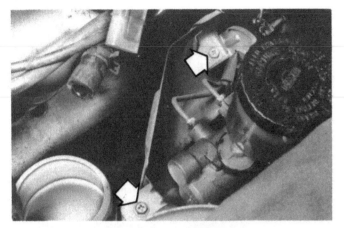

4.7a   To gain access to the oxygen sensor, first remove the two heat shield retaining screws

4.7b   After the heat shield has been removed, unplug the oxygen sensor connector

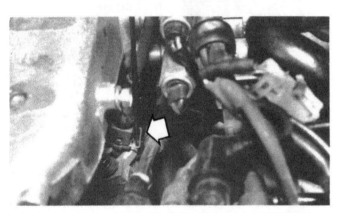

4.9   Disconnect the electrical plug from each fuel injector by inserting a screwdriver between the metal clip and the connector and prying it outward

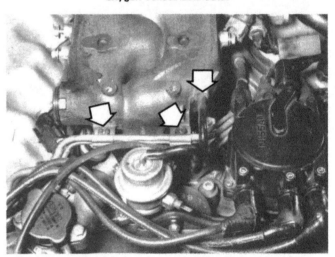

4.16   Remove the vacuum line retaining screws and the vacuum hose connected to the front of the collector (arrows)

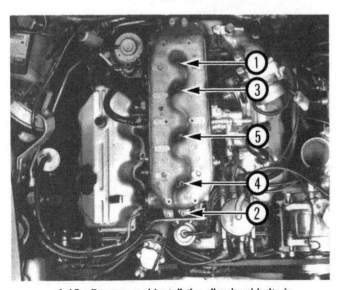

4.19   Remove and install the allen head bolts in the order shown

4.20   Carefully lift the collector and throttle chamber from the intake manifold

15 Disconnect the remaining hoses connected to the right side of the collector and the bolt attaching the EGR solenoid valve bracket to the collector.

16 At the front of the engine, remove the metal vacuum hose by removing the two retaining bolts attached to the intake collector. Disconnect the rubber vacuum hose attached to the front of the collector next to the metal hose bracket (see illustration).

17 Disconnect both vacuum lines and both fuel lines connected to the throttle chamber.

18 Disconnect the power brake vacuum line and the manifold vacuum line below it. Remove the air hose bracket attached to the intake collector.

19 Remove the five allen head attaching bolts that secure the collector to the intake manifold. **Note:** *Loosen the bolts in the numerical order shown and tighten in reverse order of removal (see illustration).*

20 Carefully lift the collector and throttle chamber as an assembly from the intake manifold (see illustration). Do not let anything fall into the intake ports.

21 Cover the intake ports with a clean shop rag to prevent dirt or gasket particles from falling into the intake.

22 Remove all of the old gasket material from the intake and engine block mating surfaces with a gasket scraper.

23 Lay a new gasket on the engine block, remove the shop rag and install the intake manifold.

24 Tighten the intake mounting bolts to the specified torque.

25 Installation of all fasteners and connectors is the reverse of removal.

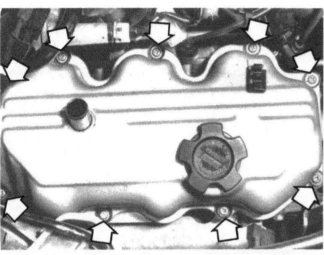

5.5   After removing the plug wires and PCV hose, remove all nine retaining screws

5.8   Remove the throttle cable and cruise control cable hold-down bracket screws and the screw holding the cruise control cable to the intake collector (arrows)

5   Rocker arm cover — removal and installation

Refer to illustrations 5.5, 5.8, and 5.13

1   Relieve the fuel pressure (Chapter 1).

2   Disconnect the negative cable at the battery. Place the cable out of the way so it cannot accidentally come in contact with the negative terminal of the battery.

Right side

3   Remove the PCV hose by sliding the hose clamp back and pulling the hose off the nipple on the rocker arm cover.

4   Remove the number 1, 3 and 5 spark plug wires from the spark plugs. Mark them clearly with either numbers or colored electrical tape to prevent confusion during installation.

5   Remove the nine rocker arm cover screws and washers (see illustration).

6   Lift off the rocker arm cover. **Caution:** *If you find it necessary to pry the cover loose, do so with extreme care. A bent rocker arm cover will not seal properly.*

Left side

7   Remove the throttle chamber air intake tube.

8   Remove both the throttle cable and cruise control cable hold-down bracket screws and the screw holding the cruise control cable to the intake collector (see illustration). Slide the cable ends out of the throttle actuator and set the cable assembly aside.

9   Remove the throttle chamber and intake collector assembly as a unit.

10 Cover the intake ports with a shop rag to prevent dirt from entering the engine.

11 Remove the two distributor cap retaining screws and lift off the distributor cap.

12 Remove the air conditioning compressor drivebelt (Chapter 1).

13 Remove the top two air conditioning compressor bolts (see illustration) and pivot the compressor back to gain access to the rocker arm cover screw.

14 Remove the nine rocker arm cover screws and lift off the rocker arm cover.

16 It is not necessary to use sealant with rubber rocker arm cover gaskets. Place the gasket in the recessed groove in the rocker arm cover and install the cover onto the head.

17 Installation of the covers, components, connectors and cables you disconected or removed is the reverse of removal.

5.13   Remove the two upper air conditioning mounting bolts (arrows), then swing the compressor out of the way to gain access to the front rocker arm cover retaining screw

6.2   To remove the rocker arm assemblies, back off the
retaining bolts in two or three stages to avoid
bending a shaft

6.3   The rocker arm shaft retaining bolts, which are
different lengths, must be installed in the same hole from
which they were removed. A small box with holes in it will
keep the bolts in order

6.4   Lift off the rocker arm shaft and rocker arm assembly

6   Rocker arm components — removal and installation

Refer to illustrations 6.2, 6.3 and 6.4

1   Remove the rocker arm cover (Section 5).
2   Back off the rocker arm shaft retaining bolts (see illustration) in
two or three stages, working your way from the ends toward the middle
of the shafts (see illustration). **Caution:** *Because some of the valves
will be open when you back out the rocker arm shaft retaining bolts,
the rocker arm shafts will be under a certain amount of valve spring
pressure. Therefore, it is imperative that the bolts are loosened gradually.
Loosening a bolt all at once near a rocker arm under spring pressure
could bend a rocker arm shaft.*
3   Remove the rocker arm shaft retaining bolts and set them aside.
The bolts must be kept in the order in which they were removed so
that they can be installed in their original location. A cardboard box
with 12 holes punched in the top can be used to store the bolts in order.
Be sure to label the front, left and right sides of the cardboard box (see
illustration).
4   Lift off the rocker arm shaft and rocker assemblies one at a time
and set them aside (see illustration). Lay them down on a nearby
workbench in the same relation to one another that they have when

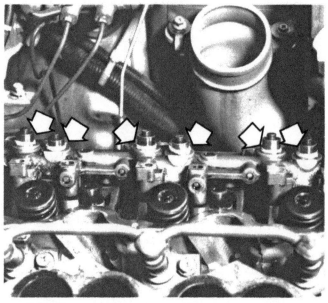

7.2   Wrap each lifter with a rubber band so the lifters
cannot fall from the lifter guides

installed. They must be installed on the same cylinder head from which
they were removed.
5   Installation is the reverse of the removal procedure. Tighten the
rocker arm shaft retaining bolts to the specified torque in a gradual
sequence similar to removal. Work from the ends of the shafts toward
the middle.

7   Lifters — removal and installation

Refer to illustrations 7.2, 7.3 and 7.5

1   Remove the rocker arm cover (Section 5) and the rocker arm shaft
assemblies (Section 6).
2   Secure the lifters by raising them slightly and wrapping a rubber
band around each one to prevent them from falling out of their respective guides (see illustration). **Note:** *If a lifter should fall out of the guide,
immediately restore it to its original guide position.*

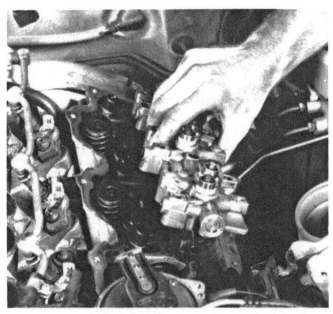

7.3 With the lifters retained by rubber bands, the lifter guide assembly can be removed

8.9 Remove the air conditioner idler pulley by removing three bolts

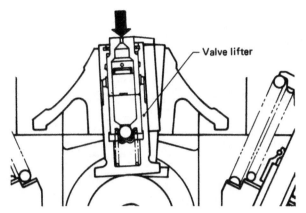

7.5 Depress the valve lifter plunger with your finger to see if it moves

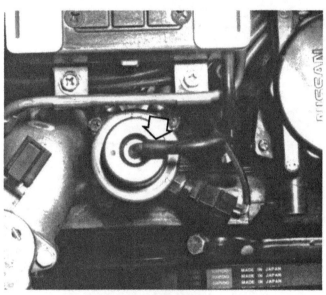

8.13 Disconnect the vacuum hose on top of the fuel pressure regulator

3 Remove the hydraulic lifter guide assembly (see illustration).
4 Remove the lifters one at a time from their bores. Keep them in order. Each lifter must be reinstalled in its original bore. **Caution:** *Try to avoid turning lifters upside down. If you turn a lifter upside down, air can become trapped inside and the lifter will have to be bled.*
5 With the lifter in its bore, push down on the lifter (see illustration). If it moves more than 1mm (0.04 in), air may be trapped inside the lifter.
6 If you think air is trapped inside a valve lifter, re-install the rocker arm and rocker arm cover.
7 Bleed air from the lifter by running the engine at 1,000 rpm under no load for about 10 minutes.
8 Remove the rocker arm cover and rocker arm again. Repeat the procedure in Step 5 once more. If there is still air in the lifter, replace it.
9 While the lifters are removed inspect them for wear. Refer to Chapter 2B for inspection procedures.
10 Upon completion of lifter inspection, place the lifters in their respective bores, install the rocker arm shaft assemblies (Section 6) and the rocker arm covers (Section 5).

8 Intake manifold — removal and installation

Refer to illustrations 8.9, 8.10, 8.13, 8.15, 8.16 and 8.17
1 Release the fuel pressure (Chapter 1).
2 Disconnect the negative cable at the battery. Place the cable out of the way so it cannot accidentally contact the negative terminal.
3 Drain the cooling system (Chapter 1).
4 Remove the throttle chamber intake air pipe.
5 Remove the intake collector with the throttle chamber attached (Section 4).
6 Remove the spark plug wires and distributor cap. Be sure to mark the spark plug wires for proper re-installation.
7 Remove the injector assembly (Chapter 4).
8 Remove the air conditioning compressor drivebelt (Chapter 1).
9 Remove the bolts securing the air conditioning idler pulley bracket (see illustration) and remove the bracket.
10 Loosen the bypass hose clamp and pry the hose from the thermostat housing.
11 Remove the bolt securing the timing belt cover to the thermostat housing.
12 Remove the upper radiator hose from the inlet pipe.
13 Disconnect the vacuum hose on top of the fuel pressure regulator (see illustration).
14 Loosen the screw securing the fuel return hose to the fuel regulator and pry off the hose.

8.15   Disconnect the bolts (arrows) securing the metal heater hose flange to the back of the intake manifold

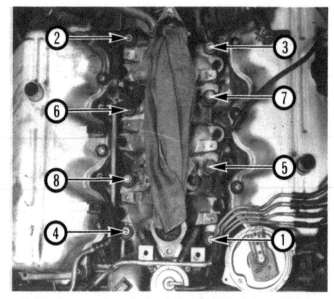

8.16   Remove the eight intake manifold hold-down bolts in the proper sequence

8.17   Lift straight up on the intake manifold to remove it

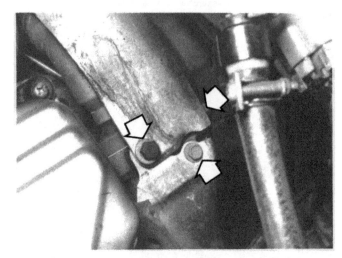

9.6   Remove the retaining bolts (arrows) attaching the heat shield to the crossover pipe

15 Disconnect the metal heater hose flange from the back of the manifold (see illustration).
16 Loosen the eight intake manifold securing bolts in numerical order (see illustration).
17 Lift straight up on the intake manifold (see illustration).
18 Clean all the old gasket material and sealant from the mounting surfaces with a gasket scraper. Don't scratch the mounting surfaces.
19 Install a new intake gasket. No sealant is necessary.
20 Installation of the intake manifold is the reverse of removal.

9   Exhaust manifold — removal and installation

Refer to illustrations 9.6, 9.8, 9.9, 9.11, 9.13, 9.17a and 9.17b
1   Relieve the fuel pressure (Chapter 1).
2   Disconnect the negative cable at the battery. Place the cable out of the way so it cannot accidentally contact the negative terminal.
3   Remove the intake collector and throttle chamber assembly (Section 4).

4   Cover the intake ports with a shop rag to prevent anything from falling into the engine.
5   Raise the vehicle and support it securely on jackstands.
6   Remove the bolts securing the crossover heat shield to the crossover pipe (see illustration).
7   Remove the crossover connecting pipe bolts from the left side of the vehicle.
8   Remove the two upper crossover pipe bolts from the right side of the vehicle (see illustration).
9   From underneath the vehicle, remove the lower bolt from the right side of the crossover connecting pipe (see illustration).

Right side

10 From the right side of the vehicle, remove the heat shield bolts securing the two heat shield halves to the right exhaust manifold.
11 Remove the exhaust manifold retaining bolts in numerical order (see illustration).
12 If your vehicle is equipped with cruise control, disconnect both cruise control bracket bolts and place the cruise control unit to one side to allow removal of the exhaust manifold.
13 Remove the exhaust manifold by lifting up on the rear of the manifold, then lift the manifold out of the engine compartment (see illustration).

9.8 Remove the crossover pipe

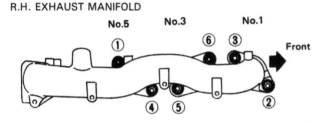

R.H. EXHAUST MANIFOLD

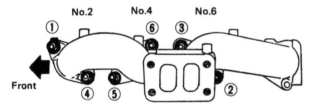

L.H. EXHAUST MANIFOLD

9.11 Loosen the exhaust manifold retaining bolts in the sequence shown

9.17a Remove the three upper bolts securing the manifold to the block

9.9 Remove the crossover connecting bolts from the left and right manifolds underneath

9.13 Raise the rear of the exhaust manifold and lift it out of the engine compartment

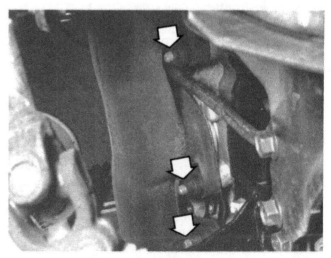

9.17b Remove the three lower bolts securing the manifold to the block

Left side

14 **Note:** *On turbo equipped models it will be necessary to remove the head and exhaust manifold as an assembly (Section 13).*
15 Remove the heat shield bolts securing the heat shield to the left exhaust manifold.
16 Remove the bolts securing the exhaust manifold to the exhaust pipe.

17 Working from above — and underneath, when necessary — remove the exhaust manifold bolts, three on top and three on bottom (see illustrations) in the proper sequence (refer to illustration 9.11). **Note:** *If you are only replacing the left side exhaust manifold gasket on normally aspirated models, removal of the manifold is not necessary. If you are removing the manifold prior to some other service procedure, remove the top two air compressor retaining bolts and the intake air tube to allow the manifold to be lifted from the engine compartment.*

Installation

18 Prior to installation of either manifold, clean all gasket mating surfaces.
19 Install a new gasket onto the studs retaining the manifold to the head. No sealant is necessary.
20 Install the manifold and tighten the nuts in the proper sequence (see illustration) to the specified torque.
21 Installation of the remaining components is the reverse of removal.

10 Vibration damper — removal and installation

Refer to illustrations 10.4 and 10.6

1 Disconnect the negative cable at the battery. Place the cable out of the way so it cannot accidentally contact the negative terminal.
2 Remove the fan shroud and fan assembly (Chapter 3).
3 Remove the drivebelts (Chapter 1).
4 Remove the pulley retaining bolts and remove the pulleys. **Caution:** *Before removing the pulleys, mark their location in relation to the vibration damper with paint in order to retain proper timing marks during installation (see illustration).*
5 Remove the starter bolts and push the starter away from the flywheel gear. Wedge a pry bar between the flywheel/driveplate teeth to prevent the crankshaft from turning while loosening the damper bolt.
6 Loosely install the damper retaining bolt to provide the gear puller with something to push against. Then install a suitable puller onto the damper and pull the damper from the end of the crankshaft (see illustration).
7 To install the damper, position the damper on the nose of the crankshaft, align the keyway and install the retaining bolt.
8 The remaining installation steps are the reverse of the removal procedure.

11 Timing belt — removal, installation and adjustment

Removal and installation

Refer to illustrations 11.13, 11.15, 11.16, 11.20a and 11.20b

1 Relieve the fuel pressure (Chapter 1).
2 Disconnect the negative cable at the battery. Place the cable out of the way so it cannot accidentally contact the negative terminal.
3 Drain the cooling system (Chapter 1).
4 Set the number 1 cylinder at TDC on its compression stroke (Section 3).
5 Remove the intake collector (Section 4).
6 Remove the rocker arm covers (Section 5).
7 Loosen the rocker arm shaft retaining bolts (Section 6).
8 Remove all of the drivebelts (Chapter 1).
9 Remove the fan and fan shroud (Chapter 3).

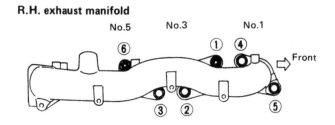

R.H. exhaust manifold

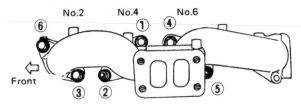

L.H. exhaust manifold

9.20 Tighten the exhaust manifold bolts in the sequence shown

10 Remove the air conditioning compressor idler pulley, if equipped (Chapter 3).
11 Remove the vibration damper (Section 10). **Note:** *Do not allow the crankshaft to rotate during removal of the damper. If the crankshaft moves, the number one piston will no longer be at TDC.*
12 Disconnect the lower radiator hose from the metal connection pipe and loosen the hose clamp at the other end of the metal connection pipe.
13 Remove the two bolts which attach the metal connection pipe to the engine block and remove the metal connection pipe (see illustration).
14 Remove the bolts securing the upper section and the lower part of the timing cover.
15 Confirm that the number 1 piston is still at TDC on its compression stroke by verifying that the timing marks on all three timing belt pulleys are aligned with their respective stationary alignment marks (see illustration). The stationary marks for the two camshaft pulleys are notches on the rear timing belt cover. The stationary mark for the crankshaft pulley is a notch in the oil pump housing. If the marks don't line up, install the crankshaft damper retaining bolt in the crankshaft and use a wrench to turn the engine over until the all timing marks are realigned. Remove the crankshaft bolt.
16 Relieve tension on the timing belt by loosening the bolt in the middle of the timing belt tensioner (see illustration).
17 Remove the timing belt from the timing gears by sliding it forward.

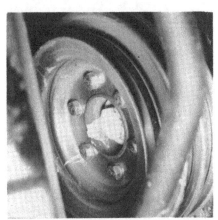

10.4 Remove the pulleys by removing the six retaining bolts, but mark the pulley relationship to the crankshaft vibration damper first to keep the timing marks in their correct alignment

10.6 Install the puller and remove the vibration damper from the end of the crankshaft

11.13 Remove the two bolts (arrows) retaining the metal coolant connection pipe to the engine block and remove the pipe

11.15 Once the timing belt cover is off, check all three timing marks for alignment and realign them if necessary

11.16 Loosen the nut (arrow) in the middle of the timing belt tensioner

Note: *If the belt is cracked, worn or contaminated with oil or coolant, replace it with a new one.*
18 Prepare to install the timing belt by turning the tensioner clockwise with an Allen wrench and temporarily tightening the locking nut.
19 Install the timing belt with the directional arrow pointing forward.
20 Align the factory white lines on the timing belt with the punch mark on each of the camshaft pulleys and the crankshaft pulley. Make sure that all three sets of timing marks are properly aligned (see illustrations). **Note:** *If one of the white lines on the timing belt is a dotted line, align the dotted line with the right-side camshaft pulley punch mark.*

Adjustment
All models
Refer to illustrations 11.21 and 11.22
21 If the tensioner was removed, reinstall it and make sure the spring is positioned properly (see illustration). Keep the tensioner steady with the Allen wrench and loosen the locking nut.
22 Using the Allen wrench, swing the tensioner 70-degrees to 80-degrees in a clockwise direction and temporarily tighten the locking nut **(see illustration)**.
23 Install all of the spark plugs.
24 Slowly turn the crankshaft clockwise two or three full revolutions, then return the number one piston to TDC on the compression stroke. **Caution:** *If resistance is felt while turning the crankshaft, it's an indication that the pistons are coming into contact with the valves. Go back over the procedure to correct the situation before proceeding.*

11.20a Align the white marks on the belt with the punch marks on the timing belt pulley and on the timing cover backing plate to insure correct timing

1984 thru 1987 models
Refer to illustrations 11.26, 11.27 and 11.31
25 Loosen the tensioner locking nut while keeping the tensioner steady with the Allen wrench.
26 Place a 0.0138-inch thick feeler gauge adjacent to the tensioner

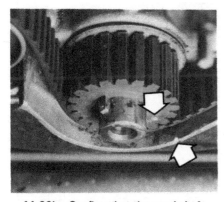

11.20b Confirm that the crankshaft timing pulley mark is aligned with the notch on the oil pump housing before releasing the timing belt tensioner

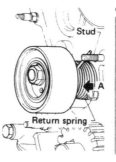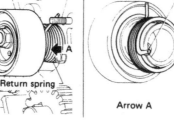

11.21 Belt tensioner spring mounting details (if the stud is removed, use Loctite on the threads during installation)

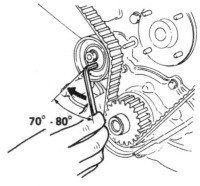

11.22 Use an Allen wrench to turn the tensioner pulley 70° to 80° in a CLOCKWISE direction

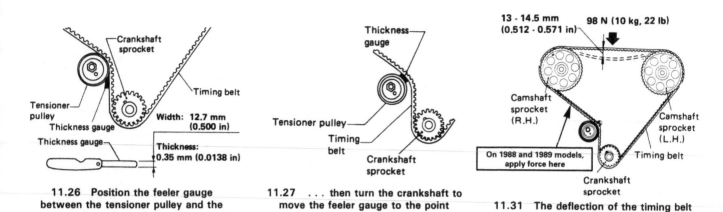

11.26   Position the feeler gauge between the tensioner pulley and the belt as shown here, . . .

11.27   . . . then turn the crankshaft to move the feeler gauge to the point shown here (it must be exact, so work carefully)

11.31   The deflection of the timing belt is checked exactly half-way between the two camshaft pulleys

pulley (see illustration).

27   Slowly turn the crankshaft clockwise until the feeler gauge is between the belt and the tensioner pulley (see illustration).

28   Tighten the tensioner locking nut, keeping the tensioner steady with the Allen wrench.

29   Turn the crankshaft to remove the feeler gauge.

30   Slowly turn the crankshaft two or three revolutions and return the number one piston to TDC.

31   Check the deflection of the timing belt by applying 22 pounds of force midway between the camshaft pulleys (see illustration). The belt should deflect 0.512 to 0.571-inch (13 to 14.5 mm). Readjust the belt if necessary.

32   Install the various components removed during disassembly, referring to the appropriate Sections in this Chapter.

1988 and 1989 models

33   Apply 22 pounds of force to the timing belt midway between the right hand camshaft sprocket and the tensioner (see illustration 11.31).

34   Follow the procedure in Steps 25 through 30 above, then reinstall the timing belt covers and other components removed to gain access to the belt.

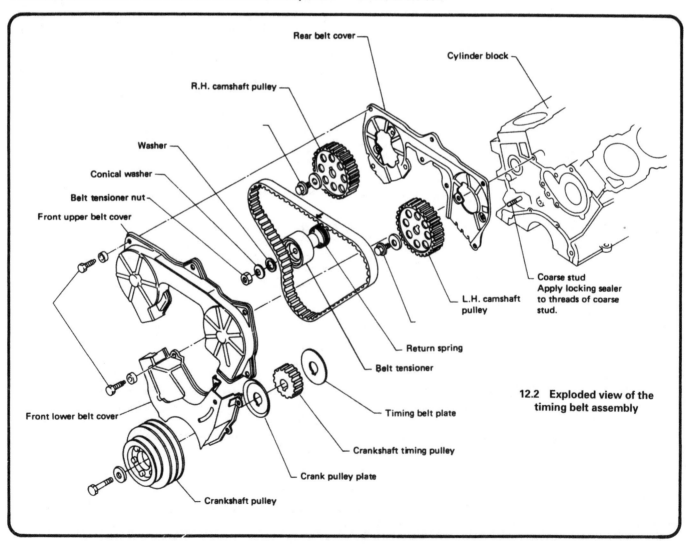

12.2   Exploded view of the timing belt assembly

12.3 Use two screwdrivers to pry the crankshaft timing gear off the end of the crankshaft

12.4 Insert the crankshaft pulley retaining bolt into the crankshaft to protect the threads and give the gear puller something to push against

12.6 Insert a screwdriver between the crankshaft and front oil seal to pry the seal out

12 Engine front oil seals — removal and installation

Refer to illustrations 12.2, 12.3, 12.4, 12.6, 12.7, 12.8a, 12.8b, 12.13, 12.14, 12.15a and 12.15b

Crankshaft oil seal

1 Disconnect the negative cable at the battery. Place the cable out of the way so it cannot accidentally contact the negative terminal.
2 Remove the fan shroud and fan assembly (Chapter 3), drivebelts (Chapter 1), pulleys (Section 10), vibration damper (Section 10) and timing belt (Section 11) (see illustration).
3 Wedge two screwdrivers behind the crankshaft timing pulley (see illustration). Carefully pry the pulley off the crankshaft. Some timing belt pulleys can be pried off easily with screwdrivers. Others are more difficult to extract because corrosion fuses them onto the nose of the crank. If the gear on your vehicle proves difficult to pry off, don't try to get it all the way off with screwdrivers. Instead, slide it just far enough to fasten the tangs of a puller.
4 Once there is enough space between the gear and the oil pump housing to install a small gear puller, thread the timing pulley attaching bolt into the nose of the crankshaft and install a puller. The bolt provides a firm base for the puller to push against and protects the threads (see illustration).
5 Turn the bolt of the puller until the pulley comes off. Remove the timing belt plate.
6 Remove the front oil seal by inserting a screwdriver between the seal and crankshaft. Pry out the old seal (see illustration).
7 Apply a thin coat of assembly lube to the inside of the seal (see illustration).
8 Fabricate a seal installation tool by cutting a short length of pipe of equal or slightly smaller outside diameter than the seal itself. File the end of the pipe that will bear down on the seal until it is free of

sharp edges. You will also need a large washer, slightly larger in O.D. than the pipe, on which the bolt head can seat (see illustration). Install the oil seal by pressing it into position with the seal installation tool (see illustration). When you see and feel the seal stop moving, don't turn the bolt any further or you will damage the seal.
9 Slide the timing belt plate onto the nose of the crankshaft.
10 Insert the Woodruff key into its groove on the crankshaft.
11 Apply a thin coat of assembly lube to the inside of the crankshaft timing pulley and slide it onto the crankshaft.

Camshaft oil seal

12 Repeat Steps 1 and 2.
13 Insert a screwdriver through the top hole in the camshaft pulley

12.7 Apply a thin coat of assembly lube to the inside of the seal before installing it onto the shaft

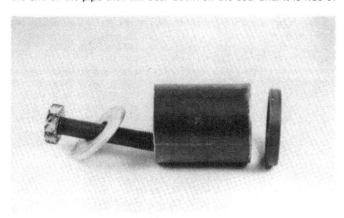

12.8a Fabricate a seal installation tool by using a piece of pipe with the same diameter as the seal and a large washer

12.8b Insert the bolt through the washer and pipe, then screw it into the crankshaft and tighten it, pressing the seal into position

12.13   Insert a screwdriver through the camshaft pulley to hold it while loosening the pulley retaining bolt

12.14   When installing the camshaft timing belt pulleys, note the R and L marks which designate the right and left cylinder heads

12.15a   Insert a screwdriver between the shaft and seal and pry it out

12.15b   The same tool you fabricated for installation of the crankshaft front seal can be used to install the camshaft seals

to lock it into place while turning the retaining bolt (see illustration).
14   Once the bolt is out, the pulley can be removed by hand. **Note:** *Each gear is marked with either an R or L (see illustration). If you are removing both camshaft pulleys, don't mix them up. They must be installed on the same cam from which they were removed.*
15   Remove the old oil seal (see illustration) and press the new seal into place (see illustration) (Step 6, 7 and 8 above). The same seal installation tool you fabricated for the crankshaft seal can be used for both camshaft seals.
16   Insert a screwdriver through the top hole in the camshaft pulley to lock it in place while you tighten the retaining bolt.
17   Installation of the remaining components is the reverse of removal.

---

13   Cylinder heads — removal and installation

Refer to illustrations 13.9, 13.12a, 13.12b, 13.14, 13.18a, 13.18b, 13.19, 13.28, 13.30 and 13.31
1   Relieve the fuel pressure (Chapter 1).
2   Disconnect the negative cable at the battery. Place the cable out of the way so it cannot accidentally contact the negative terminal.
3   Drain the engine coolant (Chapter 1).
4   Remove the intake collector (Section 4).
5   Remove the timing belt (Section 11).
6   Remove the injector assembly (Chapter 4).
7   Remove the intake manifold (Section 8).
8   Remove both cam pulleys (Section 12).

9   Remove the rear timing cover plate by removing the four bolts (see illustration).

Left cylinder head — removal
10   Remove the air conditioning compressor from its bracket and set it aside. It may be helpful to secure the compressor with wire to make sure it stays clear (Chapter 3).
11   Remove the distributor hold-down bolt and pull the distributor out of the engine (Chapter 5).
12   On normally aspirated models, remove the alternator bolts and wire the alternator out of the way (see illustrations).
13   On turbo models, disconnect the turbo from the exhaust manifold and lay it in the engine compartment out of the work area.
14   Remove the bolts attaching the air conditioner/alternator bracket to the engine and remove the bracket (see illustration).
15   Remove the bolt and two nuts attaching the crossover pipe to the left side of the exhaust (Section 9).
16   Remove the nuts from the exhaust manifold on the left side and slide the manifold away from the cylinder head (Section 9).
17   Remove the rocker arm components (Section 6) (see illustration).
18   Loosen and remove the cylinder head bolts in several stages in the proper numerical sequence (see illustration). Head bolts must be installed in their original locations. To keep them from getting mixed up, fabricate an organizer out of cardboard with holes punched in it. Insert the head bolts into the cardboard in the order they are removed. Be sure to mark the cardboard L and R and indicate the front of the engine (see illustration).

13.9 Remove the four retaining bolts and the rear timing cover plate

13.12a Remove the front retaining bolts on the alternator (arrows)

13.12b Remove the rear alternator bolt (arrow) and lower the alternator out of the work area

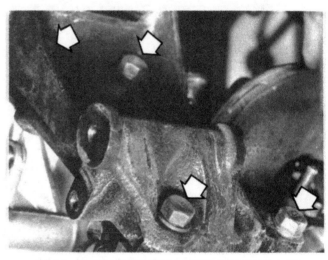

13.14 Remove four bolts (arrows) to remove the air conditioning/alternator bracket on non-turbo models or the air conditioning bracket on the turbo models

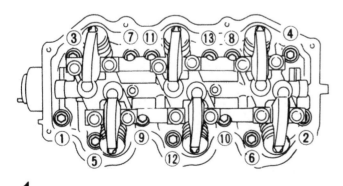

ENGINE FRONT

Loosen in numerical order.

13.18a Loosen the cylinder head bolts in the sequence shown

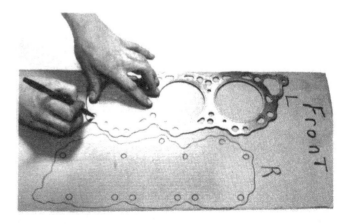

13.18b Fabricate a bolt holder from a piece of cardboard to keep the bolts in order. Be sure to indicate left, right and front on the cardboard

13.19   Remove the small bolt securing the cylinder head (arrow) before attempting to remove the head

13.28   Stuff rags into the cylinders to prevent debris from falling into the cylinders

13.30   After scraping off all old gasket material, place the new gasket on the head, making sure that the gasket seats properly on the alignment dowels (arrows)

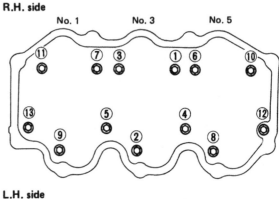

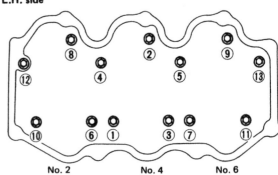

13.31   Cylinder head tightening sequence

19 Remove the small bolt, located at the lower front of the head, securing the head to the block (see illustration).
20 Lift the head off the block. If you find it necessary to pry the head from the block, don't scratch or gouge the mating surfaces.

Right cylinder head — removal

21 After Steps 1 through 9 are completed, remove the power steering pump (Chapter 10).
22 Disconnect the right side exhaust crossover connecting pipe bolts (Section 9).
23 Remove the exhaust manifold (Section 9).
24 Remove the rocker arm components (Section 6).
25 Remove the cylinder head bolts and keep them in order as described in Step 18.
26 Remove the small bolt, located at the rear of the head, which secures the cylinder head to the engine block.
27 Lift the head off the engine. If you find it necessary to pry off the head, avoid damaging the mating surfaces.

Installation

28 Insert shop rags into the cylinders to prevent gasket material from falling into the bores (see illustration).

29 Remove all traces of gasket material from both the cylinder head and engine block mating surfaces with a gasket scraper.
30 Install the new gasket. No sealant is necessary. Use the alignment dowels for proper positioning (see illustration). Remove the shop rags from the cylinders.
31 Using the alignment dowels for proper positioning, install the head. Tighten the head bolts in the proper sequence (see illustration) to the specified torque.
32 Install the lifter/guide assembly (Section 7) and rocker arm shaft assembly (Section 6). Leave the bolts loose to facilitate installation of the timing belt.
33 Installation is the reverse of the removal procedure for the remaining components.

14.3 Insert a screwdriver between the seal and crankshaft then pry it out

14.4 Apply a small amount of grease to the inside of the new seal

14.5 Push the seal onto the crankshaft, then gently tap the seal into its recessed groove in the retainer

14.6 Once the seal is flush with the retainer face, use a blunt punch to carefully drive the seal the rest of the way

14 Rear engine oil seal — replacement

Refer to illustrations 14.3, 14.4, 14.5 and 14.6

1 Remove the transmission (see Chapter 7A for manual transmissions, Chapter 7B for automatics).
2 Remove the flywheel/driveplate (Chapter 8).
3 Insert a screwdriver between the oil seal and crankshaft, then pry the seal out of its retainer (see illustration).
4 Lubricate the inside of the new oil seal with grease to facilitate its installation (see illustration).
5 Push the seal onto the crankshaft and gently tap it into its bore with a soft face hammer (see illustration).
6 Once the seal face is flush with the seal retainer and the crankshaft flange, very carefully drive the seal the rest of the way into the retainer

with the blunt end of a punch. A soft brass punch is best (see illustration). **Caution:** *Be extremely careful. Take your time and drive the seal gently and evenly into place. Damaging a new seal will result in an oil leak.*
7 Intallation is the reverse of removal.

15 Engine mounts — removal and installation

Refer to illustration 15.2

1 Engine mounts should be periodically inspected for hardening or cracking of the rubber, or separation of the rubber from the metal backing.
2 Replace the front mounts by loosening the nuts and bolts retaining

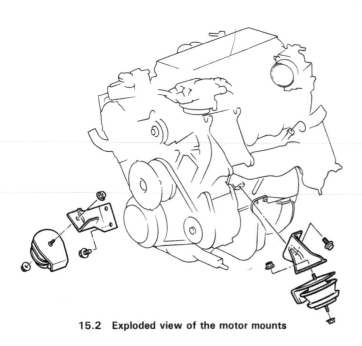

15.2 Exploded view of the motor mounts

17.12 Disconnect the hoses attached to the air compressor by removing the retaining bolts (arrows) on each line

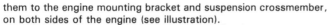

them to the engine mounting bracket and suspension crossmember, on both sides of the engine (see illustration).

3   Take the weight of the engine off the mounts by placing a jack and wooden block under the oil pan. Carefully raise the engine enough to allow removal of the mounts.

4   If you intend to use the old mounts again, mark them clearly to ensure that they are installed, right side up, on the same side as before. Remove the nuts and bolts retaining the mounts and remove the mounts, noting the correct installed position.

5   Installation is the reverse of the removal procedure.

16 Engine removal — methods and precautions

If it has been decided that an engine must be removed for overhaul or major repair work, certain preliminary steps should be taken.

Locating a suitable work area is extremely important. A shop is, of course, the most desirable place to work. Adequate work space, along with storage space for the vehicle, is very important. If a shop or garage is not available, at the very least a flat, level, clean work surface made of concrete or asphalt is required.

Cleaning the engine compartment and engine prior to removal will help keep tools clean and organized.

An engine hoist or A-frame will be necessary. Make sure that the equipment is rated in excess of the combined weight of the engine and its accessories. Safety is of primary importance, considering the potential hazards involved in lifting the engine out of the vehicle.

If the engine is removed by a novice, a helper should be available. Advice and aid from someone more experienced would also be helpful. There are many instances when one person cannot simultaneously perform all of the operations required when lifting the engine out of the vehicle.

Plan the operation ahead of time. Arrange for or obtain all of the tools and equipment you will need prior to beginning the job. Some of the equipment necessary to perform engine removal and installation safely and with relative ease are (in addition to an engine hoist) a heavy duty floor jack, complete sets of wrenches and sockets as described in the front of this manual, wooden blocks and plenty of rags and cleaning solvent for mopping up the inevitable spills. If the hoist is to be rented, make sure that you arrange for it in advance and perform beforehand all of the operations possible without it. This will save you money and time.

Plan for the vehicle to be out of use for a considerable amount of time. A machine shop will be required to perform some of the work which the do-it-yourselfer cannot accomplish due to a lack of special equipment. These shops often have a busy schedule, so it would be

wise to consult them before removing the engine in order to accurately estimate the amount of time required to rebuild or repair components that may need work.

Always use extreme caution when removing and installing the engine. Serious injury can result from careless actions. Plan ahead. Take your time and a job of this nature, although major, can be accomplished successfully.

17 Engine — removal and installation

Refer to illustrations 17.12, 17.14, 17.17, 17.18, 17.21, 17.22, 17.23, 17.24, 17.25, 17.26, 17.31, 17.32, 17.35, 17.36 and 17.37

1   If the vehicle is equipped with air conditioning, take it to a factory dealer or certified air conditioning specialist and have the system evacuated.

2   Relieve the fuel pressure, Chapter 1.

3   Disconnect the negative cable at the battery. Place the cable out of the way so it cannot accidentally contact the negative terminal.

4   Raise the vehicle and support it securely on jackstands.

5   Drain the engine oil (Chapter 1).

6   Drain the cooling system (Chapter 1).

7   Open the hood, remove the bolts securing the stay to the hood, and install the stay in the lower hole provided for the securing bolts. Install the securing bolts.

8   Remove the intake collector (Section 4).

9   Remove the engine drivebelts (Chapter 1).

10 Remove the radiator shroud and fan assembly (Chapter 3).

11 Disconnect the distributor coil wire at the distributor and tuck it out of the way behind the coil.

12 Remove the four bolts retaining the two coolant hoses to the air compressor (see illustration). **Note:** *You should have already had the pressure in the air conditioning system relieved by a professional.*

13 Remove the air compressor (Chapter 3).

14 Disconnect the upper and lower radiator hoses, the bypass hose and the inlet and outlet fuel line hoses (see illustration).

15 Unplug the power steering pump electrical connector (Chapter 10).

16 Disconnect the two hoses connected to the power steering pump.

17 Unbolt the large banjo bolt to remove the high pressure line. The low pressure hose is removed by loosening the hose clamp and pulling the hose from the nipple (see illustration). **Note:** *Place a pan under the pump to catch leaking power steering fluid.*

18 Disconnect the two heater hoses at the rear of the engine by

17.14 Disconnect the engine hoses

1) Upper radiator hose   4) Fuel outlet hose
2) Lower radiator hose   5) Radiator overflow hose
3) Fuel inlet hose

17.18 Disconnect the heater hoses by removing the hose
clamps (arrows) connecting them to the
inlet and outlet pipes

17.17 Disconnect the high pressure hose from the power
steering pump by unbolting the large bolt on the banjo
fitting, then remove the low pressure hose by unscrewing
the hose clamp and pulling the hose from the nipple

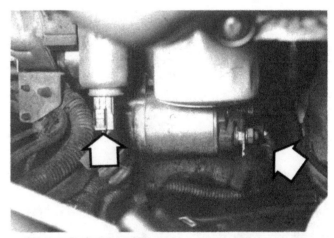

17.21 Disconnect the oil sending unit electrical connector
and both electrical wires attached to the starter (arrows)

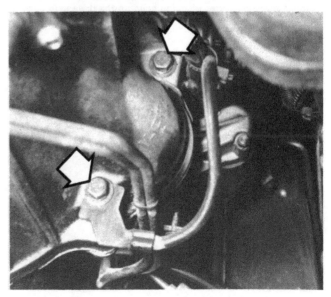

17.22 Remove the two starter retaining bolts (arrows)
and remove the starter

loosening the clamps where the rubber hoses connect to the metal
inlet and outlet tubes (see illustration).
19 Remove the bolts securing the right exhaust crossover pipe to the
exhaust manifold (Section 9).
20 Remove the nuts retaining the right exhaust manifold and remove
it (Section 9).
21 Disconnect the oil sending unit electrical connector and disconnect the electrical wires to the starter (see illustration).
22 From underneath the car, remove the two starter bolts and remove
the starter (see illustration).

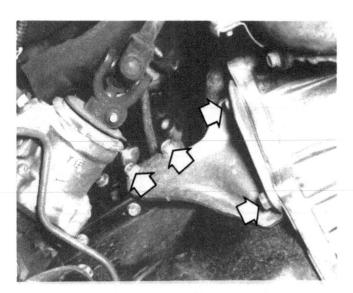

17.23   There are four retaining bolts on the left side
engine-to-transmission bracket (arrows) and three on
the other side

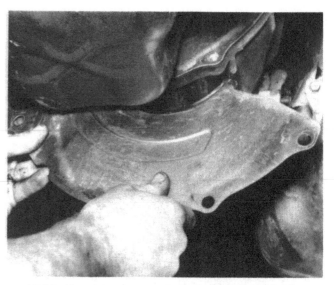

17.24   On automatic transmission equipped models, the
torque converter dust plate can be slipped out from
between the oil pan and the converter after the support
brackets have been removed

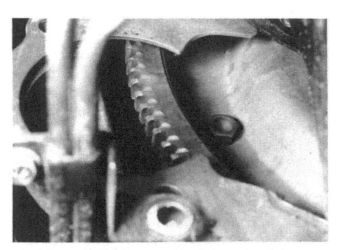

17.25   On automatic transmission equipped models, the
four torque converter retaining bolts on the driveplate can
be removed through the transmission starter tunnel

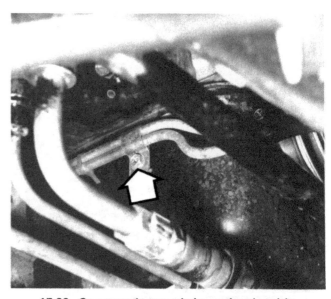

17.26   On automatic transmission equipped models,
remove the two bolts attaching the transmission cooler
lines to the engine (arrow)

23 Remove the brackets attached to the transmission housing and
to the side of the engine block. There are four retaining bolts on the
left side bracket and three bolts on the right side (see illustration).
24 After the brackets have been removed, slide the torque converter
dust plate out. It is between the torque converter and oil pan on
automatic transmision equipped models (see illustration).
25 On automatic transmission equipped models, rotate the crankshaft
and remove the four torque converter bolts located around the periphery
of the driveplate. The bolts can be accessed through the starter housing
opening in the front of the transmission (see illustration).
26 On automatic transmission equipped models, remove both bolts
securing the two transmission fluid cooling lines to the engine (see
illustration).
27 Remove the alternator (Chapter 5).
28 Disconnect the exhaust pipe from the left side exhaust manifold
by removing the three retaining nuts (Section 9).
29 Place a jack under the transmission to support the transmission.

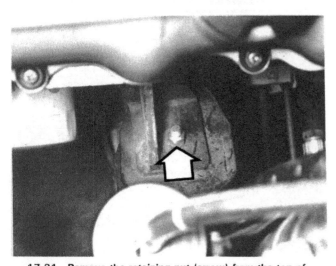

17.31   Remove the retaining nut (arrow) from the top of
each motor mount

17.32   Hook a lifting chain to the lifting brackets located on the right front and left rear of the engine and hook the chain to an engine lifting hoist

17.35   Once the engine is disconnected from the transmission, raise it and roll it away from the engine compartment

17.36   To insure proper balance, mark the flywheel's relation to the crankshaft

17.37   Remove the six flywheel retaining bolts

Place a block of wood between the jack and the transmission to protect the transmission pan.

30   Remove the remaining engine-to-transmission bolts.

31   Remove the motor mount retaining nuts from the top of each motor mount (see illustration).

32   Affix a chain to the two engine lifting brackets. One bracket is located on the right front of the engine and the other at the left rear of the engine (see illustration).

33   Push the engine hoist into position and hook the lifting chain to the hoist. Hook the chain in the middle to allow the engine to be lifted evenly.

34   Raise the engine with the lift enough to take its weight off the engine mounts, then pull it loose from the transmission.

35   Raise the engine straight up until it clears the engine well, then roll it away from the engine compartment (see illustration).

36   Mark the flywheel's relation to the crankshaft so the flywheel can be replaced in its original position (see illustration).

37   With the engine on the hoist, remove the flywheel retaining bolts and remove the flywheel (see illustration).

38   Pull the engine backing plate off the rear of the engine.

39   Install the engine on an engine stand or secure it on a work bench prior to servicing it.

Chapter 2 Part B
General engine overhaul procedures

Contents

Specifications

Head, camshaft and valve train

Head surface warpage limit	
standard	0.05 mm (0.002 in)
maximum	0.1 mm (0.004 in)
Valve stem diameter	
intake	6.965 to 6.980 mm (0.2742 to 0.2748 in)
exhaust	7.965 to 7.970 mm (0.3136 to 0.3138 in)
Valve margin	
intake	1.3 mm (0.051 in)
exhaust	1.5 mm (0.059 in)
Valve margin limit	More than 0.5 mm (0.020 in)
Valve spring free height	
outer	51.2 mm (2.016 in)
inner	44.1 mm (1.736 in)
Valve spring pressure/height	
outer	30.0 mm @ 53.4 kg (1.81 in @ 117.7 lb)
inner	25.0 mm @ 26.0 kg (0.984 in @ 57.3 lb)
Valve spring assembled height	
outer	40.0 mm @ 25.5 kg (1.575 in @ 56.2 lb)
inner	35.0 mm @ 11.0 kg (1.378 in @ 24.3 lb)
Valve spring out of square	
outer	2.2 mm (0.087 in)
inner	1.9 mm (0.075 in)
Hydraulic valve lifter	
lifter outside diameter	15.947 to 15.957 mm (0.6278 to 0.6282 in)
lifter guide inside diameter	16.000 to 16.013 mm (0.6299 to 0.6304 in)
lifter movement limit	0.1 mm (0.04 in)
Valve stem to guide clearance	
intake	
standard	0.020 to 0.053 mm (0.0008 to 0.0021 in)
maximum	0.10 mm (0.004 in)
exhaust	
standard	0.040 to 0.073 mm (0.0016 to 0.0029 in)
maximum	0.20 mm (0.008 in)
Rocker shaft and rocker arm	
shaft outer diameter	17.979 to 18.000 mm (0.7078 to 0.7087 in)
rocker arm inner diameter	18.007 to 18.028 mm (0.7089 to 0.7098 in)
Clearance between rocker arm and shaft	0.007 to 0.049 mm (0.0003 to 0.0019 in)
Intake valve seat	
seat angle	45-degrees
seat width	1.75 mm (0.0689 in)
seat diameter (upper edge)	41.6 to 41.8 mm (1.638 to 1.646 in)
Exhaust valve seat	
seat angle	45-degrees
seat width	1.7 mm (0.067 in)
seat diameter	34.4 to 34.6 mm (1.354 to 1.362 in)

Camshaft
  journal to bearing clearance
    standard .... 0.045 to 0.090 mm (0.0018 to 0.0035 in)
    maximum .... 0.15 mm (0.0059 in)
  inner diameter of camshaft bearing .... 47.00 to 47.025 mm (1.8504 to 1.8514 in)
  outer diameter of camshaft journal .... 46.935 to 46.955 mm (1.8478 to 1.8486 in)
  camshaft runout – total indicator reading .... Less than 0.04 mm (0.0016 in)
  camshaft end play .... 0.03 to 0.06 mm (0.0012 to 0.0024 in)
  cam lobe height .... 39.537 to 39.725 mm (1.5566 to 1.5640 in)
  cam wear limit .... 0.15 mm (0.0059 in)

Cylinder block

Surface flatness
  standard .... less than 0.03 mm (0.0012 in)
  limit .... 0.10 mm (0.0039 in)
Cylinder bore
  inner diameter – standard
    grade no. 1 .... 87.000 to 87.010 mm (3.4252 to 3.4256 in)
    grade no. 2 .... 87.010 to 87.020 mm (3.4256 to 3.4260 in)
    grade no. 3 .... 87.020 to 87.030 mm (3.4260 to 3.4264 in)
    grade no. 4 .... 87.030 to 87.040 mm (3.4264 to 3.4268 in)
    grade no. 5 .... 87.040 to 87.050 mm (3.4268 to 3.4272 in)
  wear limit .... 0.20 mm (0.0079 in)
  out-of-round .... less than 0.015 mm (0.0006 in)
  taper .... less than 0.015 mm (0.0006 in)
Main journal inner diameter
  grade no. 0 .... 66.645 to 66.654 mm (2.6238 to 2.6242 in)
  grade no. 1 .... 66.654 to 66.663 mm (2.6242 to 2.6245 in)
  grade no. 2 .... 66.663 to 66.672 mm (2.6245 to 2.6249 in)

Piston and rings

Piston skirt diameter
  grade no. 1 .... 86.965 to 86.975 mm (3.4238 to 3.4242 in)
  grade no. 2 .... 86.975 to 86.985 mm (3.4242 to 3.4246 in)
  grade no. 3 .... 86.985 to 86.995 mm (3.4246 to 3.4250 in)
  grade no. 4 .... 86.995 to 87.005 mm (3.4250 to 3.4254 in)
  grade no. 5 .... 87.005 to 87.015 mm (3.4254 to 3.4258 in)
Piston clearance to cylinder block .... 0.025 to 0.045 mm (0.0010 to 0.0018 in)
Piston rings
  side clearance
    top
      standard .... 0.040 to 0.073 mm (0.0016 to 0.0029 in)
      maximum .... 0.10 mm (0.004 in)
    2nd
      standard .... 0.030 to 0.063 mm (0.0012 to 0.0025 in)
      maximum .... 1.0 mm (0.04 in)
    oil rail
      standard .... 0.015 to 0.190 mm (0.0006 to 0.0075 in)
      maximum .... 1.0 mm (0.04 in)
  end gap
    top
      standard .... 0.21 to 0.44 mm (0.008 to 0.017 in) (1988 and later turbo = 0.008 to 0.012 in)
      maximum .... 1.0 mm (0.040 in)
    2nd
      standard .... 0.18 to 0.44 mm (0.007 to 0.017 in)
      maximum .... 1.0 mm (0.040 in)
    oil rail
      standard .... 0.20 to 0.76 mm (0.008 to 0.030 in)
      maximum .... 1.0 mm (0.040 in)

Crankshaft

Main journal diameter
  grade no. 0 .... 62.967 to 62.975 mm (2.4790 to 2.4793 in)
  grade no. 1 .... 62.959 to 62.967 mm (2.4787 to 2.4790 in)
  grade no. 2 .... 62.951 to 62.959 mm (2.4784 to 2.4787 in)
Rod journal diameter .... 49.961 to 49.974 mm (1.9670 to 1.9675 in)
Out-of-round .... less than 0.005 mm (0.0002 in)
End play
  standard .... 0.05 to 0.17 mm (0.0020 to 0.0067 in)
  maximum .... 0.30 mm (0.0118 in)
Main bearing clearance
  standard .... 0.028 to 0.055 mm (0.0011 to 0.0022 in)
  maximum .... 0.090 mm (0.0035 in)

Crankshaft (continued)

Rod bearing clearance
standard .... 0.010 to 0.052 mm (0.0004 to 0.0020 in)
maximum .... 0.090 mm (0.0035 in)

Torque specifications

	Kg-m	Ft-lbs
Oil pan		
1984 thru 1987	0.5 to 0.7	3.6 to 5.1
1988 on	0.7 to 0.8	5.1 to 5.8
Oil pump regulator valve	4.0 to 5.0	29 to 36
Oil pump mounting bolts		
6 mm	0.6 to 0.7	4.3 to 5.1
8 mm	1.2 to 1.6	9 to 12
Oil strainer	1.6 to 2.1	12 to 15
Rear oil seal retainer	0.6 to 0.7	4.3 to 5.1
Connecting rod	4.5 to 5.5	33 to 40
Main bearing cap	9.2 to 10.2	67 to 74
Rocker cover	0.1 to 0.3	0.7 to 2.2
Rocker shaft	1.8 to 2.2	13 to 16
Camshaft plate	8.0 to 9.0	58 to 65
Flywheel	10 to 11	72 to 80
Cylinder head – tighten in five stages		
1st stage	3.0	22
2nd stage	6.0	43
3rd stage	Loosen all bolts	
4th stage	3.0	22
5th stage	5.5 to 6.5	40 to 47
Intake manifold bolt – tighten in two stages		
1st stage	0.3 to 0.5	2.2 to 3.6
2nd stage	1.6 to 2.0	12 to14
Intake manifold nut – tighten in two stages		
1st stage	0.3 to 0.5	2.2 to 3.6
2nd stage	24 to 27	17 to 20

1 General information

Included in this portion of Chapter 2 are the general overhaul procedures for the cylinder head and internal engine components. The information ranges from advice concerning preparation for an overhaul and the purchase of replacement parts to detailed, step-by-step procedures covering removal and installation of internal engine components and the inspection of parts.

The following Sections have been written based on the assumption that the engine has been removed from the vehicle. For information concerning in-vehicle engine repair, as well as removal and installation of the external components necessary for the overhaul, see Part A of Chapter 2 and Section 2 of this Part.

The specifications included here in Part B are only those necessary for the inspection and overhaul procedures which follow. Refer to Part A for additional specifications related to this engine.

2 Engine overhaul — general information

Because a number of factors must be taken into consideration, it is not always easy to determine when, or if, an engine should be completely overhauled.

High mileage is not necessarily an indication that an overhaul is needed, while low mileage does not preclude the need for an overhaul. Frequency of servicing is probably the most important consideration. An engine that has had regular and frequent oil and filter changes, as well as other required maintenance, will most likely give many thousands of miles of reliable service. Conversely, a neglected engine may require an overhaul very early in its life.

Excessive oil consumption is an indication that piston rings and/or valve guides are in need of attention. Make sure that oil leaks are not responsible before deciding that the rings and guides are bad. Perform a cylinder compression test (Chapter 1) or have a leak-down test performed by an experienced mechanic to determine for certain the extent of the work required.

If the engine is making obvious knocking or rumbling noises, the connecting rod and/or main bearings are probably at fault. Check the oil pressure with a gauge installed in place of the oil pressure sending unit and compare it to the Specifications. If it is extremely low, the bearings and/or oil pump are probably worn out.

Loss of power, rough running, excessive valve train noise and high fuel consumption rates may also point to the need for an overhaul, especially if they are all present at the same time. If a complete tune-up does not remedy the situation, major mechanical work is the only solution.

An engine overhaul involves restoring the internal components to the same specifications as those of a new engine. During an overhaul, the piston rings are replaced and the cylinder walls are reconditioned (rebored and/or honed). If a rebore is done, new pistons are required. The main and connecting rod bearings are replaced with new ones and, if necessary, the crankshaft may be reground to restore the journals. Generally, the valves are serviced as well, since they are usually in less-than-perfect condition at this point. While the engine is being overhauled, other components, such as the carburetor/fuel injection system, distributor, starter and alternator can be rebuilt as well. The end result should be a like-new engine that will give many trouble-free miles.

Before beginning the engine overhaul, read through the entire procedure to familiarize yourself with the scope and requirements of the job. Overhauling an engine is not difficult, but it is time consuming. Plan on the vehicle being tied up for a minimum of two weeks, especially if parts must be taken to an automotive machine shop for repair or reconditioning. Check on availability of parts and make sure that any necessary special tools and equipment are obtained in advance. Most work can be done with typical shop tools, although a number of precision measuring tools are required for inspecting parts to determine if they must be replaced. Often an automotive machine shop will handle the inspection of parts and offer advice concerning reconditioning and replacement. **Note:** *Always wait until the engine has been completely disassembled and all components, especially the engine block, have been inspected before deciding what service and repair operations must be performed by an automotive machine shop.* Since the condition of the block will be the major factor to consider when determining whether to overhaul the original engine or buy a rebuilt one, never purchase parts or have machine work done on other components until the block has been thoroughly inspected. As a general rule, time is the primary cost of an overhaul, so it does not pay to install worn or sub-standard parts.

As a final note, to ensure maximum life and minimum trouble from a rebuilt engine, assemble everything with care in a spotlessly clean environment.

3 Engine rebuilding alternatives

The do-it-yourselfer is faced with a number of options when performing an engine overhaul. The decision to replace the engine block, piston/connecting rod assemblies and crankshaft depends on a number of factors, with the number one consideration being the condition of the block. Other considerations are cost, access to machine shop facilities, parts availability, time required to complete the project and experience.

Some of the rebuilding alternatives include:

Individual parts — If the inspection procedures reveal that the engine block and most engine components are in reusable condition, purchasing individual parts may be the most economical alternative. The block, crankshaft and piston/connecting rod assemblies should all be inspected carefully. Even if the block shows little wear, the cylinder bores should receive a finish hone.

Crankshaft kit — This rebuild package usually consists of a reground crankshaft and a matched set of pistons and connecting rods. The pistons will already be installed on the connecting rods. Piston rings and the necessary bearings are included in the kit. These kits are commonly available for standard cylinder bores, as well as for engine blocks which have been bored to a regular oversize.

Short block — A short block consists of an engine block with a crankshaft and piston/connecting rod assemblies already installed. All new bearings are incorporated and all clearances will be correct. Depending on where the short block is purchased, a guarantee may be included. The existing camshaft, valve train components, cylinder head and external parts can be bolted to the short block with little or no machine shop work necessary.

Long block — A long block consists of a short block plus an oil pump, oil pan, cylinder heads, rocker arm covers, camshafts and valve train components, timing sprockets and belt and timing belt cover. All components are installed with new bearings, seals and gaskets incorporated throughout. The installation of manifolds and external parts is all that is necessary. Some form of guarantee is usually included with the purchase.

Give careful thought to which alternative is best for you and discuss the situation with local automotive machine shops, auto parts dealers or your local dealership before ordering or purchasing replacement parts.

4 Engine overhaul — disassembly sequence

Refer to illustrations 4.5a, 4.5b and 4.5c

1   It is much easier to disassemble and work on the engine if it is mounted on an engine stand. These stands can often be rented for a reasonable fee from an equipment rental yard. Before the engine is mounted on a stand, the flywheel/driveplate should be removed from the engine (Chapter 8).

2   If a stand is not available, it is possible to disassemble the engine

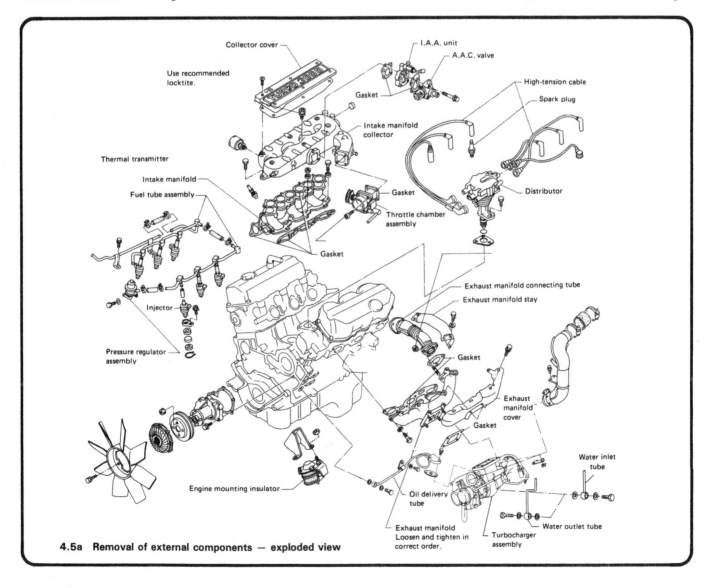

4.5a   Removal of external components — exploded view

with it blocked up on a sturdy workbench or on the floor. Be extra careful not to tip or drop the engine when working without a stand.

3   If you are going to obtain a rebuilt engine, all external components must come off your old engine first in order to be transferred to the replacement engine, just as they will if you are doing a complete engine overhaul yourself. These include:

> Alternator and bracket
> Air conditioning compressor and bracket
> Power steering pump and bracket
> Emission control components
> Distributor, spark plug wires and spark plugs
> Thermostat and housing cover
> Water pump
> Fuel injection components
> Intake/exhaust manifolds
> Oil filter
> Engine mounts
> Flywheel/driveplate

Note: *When removing the external components from the engine, pay close attention to details that may be helpful or important during installation. Note the installed position of gaskets, seals, spacers, pins, washers, bolts and other small items.*

4   If you are obtaining a short block, which consists of the engine block, crankshaft, pistons and connecting rods all assembled, then the cylinder heads, oil pan and oil pump will have to be removed as well. See *Engine rebuilding alternatives* for additional information regarding the various approaches you should consider.

5   If you are planning a complete overhaul, the engine must be disassembled and the components removed in the following order (see illustrations):

> Rocker arm covers
> Timing belt cover
> Timing belt and gears
> Cylinder heads
> Oil pan
> Oil pump
> Piston/connecting rod assemblies
> Crankshaft

6   Before beginning the disassembly and overhaul procedures, make sure the following items are available:

> Common hand tools
> Small cardboard boxes or plastic bags for storing parts
> Gasket scraper
> Ridge reamer
> Vibration damper puller
> Micrometers
> Telescoping gauges
> Dial indicator set

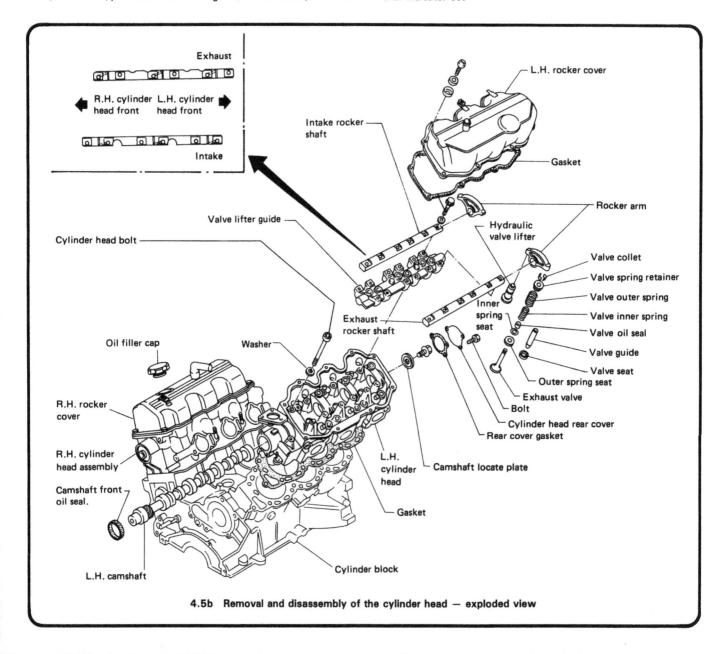

4.5b   Removal and disassembly of the cylinder head — exploded view

Valve spring compressor
Cylinder surfacing hone
Piston ring groove cleaning tool
Electric drill motor
Tap and die set
Wire brushes
Cleaning solvent

5 Cylinder head — disassembly

Refer to illustrations 5.4, 5.5a, 5.5b, 5.6, 5.7, 5.8 and 5.9

1 Remove the cylinder head from the engine (Chapter 2A).
2 Remove the rocker arm shafts (Chapter 2A).
3 Remove the hydraulic valve lifters and lifter guide (Chapter 2A).

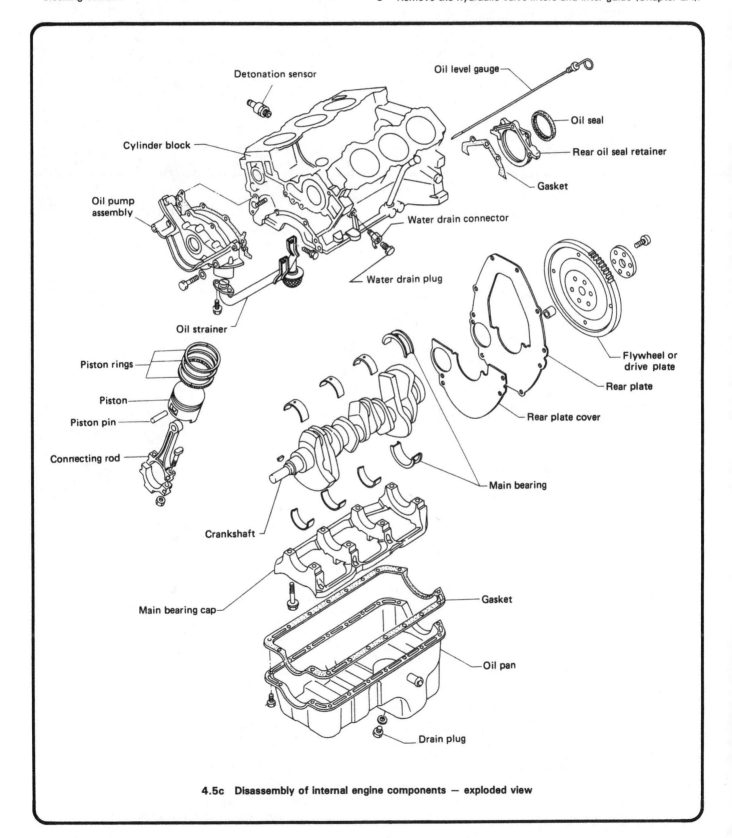

4.5c Disassembly of internal engine components — exploded view

5.4   Remove the three camshaft plate retaining bolts

5.5a   Use the holding lug located behind the number three bearing to hold the camshaft

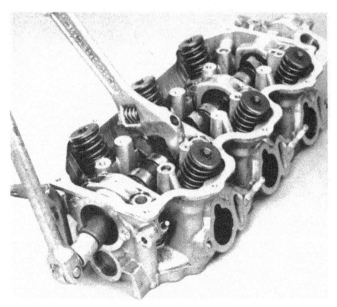

5.5b   While holding the camshaft, break loose the camshaft retaining bolt

5.6   Carefully pry out the camshaft oil seal with a small screwdriver

4   Remove the camshaft plate cover retaining bolts and gently pry it off (see illustration).
5   Use the holding lug on the camshaft (see illustration) to secure the camshaft with a wrench while loosening the camshaft plate retaining bolt (see illustration).
6   Carefully remove the camshaft oil seal at the front of the cylinder head by prying out the seal with a small screwdriver (see illustration).
7   Carefully pull the camshaft out the front of the head (see illustration). **Caution:** *Be careful. Do not scratch the cam lobes or bearing journals.*
8   Install a valve spring compressor, compress the spring, remove the keepers from the valve stem with a small magnet, then remove the retainer and valve springs (see illustration). **Caution:** *Before extracting the valve itself from the valve guide, check for mushrooming on the tip of the stem. Do not attempt to extract a valve with a mushroomed tip through its guide before deburring the tip with a small file.* Repeat this step for each valve.
9   Place the valve stem components for each valve in separate bags and mark each bag so the components can be installed in their original positions (see illustration).

5.7   Pull the camshaft out the front of the head, using both hands to support the camshaft

5.8 Use a valve spring compressor to compress the spring, then remove the keepers from the valve stem with a small magnet

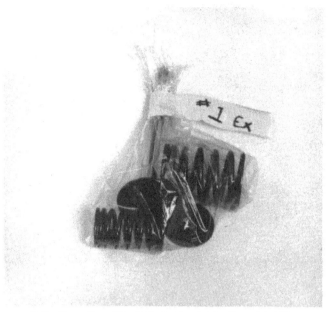

5.9 Keep the valve spring components for each valve separate so each valve assembly can be installed in its original location

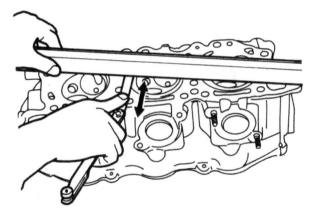

6.12a Lay a straightedge along the head and use a feeler gauge to check for warpage

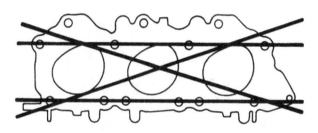

6.12b Make three more checks for head straightness by laying the straightedge at these angles

---

6 Cylinder head — cleaning and inspection

Refer to illustrations 6.12a, 6.12b, 6.14, 6.15, 6.16, 6.17, 6.18, 6.19, 6.23, 6.24 and 6.25

1 Thorough cleaning of the cylinder head and related valve train components, followed by a detailed inspection, will enable you to decide how much valve service work must be done during the engine overhaul.

Cleaning

2 Scrape away all traces of old gasket material and sealing compound from the head, intake manifold and exhaust manifold sealing surfaces with a gasket scraper.
3 Remove any scale build-up around the coolant passages.
4 Run a stiff wire brush through the oil holes to remove any deposits that may have formed in them.
5 Run an appropriate size tap into each of the threaded holes to remove any corrosion and thread sealant that may be present. If compressed air is available, use it to clear the holes of debris produced by this operation.
6 Clean the exhaust and intake manifold stud threads in a similar manner with an appropriate size die.
7 Clean the cylinder head with solvent and dry it thoroughly. Com-

pressed air will speed the drying process and ensure that all holes and recessed areas are clean. **Note:** *Decarbonizing chemicals may prove helpful when cleaning cylinder heads and valve train components. Follow the instructions on the container. Such chemicals are very caustic and should be used with caution. Make sure they're not harmful to aluminium heads.*
8 Clean the rocker arms with solvent and dry them thoroughly. Compressed air will speed the drying process and can be used to clean out the oil passages.
9 Clean all the valve springs, retainers, collets and spring seats with solvent and dry them thoroughly. Clean each valve assembly one at a time to avoid mixing up the parts.
10 Scrape off any heavy deposits that may have formed on the valves, then use a motorized wire brush to remove deposits from the valve heads and stems. Again, make sure the valves do not get mixed up.

Inspection

Cylinder head

11 Inspect the head very carefully for cracks, evidence of coolant leakage or other damage. If cracks are found, a new cylinder head should be obtained.
12 Using a straightedge and feeler gauge, check the head gasket mating surface for warpage at four angles (see illustrations). If any measurement exceeds the specified warpage, have the head resurfaced at an automotive machine shop.
13 Examine the valve seats in each of the combustion chambers. If any of them are pitted, cracked or burned, the head will require valve service that is beyond the scope of the home mechanic.

6.14   A dial indicator can be used to determine valve
stem-to-guide clearance

6.15   Inspect the cam bearing journal surfaces for pitting,
scoring or abnormal wear. If defective, the head must
be replaced

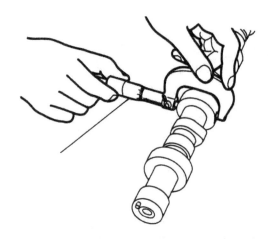

6.16   Measure the camshaft bearing journal diameter and
subtract that figure from the bearing inside diameter to
obtain the camshaft oil clearance

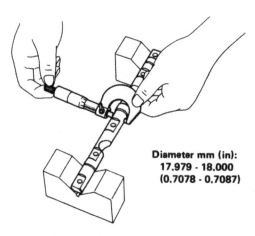

Diameter mm (in):
17.979 - 18.000
(0.7078 - 0.7087)

6.17   Measure the rocker arm shafts for wear

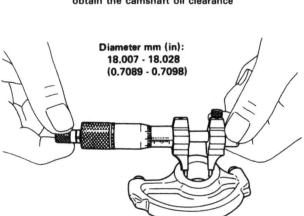

Diameter mm (in):
18.007 - 18.028
(0.7089 - 0.7098)

6.18   An inside micrometer or vernier caliper can be used
to check the rocker arms

6.19   Inspect the rocker arm pad for scuffing or wear and
check that the oil passage is free of any obstructions

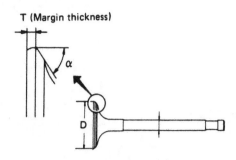

6.23 Measure the valve margin

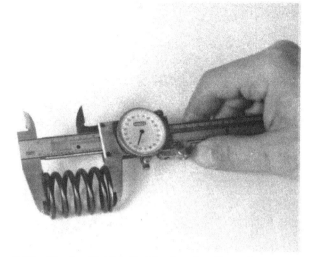

6.24 Measure the free length of each valve spring

14 Use a dial indicator to measure the lateral movement of each valve stem with the valve in the guide and approximately 1/16-inch off the seat. Mount the indicator onto the head, push the valve stem to one side and zero the dial indicator. Push the valve stem to the other side and read the indicator. Because the dial indicator reads the total side play of the valve stem, you must divide your measurement by 2 to get the actual valve stem to guide clearance (see illustration).

15 Check the camshaft bearing surfaces for pitting, scoring or any other abnormal wear (see illustration). If the bearing surfaces are damaged the head will have to be replaced.

16 Measure the inside diameter of the camshaft bearing surfaces and the outside diameter of the camshaft bearing journals. Then subtract each journal diameter from its corresponding bearing surface inside diameter to compute the journal-to-bearing oil clearances (see illustration). Compare your results to the specified clearance.

Rocker arm components

17 Check the valve rocker arms and rocker arm shafts for pits, wear and rough spots. Measure the outside diameter of each rocker shaft, at the journals on which the rocker arms pivot, with a micrometer (see illustration) and compare your measurements to the specified outside diameter for rocker arm shafts.

18 Measure the inside diameter of each rocker arm with an inside micrometer (see illustration) or a dial caliper and compare your measurements to the specified inside diameter for rocker arms.

19 Inspect the lifter pads and the pads at the valve end of the rocker arms for indications of scuffing or abnormal wear and check to make sure the oil passage is clear (see illustration). Any rocker arm with grooved pads must be replaced. Do not attempt to true the pad surface by grinding.

20 All damaged or excessively worn parts must be replaced with new components.

21 Clean all the parts thoroughly. Make sure that all oil passages are open.

Valves

22 Carefully inspect each valve face for cracks, pits and burn spots. Check the valve stem and neck for cracks. Roll the valve stem between the palms of your hands and watch the valve head. If it wobbles, the valve may be bent. Check the end of the stem for pits and excessive wear. The presence of any of these conditions indicates the need for valve service by a shop.

23 Measure the width of the valve margin (see illustration) on each valve and compare it to Specifications. Any valve with a margin narrower than specified must be replaced with a new one.

Valve train components

24 Check each valve spring for pitting and wear on the ends. Measure the length of each spring and compare it to the specified free height (see illustration). A spring that is shorter than specified has sagged and must not be used.

25 Stand each spring upright on a flat surface and check it for squareness (see illustration).

26 Check the spring retainers and keepers for obvious wear and cracks. Extensive damage can occur in the event of failure during engine operation, so any parts in marginal condition must be replaced.

27 If the inspection process indicates that the valve components are

6.25 Check each valve spring for squareness and replace any which have sagged off center

in generally poor condition and worn beyond the limits specified, often the case in an older engine being overhauled, refer to the next Section, *Valves — servicing*, before installing the valve train in the cylinder head.

28 If the inspection turns up no excessively worn parts, and if the valve faces and seats are in good condition, the valve train components can be reinstalled in the cylinder head without major servicing. Refer to Section 8 for cylinder head reassembly procedures.

---

7 Valves — servicing

---

1 Because of the complex nature of the job and the special tools and equipment needed, servicing of the valves, the valve seats and the valve guides — commonly known as a valve job — is best left to an automotive machine shop.

2 The home mechanic can remove and disassemble the head, do the initial cleaning and inspection, then reassemble and deliver the head to a dealer service department or an automotive machine shop for the actual valve servicing.

3 The dealer service department or automotive machine shop, will remove the valves and springs, recondition or replace the valves and valve seats, recondition the valve guides, check and replace the valve springs, spring retainers and keepers (as necessary), replace the valve guide seals, reassemble the valve components and make sure the in-

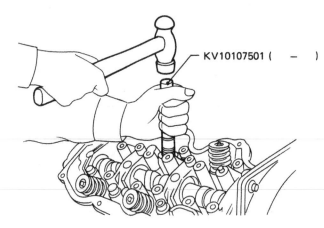

8.3 Installing intake oil seals with a seal installation tool

8.4 Measuring the valve spring installed height without
the spring installed

8.5 Make sure the outer valve spring has its narrow pitch
side (arrow) toward the cylinder head

stalled spring height is correct. The cylinder head gasket surface will also be resurfaced if it is warped.

4   After the valve job has been performed by a shop the head will be in like-new condition. When the head is returned, be sure to clean it again before installation to remove any metal particles and abrasive grit that may still be present from the valve service or head resurfacing operations. Use compressed air, if available, to blow out all the oil holes and passages.

8 Cylinder head — assembly

Refer to illustrations 8.3, 8.4, 8.5, 8.7, 8.10a, 8.10b and 8.10c

1   Regardless of whether or not the head was sent to an automotive repair shop for valve servicing, make sure it is clean before beginning reassembly.

2   If the head was sent out for valve servicing, the valves and related components will already be in place. Begin the reassembly procedure with Step 7.

3   Install new seals on each of the valve guides. To install the intake oil seals, you will need a seal installation tool (Nissan part number KV10107501) or an appropriate size deep socket. Gently tap each seal into place until it is properly seated onto the guide (see illustration). **Caution:** *Do not hammer on the guide seal once it is seated or you may damage the seal. Do not twist or cock the seals during installation or they will not seal properly on the valve stems.*

4   Install the valves, taking care not to damage the new valve guide oil seals. Slip the valve spring shim(s) around the valve guide boss and set the keepers and retainer in place. Measure the installed spring height by lifting up on the retainer until the valve is seated. Measure the distance between the upperside of the shim(s) and the underside of the retainer (see illustration). Compare your measurement to the

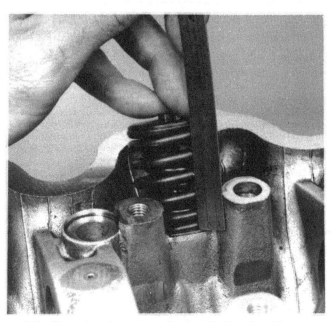

8.7 Measuring the valve spring installed height with the
spring installed

specified installed height. Add shims to bring the height within specifications.

5   Once the correct height is established, remove the keepers and retainer and install the valve springs. **Note:** *The outer spring has a graduated pitch. Install it with the narrow pitch end toward the cylinder head (see illustration).*

6   Compress the springs and retainer with a valve compressor and slip the keepers into place. Release the compressor, making sure the keepers are seated properly in the valve stem upper groove. If necessary, grease can be used to hold the keepers in place as the compressor is released.

7   Double-check the installed valve spring height of each valve. Measure the installed spring height with a small ruler and compare it to the specified installed height (see illustration). If it was correct prior to reassembly, it should still be within the specified limits. If it is not, you must install an additional valve spring seat shim(s) to bring the height within the specified limits.

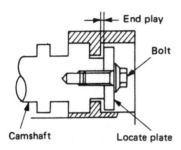

8.10a Cross-sectional view of the camshaft, plate and retaining bolt

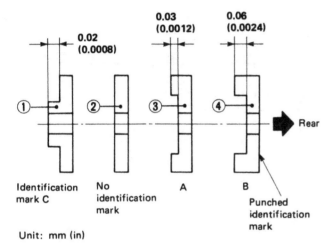

Unit: mm (in)

8.10c If the camshaft end play exceeds the specified limit, select a different cam plate to bring the end play within specification

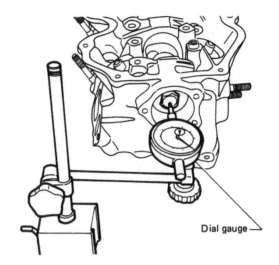

8.10b Measuring camshaft end play with a dial indicator

9.5 Apply RTV sealant to the engine block where the oil pump and rear main oil seal retainer meet the engine block (arrows)

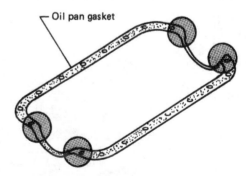

9.6 Apply sealant to the upper and lower corner surfaces of the oil pan gasket before assembly

8 Dip the camshafts in engine oil before installing them in the heads. Carefully insert the camshafts into the heads. Do not allow the surfaces of the camshaft lobes to hit the cylinder head casting or you may damage them.

9 Install the plate and camshaft retaining bolt. Tighten the bolt to the specified torque.

10 Check camshaft end play with the plate installed (see illustrations). Set up a dial indicator touching the plate retaining bolt and push the camshaft toward the rear of the cylinder head and zero the dial indicator. Then pull the camshaft forward. Compare your reading to specified end play. If the end play is excessive, obtain the next size plate at a dealer to bring end play within the specified limit (see illustration).

11 Install the cylinder heads, timing pulleys and timing belt (Chapter 2A).

12 Install the hydraulic valve lifters and lifter guide assembly onto each head. **Note:** *Install the valve lifters in their original lifter bores. Hold them in place with rubber bands during lifter guide assembly installation (Chapter 2A).*

13 Remove the rubber bands holding the lifters. Install the rocker shafts with the rocker arms attached and tighten the rocker shafts in two or three stages to the specified torque (Chapter 2A).

9 Oil pan – removal and installation

Refer to illustrations 9.5, 9.6 and 9.7

1 With the engine mounted on an engine stand, drain the oil.

2 Rotate the engine on the stand until the oil pan is facing upward. Remove all the oil pan retaining bolts.

3 It may be necessary to pry the pan from the block with a screwdriver. **Caution:** *Do not damage the gasket mounting surface or bend the oil pan.*

4 Use a small scraper to remove any sealant or gasket material from both the gasket mounting surfaces of the oil pan flange and the engine block. Thoroughly clean the oil pan in solvent.

5 Apply a small amount of RTV sealant to the four corners of the engine block gasket mating surface (see illustration).

6 Apply sealant to the upper and lower surfaces of the oil pan gasket itself at the same four corners (see illustration). **Note:** *On 1988 and later models, no oil pan gasket is used. A continuous bead of sealant 5/32-inch wide should be applied to the oil pan flange to form a seal (apply it to the inside of the bolt holes).*

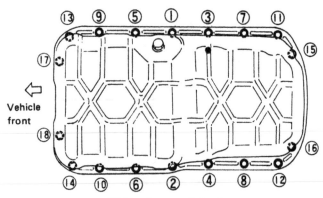

Vehicle
front

Tighten in numerical order.

9.7   Tightening sequence for the oil pan bolts

10.2   Remove the four pickup tube retaining bolts (arrows)
to remove the oil pump pickup tube

10.3   Remove the oil pump to engine retaining bolts
(arrows) to remove the pump from the engine

10.4   Use a block of wood and a hammer to gently break
the oil pump gasket seal

7   Installation of the oil pan is the reverse of the removal procedure.
Tighten oil pan flange bolts to the specified torque in numerical order
(see illustration).

10   Oil pump — removal, inspection and installation

*Refer to illustrations 10.2, 10.3, 10.4, 10.6, 10.7, 10.9, 10.10a,
10.10b and 10.13*

Removal

1   Remove the oil pan (Section 9).
2   Remove the four oil pump pickup tube retaining bolts (see illustration) and remove the oil pump pick up tube.
3   From the front of the engine remove the oil pump-to-engine block retaining bolts (see illustration).
4   Use a block of wood and small hammer to break the oil pump gasket seal (see illustration).
5   Pull forward on the oil pump and remove it from the engine block.
6   Use a small scraper to clean old gasket material and sealant from the oil pump and engine block mating surfaces (see illustration).

10.6   Clean all the old gasket material from the
mating surfaces

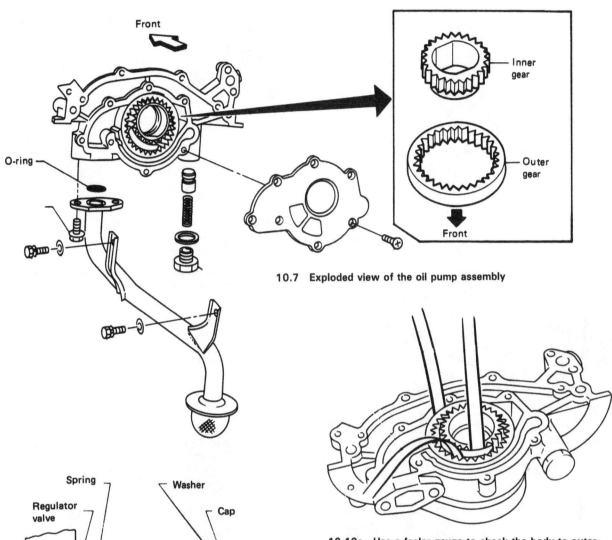

10.7   Exploded view of the oil pump assembly

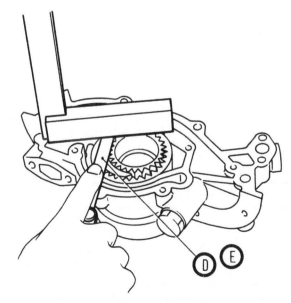

10.9   Check the oil pressure regulator valve sliding surface
and the valve spring for damage and replace if necessary

10.10a   Use a feeler gauge to check the body-to-outer
gear clearance, the clearance between the inner gear
and crescent and the clearance between the outer gear
and crescent

10.10b   Check the housing-to-inner gear clearance and the
housing-to-outer gear clearance

Inspection

7   Use a large Phillips screwdriver to remove the seven retaining
screws holding the front and rear halves of the oil pump together (see
illustration).
8   Thoroughly clean all components in solvent, then inspect them for
wear and damage.
9   Remove the oil pressure regulator cap bolt, washer, spring and
valve. Check the oil pressure regulator valve sliding surface and valve
spring (see illustration). If either the spring or the valve is damaged,
they must be replaced as a set.
10   Check the following clearances with a feeler gauge (see illustrations) and compare your measurements to the specified clearances:
   a) Body to outer gear
   b) Inner gear to crescent
   c) Outer gear to crescent
   d) Housing to inner gear
   e) Housing to outer gear
If any of these clearances are excessive, replace the entire gear set
or the entire oil pump assembly.

10.13   Before installing the oil pickup tube, replace the
rubber O-ring

11.1   A ridge reamer must be used to remove the wear
ridge at the top of each cylinder before the pistons can
be removed

11.3   Check to make sure the rods and caps are identified
before removal

11.6   To prevent damage to the crankshaft journals and
the cylinder walls, slip sections of hose over the rod bolts
before removing the pistons

11   Assemble the oil pump and tighten the retaining screws to the
specified torque. Install the oil pressure regulator valve, spring and
washer, then tighten the oil pressure regulator valve cap to the specified
torque. **Note:** *Pack the pump with petroleum jelly to prime it for startup.*

Installation

12   Apply gasket sealant to the oil pump gasket mounting surface.
13   Use new gaskets on all disassembled parts and reverse the removal
procedures for installation. Tighten all fasteners to the specified torque.
Note: *Before installing the oil pump pick up tube replace the rubber
O-ring (see illustration).*

---

11   Piston/connecting rod assembly — removal

Refer to illustrations 11.1, 11.3, 11.6 and 11.7

1   Using a ridge reamer, completely remove the ridge at the top of
each cylinder (see illustration). Follow the manufacturer's instructions
provided with the ridge reaming tool. **Caution:** *Failure to remove the
ridge before attempting to remove the piston/connecting rod assemblies
can result in broken piston rings.*
2   After all of the cylinder wear ridges have been removed, rotate
the engine until the crankshaft is facing upward.
3   Check the connecting rods and connecting rod caps for identifica-

tion marks (see illustration).
4   If they are not plainly marked, identify each rod and cap, using
a small punch to make the appropriate number of indentations to indicate the cylinders with which they are associated.
5   Loosen each of the connecting rod cap nuts approximately 1/2
turn, then remove all of them. Lift off the rod caps and bearing inserts.
Note: *Do not drop the bearing inserts out of the caps. The bearing inserts must remain paired with their respective rod caps for diagnostic
purposes.*
6   Slip a short length of plastic or rubber hose over each connecting
rod cap bolt to protect the crankshaft journal and cylinder wall during
piston/rod removal (see illustration).
7   Push the connecting rod/piston assembly out through the top of
the engine. Use a wooden tool to push on the upper bearing insert in
the connecting rod (see illustration). If you feel resistance, double-check
to make sure that all of the ridge has been removed from the cylinder.
8   Repeat this procedure for the remaining cylinders. After removal,
reassemble the connecting rod caps and bearing inserts to their respective connecting rods and install the cap nuts finger tight. Leave the
old bearing inserts in place until reassembly to prevent the connecting
rod bearing surfaces from accidental nicks or gouges.

11.7   The wooden handle of a hammer can be used to tap the piston from the cylinder block — be ready to catch the piston as it falls from the block

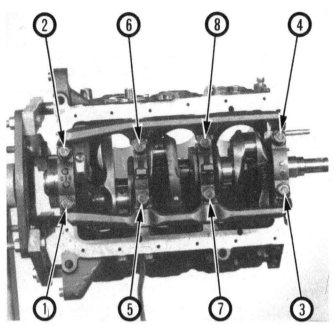

12.6   Loosen the main bearing cap assembly retaining bolts in the sequence shown

13.1a   Use a large punch to drive the freeze plugs into the block

13.1b   Grab the plugs with pliers and pull them out of the block at an angle

12   Crankshaft — removal

Refer to illustration 12.6

1   Remove the oil pan (Section 9).
2   Remove the pick up tube and oil pump (Section 10).
3   Remove the connecting rods and pistons (Section 11).
4   Remove the four oil seal retainer bolts.
5   Remove the oil seal retainer by pulling the retainer from the retaining pins.
6   Loosen each of the main bearing cap assembly retaining bolts 1/4-turn at a time, in the correct numerical order (see illustration) until the bolts can be removed by hand. Check the main bearing cap assembly to see that it is marked with an arrow pointing to the front of the engine. If it is not, mark the main bearing cap assembly with stamping dies or a center punch to insure that it is bolted back on facing the proper direction.

7   Gently tap the cap assembly with a soft face hammer, then separate it from the engine block. If necessary, use the main bearing cap bolts as levers to remove the caps. Do not drop the bearing inserts when lifting off the caps.
8   Carefully lift the crankshaft out of the engine. It is a good idea to have an assistant available, since the crankshaft is quite heavy. With the bearing inserts in place in the engine block and in the main bearing cap assembly, return the cap assembly to the engine block and tighten the bolts finger tight.

13   Engine block — cleaning

Refer to illustrations 13.1a, 13.1b, 13.2 and 13.9

1   Knock the freeze plugs into the block with a hammer and punch (see illustration), then grasp them with large pliers and pull them back through the holes (see illustration).

13.2   Use a gasket scraper to remove all traces of gasket material from the engine block

13.9   Coat the new soft plug with silicone sealant, then use a large socket to drive it into the block, being careful to keep it straight

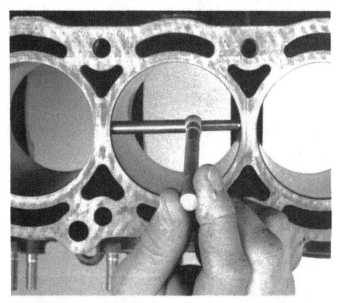

14.4a   A telescoping bore gauge is used to measure cylinder bore diameter

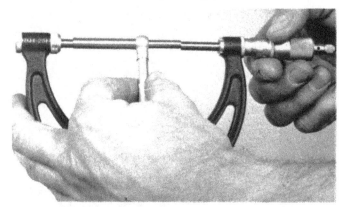

14.4b   After extending the telescoping gauge until it touches the cylinder bore, remove and measure it with a micrometer

2   Using a gasket scraper, remove all traces of gasket material from the engine block (see illustration). Be very careful not to nick or gouge the gasket sealing surfaces.

3   Remove the main bearing cap assembly and separate the bearing inserts from the caps and the engine block. Tag the bearings to indicate the main bearing cap from which they were removed and to show whether they were in the cap or the block. Set them aside.

4   If the engine is extremely dirty, take it to an automotive machine shop to be steam cleaned or hot tanked.

5   After the block is returned, clean all oil holes and oil galleries. Brushes for cleaning oil holes and galleries are available at most auto parts stores. Flush the passages with warm water until the water runs clear, dry the block thoroughly and wipe all machined surfaces with a light, rust-preventative oil. If you have access to compressed air, use it to speed the drying process and to blow out all the oil holes and galleries.

6   If the block is not extremely dirty or sludged up, you can do an adequate cleaning job with warm soapy water and a stiff brush. Take plenty of time and do a thorough job. Regardless of the cleaning method used, be very sure to thoroughly clean all oil holes and galleries, dry the block completely and coat all machined surfaces with light oil.

7   The threaded holes in the block must be clean to ensure accurate torque readings during reassembly. Run the proper size tap into each of the holes to remove any rust, corrosion, thread sealant or sludge

and to restore any damaged threads. If possible, use compressed air to clear the holes of debris produced by this operation. Thoroughly clean the threads on the head bolts and the main bearing cap assembly bolts as well.

8   Reinstall the main bearing cap assembly and snug the bolts finger tight.

9   After coating the sealing surfaces of the new freeze plugs with a good quality gasket sealer, install them in the engine block. Make sure they are driven in straight and seated properly or leakage could result. Special tools are available for this purpose, but equally good results can be obtained with a large socket of an outside diameter slightly smaller than that of the freeze plug (see illustration). Gently tap the plugs into place with a hammer.

10   If the engine is not going to be reassembled right away, cover it with a large plastic trash bag to keep it clean.

---

14   Engine block — inspection

Refer to illustrations 14.4a, 14.4b and 14.7

1   Thoroughly clean the engine block as described in the previous Section.

2   Check the block for cracks, rust and corrosion. Look for stripped threads in bolt holes. It is also a good idea to have the block checked for hidden cracks by an automotive machine shop with the special

14.7 Cylinders should always be honed before installing new rings to give the bore surface the proper finish for proper ring seating

15.2 Measure the outside diameter of each camshaft journal

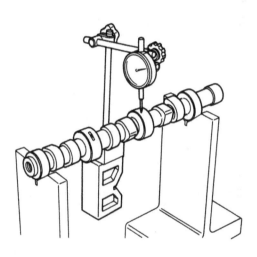

15.3 Measuring cam journal runout

equipment for this type of work. If cracks are found, the block must be repaired or replaced.

3   Check the cylinder bores for scuffing and scoring.

4   Using a cylinder bore gauge, measure the diameter at the top (just under the ridge), center and bottom of each cylinder bore, parallel to the crankshaft axis (see illustrations). Next, measure each cylinder's diameter at the same three locations at a 90 degree angle to the crank axis. Compare your results to the specified cylinder bore diameter. If the cylinder walls are badly scuffed or scored, or if they are out-of-round or tapered beyond the specified limits, have the engine block rebored and honed at an automotive machine shop. If a rebore is done, oversize pistons and rings will be required.

5   If the cylinders are not worn to the limit and if the piston-to-cylinder clearances fall within specification (Section 16), then the cylinders need not be rebored. But they must still be honed.

6   Before honing the cylinders, install the main bearing cap assembly (without the bearings) and tighten the main bearing cap retaining bolts to the specified torque.

7   To perform the honing operation you will need the proper size flexible hone with fine stones, plenty of light oil or honing oil, some clean shop rags and an electric drill motor. Mount the hone in the drill chuck, compress the stones and slip the hone into the cylinder (see illustration). Lubricate the cylinder thoroughly, turn on the drill and move the hone up and down in the cylinder at a pace which will produce a fine cross-hatch pattern on the cylinder walls (with the cross-hatch lines intersecting at approximately a 60° angle). Be sure to use plenty of lubricant. Do not withdraw the hone from the cylinder while it is running. Instead, shut off the drill and continue moving the hone up and down in the cylinder until it comes to a complete stop, then compress the stones and withdraw the hone. Wipe the oil out of the cylinder and repeat the procedure on the remaining cylinders. If you do not have the tools or do not desire to perform the honing operation, most automotive machine shops will do it for a reasonable fee.

8   After the honing job is complete, chamfer the top edges of the cylinder bores with a small file so the rings won't catch when the pistons are installed.

9   The entire engine block must be thoroughly washed again with warm, soapy water to remove all traces of the abrasive grit produced during the honing operation. Be sure to run a brush through all oil holes and galleries and flush them with running water. After rinsing, dry the block and apply a coat of light rust preventative oil to all machined surfaces. Wrap the block in a plastic trash bag to keep it clean and set it aside until reassembly.

15   Camshaft, lifters and bearing surfaces — inspection

Refer to illustrations 15.2, 15.3, 15.4, 15.5, 15.6, 15.7, 15.10 and 15.11

1   Visually check the camshaft bearing surfaces for pitting, scoring or any abnormal wear. If the bearing surfaces are damaged the head will have to be replaced (Section 6).

2   Measure the outside diameter of each camshaft bearing journal and record your measurements (see illustration). Compare your readings to the specified journal outside diameter, then measure the inside diameter of each corresponding camshaft bearing tunnel with a vernier caliper and record those measurements. Compare your readings to the specified camshaft bearing inside diameter. Subtract each cam journal outside diameter from its respective cam bearing tunnel inside diameter to compute the oil clearance for each bearing. Compare your results to the specified journal to bearing clearance. If any of these measurements fail to fall within the standard specified wear limits, either the camshaft or the head, or both, must be replaced.

3   Check camshaft runout by placing the camshaft between two V-blocks and set up a dial indicator on the center journal (see illustration). Zero the dial indicator. Turn the camshaft slowly and note the dial indicator readings. Record your readings and compare them with the specified runout. If the measured runout exceeds the specified runout, replace the camshaft.

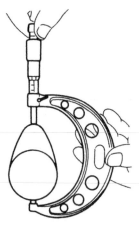

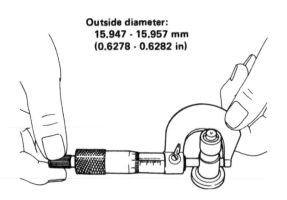

Outside diameter:
15.947 - 15.957 mm
(0.6278 - 0.6282 in)

15.4   Measuring cam lobe height

15.5   Check the contact and sliding surfaces of each lifter (arrows) for wear or scratches

15.6   Measure the outside diameter of each lifter with a micrometer

15.7   Measure the inside diameter of each lifter bore, subtract the valve lifter diameter to find the lifter-to-bore clearance and compare your measurements with the specified clearance

15.10   Measure the rocker arm shaft diameter at each journal where a rocker arm rides on the shaft

4   Check the camshaft cam lobe height for wear by measuring it with a micrometer (see illustration). Compare your measurement to the specified cam lobe height. Then subtract your measured cam lobe height from the specified height to compute wear on the cam nose. Compare this figure to the specified wear limit. If it's greater than the specified wear limit, replace the camshaft. Repeat this procedure for each cam lobe.

5   Inspect the contact and sliding surfaces of each lifter for wear or scratches (see illustration). **Note:** *If the lifter pad is worn, it's a good idea to check the corresponding camshaft lobe, because it will probably be worn too.* **Caution:** *Do not turn the lifters upside down — air can enter and become trapped inside (see Chapter 2A, Section 7).*

6   Measure the outside diameter of each lifter with a micrometer (see illustration) and compare your readings to the specified outside diameter. If any lifter is worn beyond the specified outside diameter, replace it. **Note:** *Lifters can be purchased individually. It's not absolutely necessary to replace one or two damaged lifters with an entire set.*

7   Check each lifter bore diameter in the lifter guide assembly (see illustration) and compare your reading to the specified tolerance. If any lifter bore is worn beyond the specified tolerance, the lifter guide assembly must be replaced.

8   Subtract the outside diameter of each lifter from the inside diameter of its lifter bore and compare the difference to the specified clearance. If both the lifter and the bore are within acceptable limits, this measurement should fall within tolerance as well. However, should you buy a new set of lifters alone, or a lifter guide assembly by itself, you may find that this clearance no longer falls within the specified tolerance.

9   Inspect the rocker arms and rocker shafts for abnormal wear and scratches.

10   Measure the outside diameter of the rocker arm shaft at each rocker arm journal with a micrometer (see illustration). Compare your measurements to the specified rocker arm shaft outside diameter.

11   Measure the inside diameter of each rocker arm with either an inside micrometer or a dial caliper (see illustration). Compare your measurements to the specified rocker arm inside diameter.

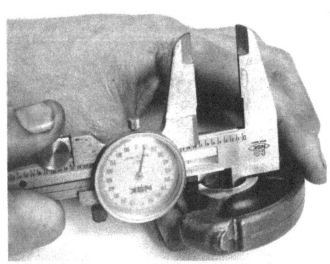

15.11   Measure the inside diameter of each rocker arm, subtract the corresponding rocker arm shaft diameter to get the oil clearance and compare your results to the specified clearance

16.4a If available, a ring groove cleaning tool can be used to remove carbon from the ring grooves

12 Subtract the outside diameter of each rocker arm shaft journal outside diameter from its corresponding rocker arm inside diameter to compute the clearance between the rocker arm shaft and the rocker arm. Compare these measurements to the specified clearance. If any of them fall outside the specified clearance, replace either the rockers or the rocker shaft, or both.

16 **Piston/connecting rod assembly — inspection**

Refer to illustrations 16.4a, 16.4b, 16.10 and 16.11

1 Using a piston ring installation tool, carefully remove the rings from the pistons. Do not nick or gouge the pistons in the process.
2 Scrape all traces of carbon from the crown (top) of the piston. A hand-held wire brush or a piece of fine emery cloth can be used once the majority of the deposits have been scraped away.
3 **Caution:** *Do not, under any circumstances, use a wire brush mounted in a drill motor to remove deposits from the pistons. The piston material is soft and will be eroded away by the wire brush.*
4 Use a piston ring groove cleaning tool to remove any carbon deposits from the ring grooves (see illustration). If a tool is not available, a piece broken off the old ring will do the job (see illustration). **Caution:** *Do not remove any metal and do not nick or scratch the sides of the ring grooves.*
5 Once all carbon deposits have been removed from the ring grooves, clean the piston/rod assemblies with solvent and dry them thoroughly.
6 If the pistons are not damaged or worn excessively, and if the engine block isn't rebored, new pistons won't be necessary. Normal piston wear appears as even vertical wear lines on the piston thrust surfaces and slight looseness of the top ring in its groove.
7 Carefully inspect each piston for cracks around the skirt, at the pin bosses and at the ring lands.
8 Look for scoring and scuffing on the thrust faces of the skirt, holes in the piston crown and burned areas at the edge of the crown. If the skirt is scored or scuffed, the engine may have been suffering from overheating and/or abnormal combustion, either of which may have caused inordinately high operating temperatures. The cooling and lubrication systems should be thoroughly inspected. A hole in the piston crown is an indication that preignition was occurring. Burned areas at the edge of the piston crown are usually evidence of detonation (spark knock). If any of the above problems exist, the causes must be corrected or the damage will occur again.
9 Corrosion of the piston (evidenced by pitting) indicates that coolant is leaking into the combustion chamber and/or the crankcase. Again, the cause must be corrected or the problem may persist in the rebuilt engine.
10 Measure the piston ring side clearance by laying a new piston ring in each ring groove and slipping a feeler gauge in between the ring

16.4b Use a piece of broken ring to remove carbon deposits from the ring grooves, but don't chip any piston material with it

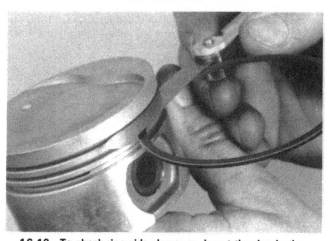

16.10 To check ring side clearance, insert the ring in the groove then use a feeler gauge to measure the remaining gap

16.11 Measure the diameter of each piston at 90° to the wrist pin

and the edge of the ring groove (see illustration). Check the clearance at 90 degree increments around each groove. The rings differ in thickness, so be sure to use the correct ring for each groove. Compare your readings to the specified ring side clearance for each ring. If the side clearance is greater than specified, new pistons will have to be used.
11 Check the piston-to-bore clearance by measuring the cylinder bore diameter (Section 6) and the piston diameter. Make sure that the pistons

and bores are correctly matched. Measure the piston across the skirt (see illustration). Subtract the piston diameter from the bore diameter for each cylinder to compute the clearance. Compare your results with the specified piston-to-bore clearance. If it is greater than specified, the block will have to be rebored and new oversize pistons and rings installed.

12 Check the piston-to-rod clearance by twisting the piston and rod in opposite directions. Any noticeable play indicates that there is excessive wear and the piston/connecting rod assemblies may have to be replaced.

13 Check the connecting rods for cracks and other damage. Temporarily remove the rod caps, lift out the old bearing inserts, wipe the rod and cap bearing surfaces clean and inspect them for nicks, gouges or scratches. After checking the rods, replace the old bearings, slip the caps into place and tighten the nuts finger tight.

17   Crankshaft — inspection

Refer to illustration 17.2

1   Clean the crankshaft with solvent and dry it thoroughly. Be sure to clean the oil holes with a stiff brush and flush them with solvent. Check the main and connecting rod bearing journals for uneven wear, scoring, pitting or cracks. Check the remainder of the crankshaft for cracks and damage.

2   Using a micrometer, measure the diameter of the main and connecting rod journals (see illustration). Compare your measurements to the specified main and connecting rod journal outside diameters. By measuring the diameter at a number of points around each journal's circumference, you will be able to determine whether the journal is out of round. Take the measurement at each end of the journal, near the crank counterweights, to determine whether the journal is tapered.

3   If the crankshaft journals are damaged, tapered, out of round or worn beyond the specified limits, the crankshaft must be reground by an automotive machine shop. Be sure to use the correct oversize bearing inserts if the crankshaft is reconditioned.

18   Main and connecting rod bearings — inspection

1   Even though the main and connecting rod bearings should be replaced with new ones during the engine overhaul, the old bearings can reveal valuable information about the condition of the engine.

2   Bearing failure usually occurs because of lack of lubrication, the presence of dirt or other foreign particles, engine overloading and/or corrosion. Therefore, the causes of bearing failure must be diagnosed and corrected before the engine is reassembled to prevent it from happening again.

3   When examining the bearings, remove the main bearing caps, the connecting rods and the rod caps. Lay them out on a clean surface in the same position as their location in the engine so that you can trace any problems to the corresponding crankshaft journal.

4   Dirt and contaminantion gets into the engine in a variety of ways. Sometimes it's left in the engine during assembly. It can also pass through filters or breathers. Or it can get into the oil, and from there into the bearings. Metal chips from machining operations and normal engine wear are also carried throughout the engine by oil. Abrasives are sometimes left in engine components after reconditioning, especially when parts are not thoroughly cleaned. Regardless of the means by which they enter the engine, these foreign particles often end up embedded in the soft material of the bearings. Larger particles will not embed in the bearings — instead they will score or gouge both bearings and shafts. The best prevention is thorough cleaning of parts and spotlessly clean engine assembly procedures. Frequent and regular engine oil and filter changes are also recommended.

5   Poor lubrication has a number of interrelated causes. Excessive heat thins the oil. Overloading squeezes the oil from bearing faces. Excessive bearing clearances cause a phenomenon known as "throw-off." A worn oil pump lowers oil pressure. High engine speeds can cause lubrication breakdown. Blocked oil passages, usually the result of misaligned oil holes in a bearing shell, can oil-starve and destroy a bearing. When lack of lubrication is the cause of bearing failure, the bearing material is wiped, or extruded, from the steel backing of the bearing. Extreme temperatures can turn the steel backing a bluish color.

17.2   Measure the diameter of each crankshaft journal at several places to detect taper and out-of-round conditions

6   Driving habits can have an effect on bearing life. Lugging the engine puts very high loads on bearings, squeezing out the thin oil film. Such loads also cause the bearings to flex, producing fatigue failure (fine cracks in the bearing face). Eventually the bearing material will loosen and tear away in pieces from the steel backing. Short trips lead to bearing corrosion because insufficient engine heat is produced to dissipate condensed water and corrosive gases. These products collect in the engine oil, forming acid and sludge. This contaminated oil carries these acids to the bearing material, which they attack and corrode.

7   Incorrect bearing installation during engine assembly will lead to bearing failure as well. Tight-fitting bearings leave insufficient bearing oil clearance and result in oil starvation. Dirt or foreign particles trapped behind a bearing insert can result in high spots on the bearing which lead to failure.

19   Piston rings — installation

Refer to illustrations 19.3a, 19.3b, 19.9 and 19.12

1   Before installing new piston rings, the ring end gaps must be checked. It is assumed that the piston ring side clearance has been checked and verified correct (Section 16).

2   Lay a new ring set next to each piston/connecting rod assembly prior to measuring the end gap for each set of rings. Once end gap has been measured, these rings must remain matched with the same piston and cylinder throughout reassembly.

3   Insert the top (number one) ring into the first cylinder and square it up with the cylinder walls by pushing it in with the top of the piston (see illustration). To measure the end gap, slip a feeler gauge between the ends of the ring (see illustration). Record your measurements for each ring and compare them to the specified end gap.

4   If any gap is larger or smaller than specified, double-check to make sure that you have the correct rings before proceeding.

5   If a gap is too small, it must be enlarged. If it's too small, the ring ends can come in contact with each other during engine operation, causing serious damage to the engine. The end gap is increased by filing the ring ends very carefully with a fine-toothed file. Mount the file in a vise equipped with soft jaws, slip the ring over the file with the ends contacting the file face and slowly move the ring against the file to remove material from the ends. When performing this operation, file only from the outside in.

6   End gap can also be too big. If your measurements indicate an excessive end gap when compared to the specified maximum end gap, you probably purchased the wrong size rings. Double check them to make sure they are correctly matched to the piston and cylinder.

7   Repeat the procedure above for each ring to be installed in the first cylinder, then go on to the next cylinder. Remember to keep rings, pistons and cylinders matched up.

19.3a Before checking piston ring end gap, the ring must be squared in the cylinder bore by pushing it down into the cylinder with the top of a piston

19.3b With the ring square in the cylinder measure the end gap with a feeler gauge

19.9 Roll the oil control ring side rails into place above and below the oil control ring expander — don't use a ring expander

19.12 Use a ring expander to install the compression rings. Be sure that the top and middle rings aren't mixed up and that the identification mark (usually one or two dots) is facing up

8   Once the ring end gaps have been checked and corrected, the rings can be installed on the pistons.
9   The three-part oil control ring (lowest one on the piston) is installed first. Slip the spacer/expander into the groove, then install the lower side rail (see illustration). Do not use a piston ring installation tool on the oil ring side rails — it could damage them. Instead, place one end of the side rail into the groove between the spacer/expander and the ring land, hold it firmly in place and slide a finger around the piston while pushing the rail into the groove. Next, install the upper side rail in the same manner.
10   After the three oil ring components have been installed, check to make sure that both the upper and lower side rails can be turned smoothly in the ring groove.
11   The middle ring is installed next. It should be stamped with a mark so it can be readily distinguished from the top ring. Always follow the instructions printed on the ring package or box — different manufacturers may require different approaches. **Note:** *Do not mix up the top and middle rings — they have different cross sections.*
12   Make sure that the identification mark is facing upward, then slip the ring into the middle groove on the piston with a piston ring installation tool (see illustration). Do not expand the ring any more than necessary to slide it over the piston.
13   Install the top ring in the same manner. Make sure the identification mark faces up.
14   Repeat this procedure for the remaining pistons and rings. Remember that the middle and top rings are different and must not be switched.

20 Crankshaft oil clearance — check

Refer to illustrations 20.7, 20.8, 20.9a, 20.9b and 20.9c

1   The running oil clearance of the main bearings should be checked anytime the crankshaft is removed. A clearance check will determine whether old bearings are excessively worn and will tell you if new bearings are the correct size. Using Plastigage, readily available at any automotive parts store, is the easiest way to measure oil clearance. However, three precautions must be taken when working with Plastigage:
   a) All traces of oil and grease must be completely removed from the crankshaft and bearing surfaces before taking any measurements.

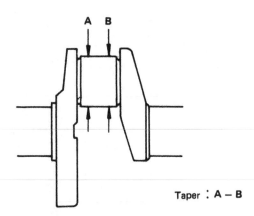

Taper : A − B

20.7   To calculate crankshaft journal taper, find the difference between the width of Plastigage (the oil clearance) at points A and B and compare the results to the specified allowable taper

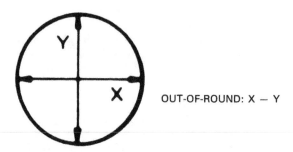

OUT-OF-ROUND: X − Y

20.8   To compute out-of-round, find the difference between diameter X and diameter Y and compare the result with the specified allowable out-of-round

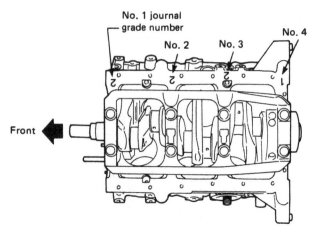

20.9a   The cylinder block main journal grade numbers are stamped into the oil pan mating surface of the block

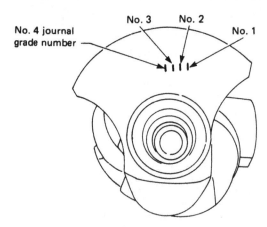

20.9b   The crankshaft main journal grade numbers are stamped on the crank counterweight

		Main journal grade number		
		0	1	2
		Main bearing grade number		
		0	1	2
Crankshaft journal grade number	0	0	1	2
	1	1	2	3
	2	2	3	4

For example:

Main journal grade number: 1

Crankshaft journal grade number: 2

Main bearing grade number = 1 + 2
= 3

20.9c   Once you know the correct main journal and crankshaft journal grade numbers, use this table to determine the correct size main bearing inserts

b)   The crankshaft must not be rotated while Plastigage is being used.

c)   All traces of Plastigage must be removed when testing is complete. Avoid scratching or nicking the crank journal or bearing surfaces when removing Plastigage. Do not use sharp tools or abrasive cleaners. Instead, remove the Plastigage with your fingernail or a small blunt piece of wood.

2   Following removal of the crankshaft and bearings, clean them thoroughly with fresh solvent, then allow them to dry.

3   After they are thoroughly dry, reinstall the main bearings into their respective positions in the engine block and main bearing caps. Then carefully lay the crankshaft into position in the engine block.

4   Cut a piece of Plastigage approximately the width of the bearing, or a little shorter, and lay it on the crank journal parallel with the crankshaft axis. Do this for each crank journal.

5   Carefully install the main bearing cap assembly and tighten the bolts in the proper sequence (Section 21) to the specified torque.

6   Remove the bearing cap assembly and measure the width of the flattened Plastigage strips. A scale is provided on the Pastigage envelope for measuring the width of the flattened strip, which is analogous to bearing clearance.

7   If the Plastigage is wider (flatter) at one end of a journal than it is at the other, the journal is tapered. To calculate taper, find the difference between the width of the Plastigage (the oil clearance) at points A and B (see illustration). Compare your measurement to the specified taper limit. Be sure to note the amount of taper, if any, for each journal.

8   To determine whether the journals are out-of-round, remove all traces of Plastigage and rotate the crankshaft 90 degrees. Lay down

new strips of Plastigage, torque the main bearing cap retaining bolts (Section 21) and check each journal-to-bearing clearance again. Compare these measurements with your initial readings. Compute the difference between the two (see illustration) and compare it to the specified out-of-round.

9   To select new main bearings, select the proper thickness as follows:

a)   The "grade number" of each cylinder block main journal is punched on the respective cylinder block (see illustration).

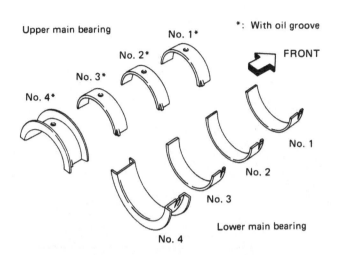

21.5 The main bearing insert must be installed in this order, with the thrust bearing at the fourth main journal and oil grooves in the upper main bearing halves

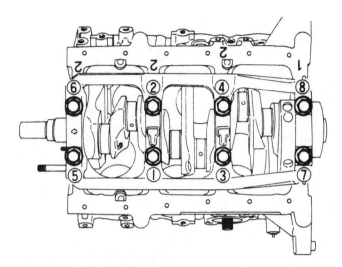

21.10 Tighten the main bearing cap retaining bolts in this sequence

b) The grade number of each crankshaft main journal is punched onto the respective crank counterweight (see illustration).
c) Select the appropriate thickness of main bearing in accordance with the following table (see illustration).
10 Once you have obtained new bearings of the proper thickness, install them and calculate oil clearance, taper and out-of-round in accordance with the procedures outlined above.

Connecting rod bearing oil clearance

11 The oil clearance of the connecting rod bearings is measured using the same method as the procedure outlined above for the main bearings. Before you can check the rod bearing clearances, however, you must tighten the main bearing caps to the specified torque and temporarily install the connecting rods on their respective journals.

---

21 Crankshaft and rear main seal — installation

Refer to illustrations 21.5, 21.10, 21.13, 21.14, 21.16, 21.17, 21.18, 21.19, 21.20 and 21.21
1 Crankshaft installation is generally the first step in engine reassembly. It is assumed at this point that the engine block and crankshaft have been cleaned, inspected and reconditioned. It is also assumed that you have assembled all the correct size main bearings and have checked the main journal-to-bearing oil clearance outlined in the previous Section.
2 Rotate the engine on its stand until the bottom of the block is facing upward.
3 Remove the main bearing cap assembly retaining bolts and lift out the cap assembly.
4 Remove the bearing inserts from the block and the main bearing cap assembly. Wipe the main bearing surfaces of the block and caps with a clean, lint-free cloth. Everything must be kept spotlessly clean from this point.
5 Clean the back sides of the new main bearing inserts and lay one bearing half in each main bearing saddle in the block. Lay the other bearing half from each bearing set in the corresponding main bearing cap (see illustration). Make sure the tab on the bearing insert fits into the recess in the block or cap. The oil hole in the block must line up with the oil hole in the bearing insert. Do not hammer the bearing into place and do not nick or gouge the bearing faces. No lubrication should be used at this time.
6 Clean the faces of the bearings in the block and the crankshaft main bearing journals with a clean, lint-free cloth. Check or clean the oil holes in the crankshaft. Any dirt here can only go one way — straight through the new bearings.
7 Apply a uniform layer of moly-based grease or engine assembly

21.13 Crankshaft end play can be checked with a dial indicator on the front counterweight

lube to each of the bearing surfaces. Be sure to coat the thrust flange faces as well as the journal face of the number 4 bearing.
8 Make sure the crankshaft journals are clean, then gently lay the crankshaft into place in the block.
9 Clean the bearing faces in the main journal caps, then apply a uniform layer of moly-based grease to each of them. Install the main cap assembly with the arrow pointing toward the front of the engine.
10 Install and tighten all main bearing bolts finger tight. Then tighten all bolts by starting at the middle main journal and working outward (see illustration). Work up to the final specified torque in three steps.
11 Rotate the crankshaft a number of times by hand and check for any obvious binding.
12 Check crankshaft free end play with a feeler gauge or a dial indicator.
13 Set a dial indicator on the engine block with the probe touching the number 4 journal. Use a screwdriver to push the crankshaft forward. Zero in the dial indicator and use a screwdriver to push the crankshaft back (see illustration). Record your reading and compare it to the specified end play.

21.14   Crank end play can also be measured with a feeler gauge on the number 4 crank throw

21.16   Place the retainer between two blocks of wood and drive the seal out of the retainer from the rear

21.17   Install the oil seal and gently tap it into place

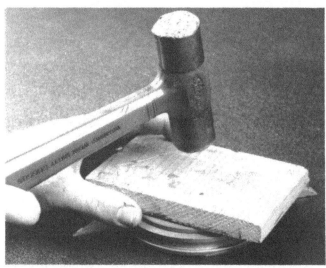

21.18   Place a block of wood over the seal and drive it down evenly by working the hammer around the edges of the seal

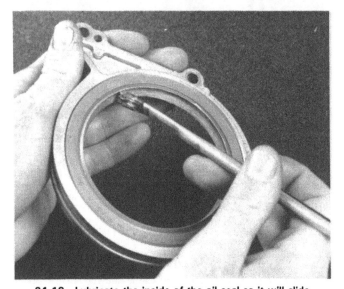

21.19   Lubricate the inside of the oil seal so it will slide smoothly onto the crankshaft

14   If a dial indicator isn't available, use a feeler gauge. Gently pry or push the crankshaft all the way to the front of the engine. Slip a feeler gauge between the crankshaft and the front face of the number 4 bearing, the rear main bearing (see illustration). The measured clearance is the end play. Compare your measurement to the specified free end play.

15   Should your measurement fail to fall within the specified crankshaft free end play, double check your reading, the bearings for correct size and the crankshaft bearing retaining bolts for proper torque. **Note:** *The thrust flange on the number 4 main bearing is the primary determinant of free end play, so if your measurement falls outside the specified end play, the number 4 bearing is probably the wrong size. If you determine that it is the right size, you may need a new crankshaft.*

16   Place the oil seal retainer, face down, atop two wooden blocks on a flat surface to raise it enough to allow removal of the seal. Use a hammer and punch to drive the seal out of the retainer (see illustration).

17   Remove the wood blocks, flip the retainer over and lay the new seal in place. Gently tap it into place with a small hammer (see illustration).

18   Place one of the wood blocks on top of the new seal and drive the new seal into place by working the hammer around the edges of

21.20 Apply sealant to the engine block mating surface

21.21 Tighten the oil seal retainer assembly bolts evenly, drawing the retainer to the block

the wood block (see illustration).

19 Lubricate the inside of the seal to allow it to slide freely onto the crankshaft (see illustration).

20 Apply gasket sealer to the engine block mating surface (see illustration).

21 Slide the rear main seal onto the rear of the crankshaft and tighten the seal retainer bolts to the specified torque in a criss cross pattern (see illustration).

22 Piston/connecting rod assembly — installation and bearing oil clearance check

Refer to illustrations 22.5, 22.6, 22.10, 22.12, 22.13, 22.14, 22.17, 22.19a and 22.19

1 Before installing the piston/connecting rod assemblies the cylinder walls must be perfectly clean, the top edge of each cylinder must be chamfered, and the crankshaft must be in place.

2 Remove the connecting rod cap from the end of the number one connecting rod. Remove the old bearing inserts and wipe the bearing surfaces of the connecting rod and cap with a clean, lint-free cloth. Everything must be kept spotlessly clean.

3 Clean the back side of the new upper bearing half, then lay it in place in the connecting rod. Make sure that the tab on the bearing fits into the recess in the rod. Do not hammer the bearing insert into place and be very careful not to nick or gouge the bearing face. Do not lubricate the bearing at this time.

4 Clean the back side of the other bearing insert and install it in the rod cap. Again, make sure the tab on the bearing fits into the recess in the cap, and do not apply any lubricant. It is imperative that the mating surfaces of the bearing and connecting rod be perfectly clean and oil-free when they are assembled.

5 Position the piston ring gaps correctly (see illustration), then slip a section of plastic or rubber hose over the connecting rod cap bolts.

6 Lubricate the piston and rings with clean engine oil and attach a piston ring compressor to the piston. The piston skirt should protrude about one-quarter inch to guide the piston into the cylinder (see illustration). The rings must be compressed as far as possible.

7 Rotate the crankshaft until the number one connecting rod journal is as far from the number one cylinder as possible (bottom dead center), and apply a uniform coat of engine oil to the cylinder walls.

8 With the dot on top of the piston facing to the front of the engine, gently place the piston/connecting rod assembly into the number one cylinder bore and rest the bottom edge of the ring compressor on the engine block. Tap the top edge of the ring compressor to make sure

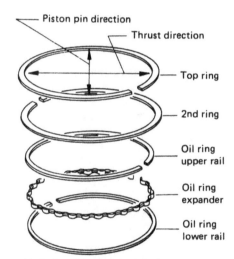

Mark should be facing upward.

22.5 Position piston ring gaps as shown

22.6 When installing the pistons and rings, allow a portion of the piston skirt to extend below the ring compressor (arrow) to center the piston in the cylinder

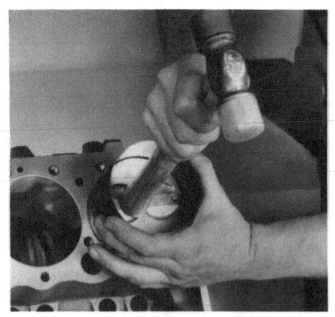

22.10   Gently drive the piston into the cylinder bore with
the end of a wooden hammer handle

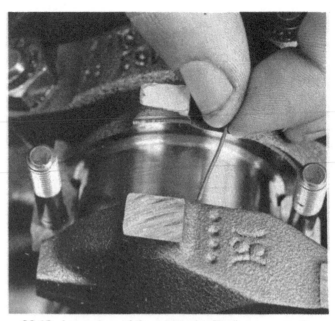

22.12   Lay a piece of the proper size Plastigage across the
rod journal before installing the rod cap

22.13   Match the connecting rod and caps to the cylinder
by matching the numbers stamped on the rod and cap

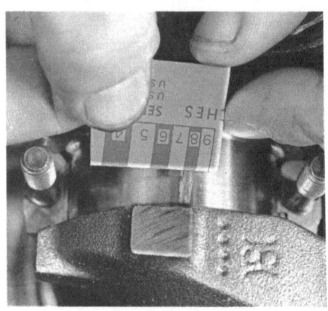

22.14   After removing the rod cap, compare the width of
the crushed Plastigage to the scale on the package to
determine oil clearance

it is contacting the block around its entire circumference.
9   Clean the number one connecting rod journal on the crankshaft
and the bearing faces in the rod.
10 Carefully tap on the top of the piston with the end of a wooden
hammer handle (see illustration) while guiding the end of the connecting rod into place on the crankshaft journal. The piston rings may try
to pop out of the ring compressor just before entering the cylinder bore,
so keep some downward pressure on the ring compressor. Work
slowly, and if any resistance is felt as the piston enters the cylinder,
stop immediately. Find out what is hanging up and fix it before proceeding. Do not, for any reason, force the piston into the cylinder or
you might break a ring and/or the piston.
11 Once the piston/connecting rod assembly is installed, the connecting rod bearing oil clearance must be checked before the rod cap is
permanently bolted in place.
12 Cut a piece of the the appropriate size Plastigage slightly shorter
than the width of the connecting rod bearing and lay it in place on the
number one connecting rod journal, parallel with the journal axis (it

must not cross the oil hole in the journal) (see illustration).
13 Clean the connecting rod cap bearing face, remove the protective
hoses from the connecting rod bolts and gently install the rod cap in
place. Make sure the mating mark on the cap is on the same side as
the mark on the connecting rod (see illustration). Install the nuts and
tighten them to the specified torque, working up to it in three steps.
Do not rotate the crankshaft at any time during this operation.
14 Remove the rod cap, being very careful not to disturb the
Plastigage. Compare the width of the crushed Plastigage to the scale
printed on the Plastigage container to obtain the oil clearance (see illustration). Compare it to the specified clearance. If the clearance is
not correct, double-check to make sure that you have the correct size
bearing inserts. Also, recheck the crankshaft connecting rod journal
diameter and make sure that no dirt or oil was between the bearing
inserts and the connecting rod or cap when the clearance was
measured.
15 Carefully scrape all traces of the Plastigage material off the rod
journal and/or bearing face. Be very careful not to scratch the bearing

22.17   The dot stamped on the piston should point to the front of the engine after installation

— use your fingernail or a piece of hardwood. Make sure the bearing faces are perfectly clean, then apply a layer of moly-based grease or engine assembly lube to both of them. You will have to push the piston into the cylinder to expose the face of the bearing insert in the connecting rod. Be sure to slip the protective hoses over the rod bolts first.

16   Slide the connecting rod back into place on the journal, remove the protective hoses from the rod cap bolts, install the rod cap and tighten the nuts to the specified torque. Again, work up to the torque in three steps.

17   Repeat the entire procedure for the remaining piston/connecting rod assemblies. Keep the back sides of the bearing inserts and the inside of the connecting rod and cap perfectly clean during reassembly. Make sure you have the correct piston for the cylinder and that the dot on the piston faces to the front of the engine when the piston is installed (see illustration). Remember, use plenty of oil to lubricate the piston before installing the ring compressor. Also, when installing the rod cap assembly for the final time, be sure to lubricate the bearing faces adequately.

18   After all the piston/connecting rod assemblies have been properly installed, rotate the crankshaft a number of times by hand and check for any obvious binding.

19   The connecting rod end play must be checked. There are two ways to do it:

   a) Slip a feeler gauge(s) between the connecting rod and the crankshaft throw (see illustration). End play is equal to the thickness of the feeler gauge(s). Compare your measurement to the specified end play.

   b) Set up a dial indicator with its probe touching the rod cap (see illustration). Force the rod to the rear by prying on it with a screwdriver between the rod cap and the main bearing cap. Zero the dial indicator. Then pry the rod forward until it stops. The indicated reading is end play. Compare this measurement to the specified end play.

20   If, using either of the above methods, you find that your measurement does not agree with specified end play, double check your reading. Then check the bearing inserts to make sure that they are the right ones. If they are, you will have to buy either a new set of rods, or a new crankshaft, or both.

---

23   Initial start-up and break-in after overhaul

1   Once the engine has been properly installed in the vehicle (Chapter 2A), double check the engine oil and coolant levels.

2   With the spark plugs out of the engine and the coil high-tension lead grounded to the engine block, crank the engine over until oil pressure registers on the gauge (if so equipped) or until the oil light goes off.

22.19a   A feeler gauge slipped between the side of the rod and the machined face of the crank throw measures the rod side clearance

22.19b   A dial indicator can be used to measure rod side clearance

3   Install the spark plugs, hook up the plug wires and the coil high tension lead.

4   It may take a few moments for the gasoline to reach the engine, but the engine should start without a great deal of effort.

5   As soon as the engine starts it should be set at a fast idle (to ensure proper oil circulation) and allowed to warm up to normal operating temperature. While the engine is warming up, make a thorough check for oil and coolant leaks.

6   Shut the engine off and recheck the engine oil and coolant levels. Restart the engine and check the ignition timing and the engine idle speed (Chapter 1). Make any necessary adjustments.

7   Drive the vehicle to an area with minimum traffic, accelerate at full throttle from 30 to 50 mph, then allow the vehicle to slow to 30 mph with the throttle closed. Repeat the procedure 10 or 12 times. This will load the piston rings and cause them to seat properly against the cylinder walls. Check again for oil and coolant leaks.

8   Drive the vehicle gently for the first 500 miles (no sustained high speeds) and monitor the oil level. It is not unusual for an engine to use oil during the break-in period.

9   At approximately 500 to 600 miles, change the oil and filter and retorque the cylinder head bolts.

10   For the next few hundred miles, drive the vehicle normally. Do not pamper or abuse it.

11   After 2000 miles, change the oil and filter again and consider the engine fully broken in.

Chapter 3
Cooling, heating and air conditioning systems

Contents

Specifications

Cap relief pressure	9 to 14 psi	
Leakage test pressure..............................	23 psi	
Thermostat opening temperature......................	170°F	
Coolant capacity		
non-turbo model	11-1/8 qt	
turbo model	11-5/8 qt	
reservoir	7/8 qt	
Torque specifications	**Kg-m**	**Ft-lbs**
Water pump bolts.................................	1.6 to 2.0	12 to 15
Thermostat housing bolts...........................	1.6 to 2.0	12 to 15
Water inlet bolts.................................	1.6 to 2.0	12 to 15
Water outlet bolts	1.6 to 2.0	12 to 15
Coolant filler housing bolt	0.3 to 0.4	2.2 to 2.9
Radiator bolts...................................	0.3 to 0.4	2.2 to 2.9
Radiator hose clamps	0.3 to 0.4	2.2 to 3.6
Midway pipe to body bolt	0.3 to 0.4	2.2 to 2.9
Water suction pipe bolt	0.8 to 1.1	5.8 to 8.0
Cooling fan bolts	0.6 to 1.0	4.3 to 7.2
Fan coupling bolts	0.6 to 1.0	4.3 to 7.2
Electric cooling fan bolts and nuts	0.3 to 0.4	2.2 to 2.9
Coolant delivery tube	3.2 to 4.1	23 to 30

1 General information

Refer to illustration 1.1

The components of the cooling system are the radiator, top and bottom water hoses, water pump, thermostat, radiator cap with pressure relief valve and heater hoses (see illustration).

The principle of the system is that cold water in the bottom of the radiator circulates upwards through the lower radiator hose to the water pump, where the pump impeller pushes the water around the cylinder block and heads through the various cast-in passages to cool the cylinder bores, combustion surfaces and valve seats. When sufficient heat has been absorbed by the cooling water, and the engine has reached an efficient working temperature, the water moves from the cylinder head past the now open thermostat into the top radiator hose and into the radiator header tank. The water then travels down the radiator tubes where it is rapidly cooled by the natural flow of air as the vehicle moves down the road. A multi-blade fan, mounted on the

water pump pulley, assists this cooling action. The water, now cooled, reaches the bottom of the radiator and the cycle is repeated.

When the engine is cold the thermostat remains closed until the coolant reaches a pre-determined temperature (see the Specifications). This assists rapid warming-up.

The system is pressurized by means of a spring-loaded radiator filler cap, which prevents premature boiling by increasing the boiling point of the coolant. If the coolant temperature goes above this increased boiling point, the extra pressure in the system forces the radiator cap internal spring-loaded valve off its seat and exposes the overflow pipe, down which displaced coolant escapes into the coolant recovery reservoir.

The coolant recovery system consists of a plastic reservoir into which the overflow coolant from the radiator flows when the engine is hot. When the engine cools, coolant is drawn back into the radiator from the reservoir and thus maintains the system at full capacity.

Aside from cooling the engine during operation, the cooling system also provides the heat for the car interior heater and heats the inlet manifold. On vehicles equipped with an automatic transmission, the transmission fluid is cooled by a cooler attached to the base of the radiator.

On cars equipped with air conditioning systems, a condenser is placed ahead of the radiator. On turbo models the condenser has an electric fan for additional cooling.

The radiator cooling fan incorporates either a fluid coupling or a fluid/temperature controlled coupling. The latter device comprises an oil-operated clutch and is a coiled bi-metallic thermostat which functions to permit the fan to slip when the engine is below normal operating temperature level and does not require the supplementary air flow provided by the fan at normal running speed. At higher engine operating temperature, the fan is locked and rotates at the speed of the water pump pulley. The fan coupling is a sealed unit and requires no periodic maintenance.

Turbo equipped models have electric cooling fans which are controlled by the water temperature sending unit. The unit acts as a switch to turn the fan on or off as needed to maintain a proper operating temperature. **Warning:** *The radiator cap should not be removed while the engine is hot. The proper way to remove the cap is to wrap a thick cloth around it, rotate the cap slowly counterclockwise to the detent and allow any residual pressure to escape. Do not press the cap down until all hissing has stopped, then push down and twist off.*

2   Antifreeze

1   It is recommended that the cooling system be filled with a water/ethylene glycol based antifreeze solution which will give protection down to at least –20 °F. This provides protection against corrosion and increases the coolant boiling point. When handling antifreeze, take care that it is not spilled on the vehicle paint, since it will cause damage if not removed immediately.

2   The cooling system should be drained, flushed and refilled at least every alternate Fall. The use of antifreeze solutions for periods of longer than two years is likely to cause damage and encourage the formation of rust and scale due to the corrosion inhibitors gradually losing their efficiency.

3   The exact mixture of antifreeze to water which you should use depends upon the relative weather conditions. The mixture should contain at least 50 percent antifreeze, but under no circumstances should the mixture contain more than 70 percent antifreeze.

3   Thermostat — removal and installation

Refer to illustrations 3.1, 3.6, 3.12, and 3.14

1   The thermostat is a restriction valve which is actuated by a heat sensitive element. It is mounted inside a housing on the right-front side of the engine (see illustration) and is designed to open and close at predetermined temperatures to allow coolant to warm up or cool the engine.

2   To remove the thermostat for replacement or testing, begin by draining the cooling system (refer to Chapter 1).

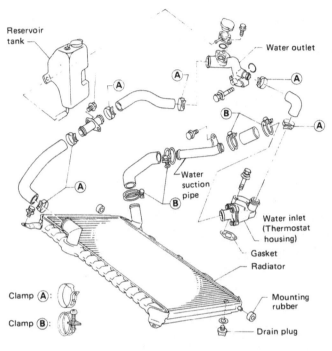

1.1   Exploded view of the cooling system

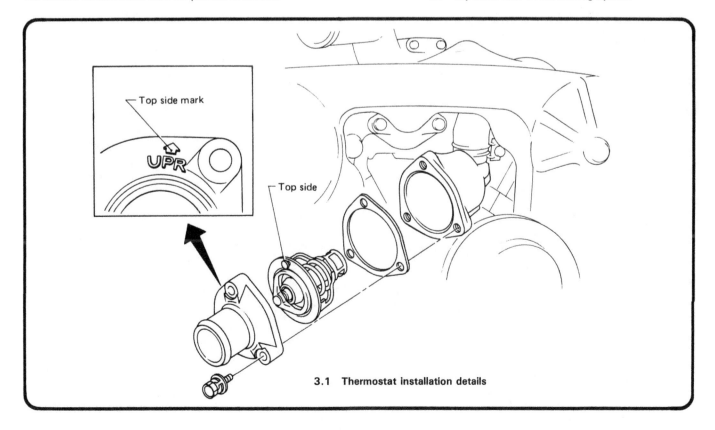

3.1   Thermostat installation details

3   Remove the fan and shroud as described in Section 11.

4   Remove the drivebelts, referring to Chapter 1.

5   Jack up the vehicle and support it securely on jackstands.

6   Remove the lower radiator hose from the metal water suction tube by loosening the hose clamps and support bracket bolts (see illustration).

7   Loosen the remaining hose clamp and remove the water suction tube from the vehicle.

8   Remove the three thermostat housing retaining bolts.

9   Remove the housing from the engine. If necessary, use a soft-face hammer to tap the housing in order to break the gasket seal.

10   Lift the thermostat from its recessed cavity, carefully noting which end is up.

11   Scrape off all traces of gasket and sealant from the gasket mounting surfaces.

12   Insert the new thermostat into the housing with the spring side in the block. The thermostat should seat into the recessed cavity (see illustration).

13   Apply gasket sealer to both sides of the new gasket and install the gasket onto the the thermostat housing.

14   Install the thermostat housing onto the block, making sure the arrow on the cover is pointing up (see illustration).

15   Tighten the housing retaining bolts to the proper specification.

16   Reverse the remaining removal procedures, installing new hose clamps if necessary.

3.6   Remove the two bolts (arrows) and hose clamps from the metal water suction tube and remove the tube from the vehicle

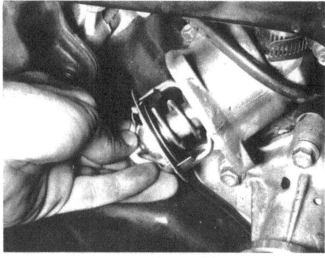

3.12   Install the new thermostat with the spring side inside the engine

3.14   Note the arrow on the housing which points up for installation

4.6   Remove the upper radiator hose by loosening the hose clamps at the radiator and at the mid-way pipe connection

4.7   Remove the air intake assembly by loosening and removing the fasteners indicated by the arrows

4 Radiator — removal and installation

Refer to illustrations 4.6 and 4.7

1 Disconnect the negative cable at the battery. Place the cable out of the way so it cannot accidentally come in contact with the negative terminal of the battery, as this would once again allow power into the electrical system of the vehicle.
2 Jack up the vehicle and support it securely on jackstands.
3 Drain the cooling system (Chapter 1).
4 Remove the front bumper fascia (Chapter 11).
5 Remove the fan shroud (Section 11).
6 Disconnect the upper radiator hose from the radiator and from the mid-way pipe connection (see illustration).
7 Remove the air intake assembly by removing the bolt securing it to the upper crossmember and loosening the large clamp securing the assembly to the air flow meter tube (see illustration).
8 Loosen the large retaining clamp and remove the flexible hose connected to the air flow meter inlet tube.
9 From underneath the vehicle, disconnect the lower radiator hose by loosening the clamp and pulling the hose from the radiator.

10 If equipped with an automatic transmission, disconnect the transmission cooling lines from the radiator by loosening the small hose clamps and pulling both the hoses from the bottom of the radiator. Plug the hoses to prevent fluid from draining out of the transmission and to prevent contamination.
11 Remove the two top radiator retaining bolts.
12 Pull the radiator out through the front of the vehicle.
13 Installation of the radiator is the reverse of the removal procedure. However, note that the radiator rubber mounting grommets must seat in their original positions.

5 Electric cooling fan — removal and installation

Refer to illustration 5.4

1 Disconnect the negative cable at the battery. Place the cable out of the way so it cannot accidentally come in contact with the negative terminal of the battery, as this would once again allow power into the electrical system of the vehicle.
2 Remove the front bumper fascia (Chapter 11).
3 Disconnect the cooling fan electrical connector.
4 Remove the four nuts retaining the electric fan to the mounting bars and lift the fan and shroud off as an assembly (see illustration).
5 Reverse the removal procedure for installation.

6 Water pump — check

Refer to illustrations 6.4 and 6.5

1 A failure in the water pump can cause serious engine damage due to overheating.
2 There are three ways to check the operation of the water pump while it is installed on the engine. If the pump is suspect, it should be replaced with a new or rebuilt unit.
3 With the engine running and warmed to normal operating temperature, squeeze the upper radiator hose. If the water pump is working properly, a pressure surge should be felt as the hose is released.
4 Water pumps are equipped with weep or vent holes. If a failure occurs in the pump seal, small amounts of water will leak from this hole. In most cases it will be necessary to use a flashlight to find the hole on top of the water pump by looking through the space just below the timing belt covers to see evidence of leakage (see illustration).
5 If the water pump shaft bearings fail there may be a squealing sound at the front of the engine while it is running. Shaft wear can be felt if the water pump pulley is forced up and down (see illustration). Do not mistake drivebelt slippage, which also causes a squealing sound, for water pump failure.

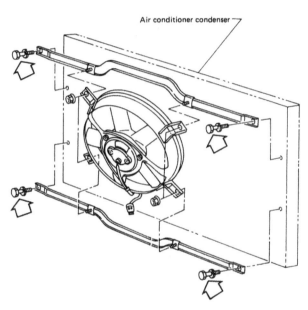

5.4 Remove the four bolts securing the fan to the condenser mounting brackets (arrows)

6.4 If water is seeping through the weep hole it is an indication that the water pump seal is leaking and the pump will have to be replaced

6.5 Check the water pump for excessive play and rough operation. The pump should turn smoothly with no play in the shaft

7   Water pump — removal and installation

Refer to illustration 7.9

1   Disconnect the negative cable at the battery. Place the cable out of the way so it cannot accidentally come in contact with the negative terminal of the battery, as this would once again allow power into the electrical system of the vehicle.
2   Jack up the vehicle and support it securely on jackstands.
3   Drain the cooling system, including both cylinder block petcocks (Chapter 1).
4   Refer to Section 11 to remove the fan and shroud.
5   Refer to Chapter 1 and remove the drivebelts.
6   Remove the water pump pulley.
7   Refer to Chapter 2A and remove the vibration damper from the crankshaft.
8   Remove the upper and lower timing belt covers (refer to Chapter 2A).
9   Remove the six water pump retaining bolts (see illustration).
10   Remove the water pump from the engine block and scrape all gasket and sealant material from the mounting surfaces.
11   Upon installation, apply gasket sealant to both sides of the new gasket and install it and the water pump to the engine block.

12   Tighten all the water pump bolts, a little at a time, to specification.
13   After running the engine for a short time, check carefully for leaks around the water pump.

8   Coolant temperature sending unit — check and replace

Refer to illustration 8.1

1   On turbo models a water temperature switch is located on the left cylinder head. This unit controls the electric cooling fan (see illustration).
2   To remove the switch, unplug the electrical connector.
3   Use the correct size socket and remove the switch from the head.
4   Installation is the reverse of the removal procedure.
5   To check the switch, use an ohmmeter and connect one of the leads to the switch wire and the other grounded to the metal head of the switch.
6   Place the switch in a pan of water heated to 212°F (boiling) and the switch should activate the ohmmeter. If the ohmmeter doesn't move, replace the switch.

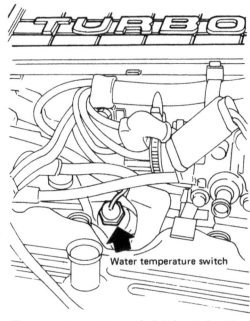

8.1   The water temperature switch is located on the top of the left cylinder head

7.9   Remove the six bolts retaining the water pump to the engine block (arrows)

9.3   Remove the ashtray and remove the facia plate retaining screws (arrows)

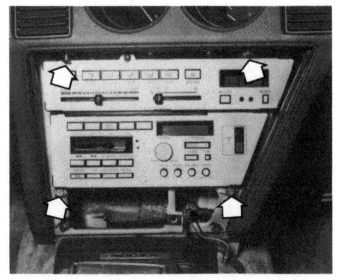

9.4   Remove the radio-air conditioner/heater control assembly retaining screws (arrows)

9 Heater and A/C control assembly — removal and installation

Refer to illustrations 9.3, 9.4, 9.5, 9.7, and 9.9

1 Disconnect the negative cable at the battery. Place the cable out of the way so it cannot accidentally come in contact with the negative terminal of the battery, as this would once again allow power into the electrical system of the vehicle.
2 Pull the ashtray out and let the front tilt down to remove it from the dash.
3 Remove the four screws retaining the facia plate to the dash (see illustration).
4 Remove the four screws retaining the radio/air conditioner/heater assembly to the dash (see illustration).
5 Pull the assembly forward and disconnect the electrical connector by pressing down on the connector tab, then pulling the connector apart (see illustration).
6 Gently pry the a/c-heater temperature control and fan buttons off the levers.
7 Remove the air conditioning/heater facia attaching screws from the back side of the assembly (see illustration).

8 Remove the two screws from the back of the clock.
9 Remove the two screws on each side and pull the air conditioning/heater assembly from the case (see illustration).
10 Reverse the removal procedure for installation. Do not overtighten the various facia screws, as the material can crack easily.

10 Blower motor — removal and installation

Refer to illustrations 10.3, 10.5, and 10.6

1 Disconnect the negative cable at the battery. Place the cable out of the way so it cannot accidentally come in contact with the negative terminal of the battery, as this would once again allow power into the electrical system of the vehicle.
2 From inside the vehicle, on the right side, remove the facia/kick panel (Chapter 11).
3 Remove the two retaining bolts securing the computer electrical connector to the kick panel (see illustration) and let the connector hang loose.
4 Remove the three blower motor retaining screws.

9.5 Reach behind the radio and press down on the electrical connector tab, then pull the connector apart

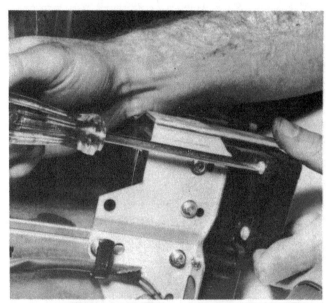

9.7 Working from the backside of the air conditioner/heater assembly, remove the facia plate retaining screws

9.9 Remove the two retaining screws on each side to remove the assembly from its bracket

10.3 To gain clearance to remove the blower motor, first remove the two bolts (arrows) retaining the computer module electrical connector from the kick panel

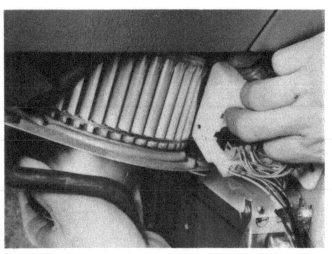

10.5 After the retaining bolts have been removed the blower motor can be removed by working it down around the components attached next to it

5   Work the blower motor out around the computer electrical connector and heater box (see illustration).

6   Disconnect the blower motor electrical connector (see illustration).

7   If the motor is being replaced with a new one, remove the cage from the blower motor shaft and reinstall on the new motor.

8   Install the blower motor assembly in the reverse order of removal.

11   Fan and fan shroud — removal and installation

Refer to illustrations 11.2, 11.3a, 11.3b, 11.4, and 11.5

1   Disconnect the negative cable at the battery. Place the cable out of the way so it cannot accidentally come in contact with the negative terminal of the battery, as this would once again allow power into the electrical system of the vehicle.

2   Remove the radiator shroud retaining bolt securing the shroud to the upper crossmember (see illustration).

3   From under the vehicle, use a small screwdriver to remove the three retaining clips attaching the shroud to the radiator. Remove the clips by prying up on the retaining clip until the clip can be pried over its plastic retaining tab (see illustrations). At this point the shroud can be lifted out of the engine compartment.

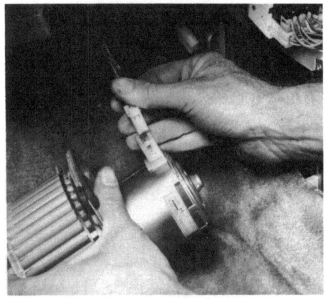

10.6   Disconnect the electrical connector

11.2   Remove the radiator shroud retaining bolt securing the shroud to the upper crossmember (arrow)

11.3a Pry off the three clips that attach the shroud to the radiator

11.3b Once the clip is raised over its retaining tab it can be pried off

4   Remove the fan and fan coupling from the water pump by removing the four retaining nuts (see illustration).
5   To separate the fan from the fluid coupling, remove the four retaining bolts (see illustration).
6   The fan assembly is installed by reversing the above steps. Tighten all fasteners to the proper torque.

12   Air conditioning system — general information

The air conditioning system used in the Nissan 300ZX maintains proper temperature by cycling the compressor on and off according to the pressure within the system, and by maintaining a mix of cooled, ambient and heated air, using basically the same blower, heater core and outlet duct system that the heating system uses.

In addition, the system is sensitive to the temperature of the air being taken in. If the intake air is below approximately 32 °F, the compressor will not operate.

The main components of the system include a belt-driven compressor, a condenser (mounted in front of the radiator), a receiver, drier and an evaporator.

The system operates by air (outside or recirculated) entering the evaporator core by the action of the blower, where it receives maximum cooling if the controls are set for cooling. When the air leaves the evaporator, it enters the heater/air conditioner duct assembly and, by means of a manually controlled deflector, it either passes through or bypasses the heater core in the correct proportions to provide the desired vehicle temperature.

Distribution of this air into the car is regulated by a manually-operated deflector, and is directed either to the floor vents, dash vents or defroster vents according to the settings.

Some models have an automatic air temperature control system which uses a variety of electrical and vacuum-linked sensors, switches and amplifiers that automatically control all the functions of the A/C and heating systems so the interior temperature is kept at a constant level. Once the control lever is set to a desired temperature range, this system automatically selects the optimum combination of air flow, air temperature and outlet location. Because of the complexity of this system, we recommend that you allow a Nissan dealer or other qualified mechanic to diagnose and repair any malfunctions of the system. **Warning:** *No part of the pressure system should be disconnected by the home mechanic. Due to the need for the specialized evacuating and charging equipment, such work should be left to your dealer or a refrigeration specialist.*

13   Air conditioning system — checks and maintenance

1   The following maintenance steps should be performed on a regular basis to ensure that the air conditioner continues to operate at peak efficiency.
   a) Check the tension of the A/C compressor drivebelt and adjust if necessary. Refer to Chapter 1.
   b) Visually inspect the condition of the hoses, checking for any cracking, hardening or other deterioration. **Warning:** *Do not replace hoses without first having the system discharged by a Nissan dealer or a qualified specialist.*
   c) Check that the fins of the condenser are not covered with foreign material, such as leaves or bugs. A soft brush and compressed air can be used to remove them.
2   The A/C compressor should be run for about 10 minutes at least once every month. This is especially important to remember during the winter months because long-term non-use can cause hardening of the internal seals.
3   Due to the complexity of the air conditioning system and the special equipment required to effectively work on it, accurate troubleshooting and repair of the system should be left to a professional mechanic. If the system should lose its cooling action, some causes can be diagnosed by the home mechanic. Look for other symptoms of trouble, such as those in the following list. In all cases, it's a good idea to have the system serviced by a professional.
   a) If bubbles appear in the sight glass (located on top of the receiver-drier) this is an indication of either a small refrigerant leak or air in the refrigerant. If air is in the refrigerant, the receiver-drier is suspect and should be replaced.
   b) If the sight glass takes on a mist-like appearance or shows many bubbles, this indicates a large refrigerant leak. In such a case, do not operate the compressor at all until the fault has been corrected.
   c) Sweating or frosting of the expansion valve inlet indicates that the expansion valve is clogged or defective. It should be cleaned or replaced as necessary.
   d) Sweating or frosting of the suction line (which runs between the suction throttle valve and the compressor) indicates that the expansion valve is stuck open or defective. It should be corrected or replaced as necessary.
   e) Frosting on the evaporator indicates a defective suction throttle valve, requiring replacement of the valve.
   f) Frosting of the high pressure liquid line (which runs between the condenser, receiver-drier and expansion valve) indicates that either the drier or the high pressure line is restricted. The

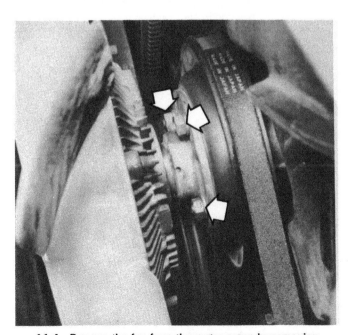

11.4   Remove the fan from the water pump by removing the retaining bolts (arrows)

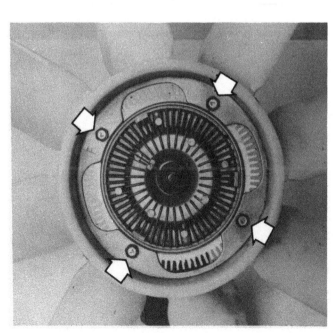

11.5   If the fan coupling is to be replaced, remove the four bolts that attach the fan and coupling (arrows)

line will have to be cleared or the receiver-drier replaced.

g) The combination of bubbles in the sight glass, a very hot suction line and, possibly, overheating of the engine is an indication that either the condenser is not operating properly or the refrigerant is overcharged. Check the tension of the drivebelt and adjust if necessary (Chapter 1). Check for foreign matter covering the fins of the condenser and clean if necessary. If no fault can be found in these checks, the condenser may need to be replaced.

14   Air conditioning compressor — removal and installation

Refer to illustration 14.7

Warning: *Prior to disconnecting any air conditioning lines anywhere in the system the car should be taken to a Nissan dealer or another qualified automotive air conditioning repair shop to have the system depressurized.*

1   Disconnect the negative cable at the battery. Place the cable out of the way so it cannot accidentally come in contact with the negative terminal of the battery, as this would once again allow power into the electrical system of the vehicle.

2   Disconnect the coil wire from the distributor and move it out of the work area.

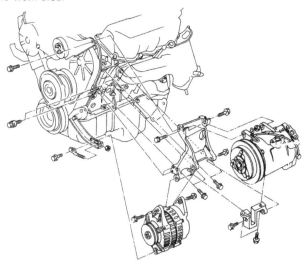

14.7   Air conditioning compressor and mounting bracket — exploded view

3   Jack up the front of the car and support it securely on jackstands. Remove the splash shield.

4   Remove the drivebelt from the compressor (Chapter 1).

5   Disconnect both hoses leading to the compressor, and immediately plug all openings.

6   Disconnect the electrical connector to the compressor by pressing down on the tab and pulling the connector apart.

7   Remove the four nuts and bolts retaining the compressor to its bracket and lift off the compressor (see illustration).

8   Turn the compressor so the clutch is facing up and remove the compressor from the engine compartment. **Caution:** *The compressor should not be left on its side or upside down for more than 10 minutes at a time, as compressor oil could enter the low pressure chambers and cause internal damage. If this should happen, the oil can be expelled from the chambers by positioning the compressor right side up and hand cranking it several times.*

9   Installation is the reverse of removal with the following notes:
   a) When reconnecting the hoses to the compressor, be sure to use new O-rings at each connection.
   b) Check the tightness of the A/C compressor bracket bolts while the compressor is removed, to be sure that none are loose.
   c) After installing the compressor, turn the compressor clutch a few times by hand.
   d) Adjust the compressor drivebelt tension. Refer to Chapter 1, if necessary.

10   Once the compressor and all A/C lines have been securely connected, the car must be taken to a Nissan dealer or other qualified shop to have the system charged.

15   Condenser — removal and installation

Refer to illustrations 15.3 and 15.4

Warning: *Prior to disconnecting any air conditioning lines anywhere in the system the car should be taken to a Nissan dealer or other qualified automotive air conditioning repair shop to have the system depressurized.*

1   Raise the vehicle and support it securely on jackstands.

2   Remove the front bumper fascia (Chapter 11).

3   Disconnect the refrigerant lines from the condenser (see illustration). Be sure to use two wrenches when loosening the flare nuts so as not to damage the lines. Immediately after disconnection, plug all openings in the lines and condenser to prevent dirt and moisture from entering. It may be necessary to disconnect the refrigerant line at its connection located near the water suction pipe attached to the engine.

4   Disconnect the refeerigerant line brackets from the condenser and then remove the two condenser mounting bolts (see illustration).

5   Lower the condenser out of the vehicle.

15.3   Disconnect the refrigerant lines shown by arrows. Use a back up wrench to avoid damaging the metal lines

15.4   Disconnect the condenser hold-down bolts as indicated by arrows (non-turbo model shown)

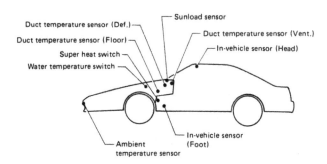

Duct temperature sensor (Def.)
Duct temperature sensor (Floor)
Super heat switch
Water temperature switch
Sunload sensor
Duct temperature sensor (Vent.)
In-vehicle sensor (Head)
In-vehicle sensor (Foot)
Ambient temperature sensor

16.1 Automatic air sensor and switch locations

6   If the condenser fins or air passages are clogged with foreign material, such as dirt, insects or leaves, use compressed air or cold water and a soft brush to clean the condenser. If the condenser is in need of other repairs, have a professional radiator shop or Nissan dealer perform the work.

7   Installation is the reverse of the removal procedure. When connecting the refrigerant lines, always use new O-rings at each connection.

8   Once all A/C lines have been securely connected, the car must be taken to a Nissan dealer or other qualified shop to have the system charged.

16   Automatic Temperature Control system — component description

Refer to illustration 16.1

The Automatic Temperature Control system works in conjunction with both the heating and air-conditioning system to maintain the interior temperature of the car at a desired level.

This section is to familiarize you with the basic function of the system and the location of its components.

Because of the complexity of the system, we recommend that any malfunctioning of the system be diagnosed and repaired by a Nissan dealer or other qualified mechanic.

The Automatic Temperature Control system is divided into nine sub-systems.

System Starting Control

Automatic amplifier (with microcomputer)

The amplifier has a built-in microcomputer, which enables it to deal with the data given to it by sensors and to provide precise control, which is necessary for operating the automatic air conditioning.

Temperature setting switch

This switch is used for initially setting the in-vehicle temperature and for memorizing the setting by the microcomputer in the automatic amplifier.

In-vehicle sensor (head and foot)

This sensor converts the temperature value of the inside air, which is drawn through the aspirator fan, into a resistance valve, which is then input into the automatic amplifier. The sensors are used to detect the typical temperature at either head or feet and sends the information to the microcomputer to compensate for the differences.

Ambient temperature sensor

This sensor transforms the value of ambient temperature into a resistance value that the automatic amplifier can read and compensate for. From this value and the values that the automatic amplifier receives from the defroster, ventilator and floor duct sensors, the amplifier knows how hard to work to correct for the temperature outside.

Sunload sensor (photo diode)

This sensor transforms sunload into current value, which is put into the automatic amplifier. The sensor is located on the defroster grill.

All the sensors send signals to the auto amplifier, and when the amplifier receives the input signals it calculates the desirable air mix door position and causes the air mix door to move to the calculated position.

Chapter 4 Fuel and exhaust systems

Contents

Specifications

Fuel pump cut-off discharge pressure	61 to 71 psi
Regulator pressure	36.3 psi
Air regulator air flow (at 20-degrees C/68-degrees F)	971 cu ft/hr (512 for 1987 and later models)
Turbocharger maximum pressure	320 to 380 mm Hg (12.60 to 14.96 in Hg)
Fuel pressure at idle	30 psi (approximate)
Fuel pressure as throttle pedal is depressed	
VG30E	37 psi (approximate)
VG30ET	44 psi (approximate)
Air flow meter voltage between terminals B and D	
1984 thru 1986	$1.6 \pm 0.5V$
1987 on	2 to 4V
Cylinder head and fuel temperature sensor thermistor resistance (1984 thru 1986)	
at -10-degrees C (14-degrees F)	7.0 to 11.4 K-ohms
at 20-degrees C (68-degrees F)	2.1 to 2.9 K-ohms
at 50-degrees C (122-degrees F)	0.68 to 1.0 K-ohms
Cylinder head and fuel temperature sensors thermistor resistance (1987 on)	
at 20-degrees C (68-degrees F)	2.3 to 2.7 K-ohms
at 50-degrees C (122-degrees F)	0.77 to 0.87 ohm
at 80-degrees C (176-degrees F)	0.30 to 0.33 ohm
Exhaust gas sensor heater resistance	5 to 6 K-ohms
Turbocharger axial play	0.013 to 0.091mm (0.0005 to 0.0036 in)

Torque specifications	Kg-m	Ft-lbs
Accelerator cable adjusting nut	0.8 to 1.0	5.8 to 7.2
Throttle chamber securing bolt	1.8 to 2.2	13 to 16
Intake collector bolt	1.8 to 2.2	13 to 16
Cylinder head temperature sensor	1.2 to 1.6	9 to 12
Exhaust gas sensor		
1984 thru 1986/1987 and later non-turbo	4.1 to 5.1	30 to 37
1987 and later turbo	1.8 to 2.4	13 to 17
EGR control valve	1.8 to 2.3	13 to 17
EGR tube	3.5 to 4.5	25 to 33
Detonation sensor	2.5 to 3.5	18 to 25
Fuel hose clamp	0.10 to 0.15	0.7 to 1.1
Throttle pedal hold down bolt	0.4 to 0.5	2.9 to 3.6
Fuel pump retaining bolt	0.23 to 0.31	1.7 to 2.2
Fuel tank gauge retaining bolt	0.23 to 0.31	1.7 to 2.2
Filler neck-to-tank retaining bolt	0.23 to 0.31	1.7 to 2.2
Fuel tank strap retaining bolt	3.2 to 4.3	23 to 31
Heat shield retaining bolts and nuts	0.4 to 0.5	2.9 to 3.6
Exhaust manifold to header pipe retaining nuts	4.7 to 6.2	34 to 45
Header pipe to catalytic converter	3.2 to 4.3	23 to 31
Exhaust pipe to catalytic converter	3.2 to 4.3	23 to 31
U-bolt retaining nut	2.3 to 3.7	17 to 27
Exhaust tube mounting bolts and nuts	0.93 to 1.2	6.7 to 8.7
Exhaust pipe to muffler inlet pipe	1.8 to 2.8	13 to 20
Turbo-to-exhaust manifold nut	4.5 to 5.5	33 to 40
Turbo oil drain pipe to turbo bolt	1.0 to 1.2	7 to 9
Turbo coolant delivery tube bolt	3.2 to 4.2	23 to 30
Oil delivery tube to block bolt	1.5 to 2.0	11 to 14
Oil delivery tube to turbo bolt	1.0 to 2.2	7 to 16

127

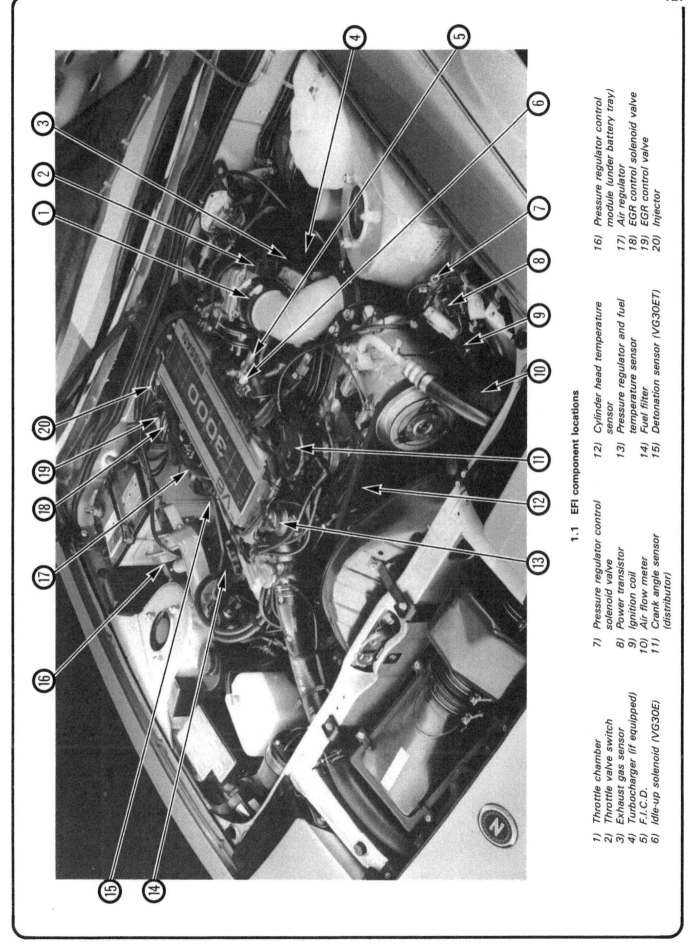

1.1 EFI component locations

1) Throttle chamber
2) Throttle valve switch
3) Exhaust gas sensor
4) Turbocharger (if equipped)
5) F.I.C.D.
6) Idle-up solenoid (VG30E)
7) Pressure regulator control solenoid valve
8) Power transistor
9) Ignition coil
10) Air flow meter
11) Crank angle sensor (distributor)
12) Cylinder head temperature sensor
13) Pressure regulator and fuel temperature sensor
14) Fuel filter
15) Detonation sensor (VG30ET)
16) Pressure regulator control module (under battery tray)
17) Air regulator
18) EGR control solenoid valve
19) EGR control valve
20) Injector

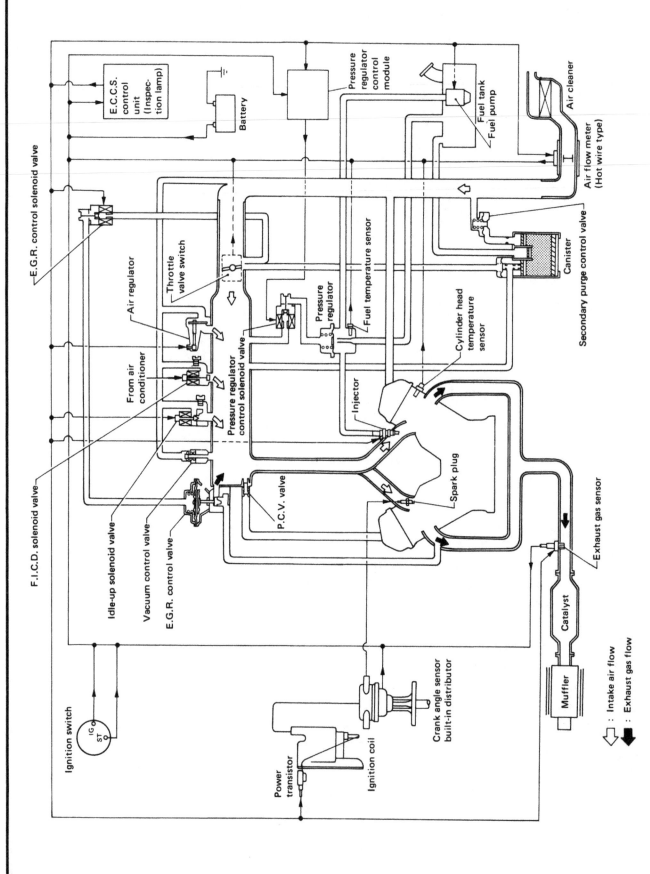

1.2a Use this ECCS diagram for the VG30E to understand how the air/fuel/exhaust system interacts

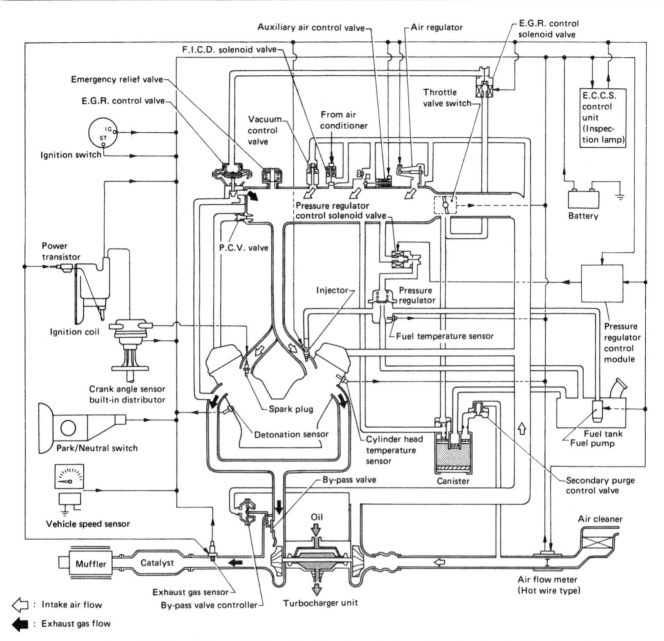

1.2b On the VG30ET model, the air is introduced into the turbocharger first, then force fed into the engine

Key:

⇦ : Intake air flow

⬅ : Exhaust gas flow

1 EFI (Electronic Fuel Injection) system — general information

Refer to illustrations 1.1, 1.2a and 1.2b

The VG30E and VG30ET engines uses an Electronic Fuel Injection (EFI) fuel system in place of the conventional carburetor (see illustrations). Electronic fuel injection provides optimum mixture ratios, and this, together with the immediate response characteristics of the fuel injection, permits the engine to run on the weakest possible fuel/air mixture. This vastly reduces the exhaust gas toxic emission. The fuel system is interrelated with and works in conjunction with the emissions control and exhaust systems covered in Chapter 6. Thus, some elements that relate directly to the fuel system are covered in that Chapter.

The EFI system consists of three subsystems: The fuel flow system, the air flow system and the electrical signaling system (see illustration). The various components that make up the entire EFI system are detailed in Section 2.

Fuel from the tank is delivered under pressure by an electric fuel pump. The amount of fuel to be injected is determined by the injection pulse duration as well as by a pressure difference between fuel pressure and intake manifold vacuum pressure. The Electronic Concentrated Control System (ECCS) control unit controls only the injection pulse duration. For this reason, the pressure difference between the fuel pressure and intake manifold vacuum pressure must be maintained at a constant level. Since the intake manifold vacuum pressure varies with engine operating conditions, a pressure regulator is placed in the fuel line to regulate the fuel pressure in response to changes in the intake manifold pressure. Where manifold conditions are such that the fuel pressure could be beyond that specified, the pressure regulator returns surplus fuel to the tank.

An injection of fuel occurs once every rotation of the engine. Because the injection signal comes from the control unit, all six injectors operate simultaneously and independent of the engine stroke. Each injection supplies half the amount of fuel required by the cylinder, and the length of the injection period is determined by information fed to the control unit by various sensors included in the system.

Elements affecting the injection duration include: Engine rpm, quantity and temperature of the intake air, throttle valve opening, temperature of the engine coolant, intake manifold vacuum pressure and amount of oxygen in the exhaust gases.

Because the EFI system operates at high fuel pressure, any leak can affect system efficiency and present a serious fire risk. Also, since the

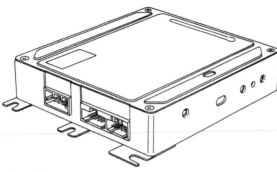

2.1 The ECCS unit controls the amount of fuel that is injected, ignition timing, idle speed, fuel pump operation and feedback of the mixture ratio

intake air flow is critical to the operation of the system, even a slight air leak will cause an incorrect air/fuel mixture. **Note:** *Certain precautions should be observed when working on the EFI system:*

a) Do not disconnect either battery cable while the engine is running.

b) Prior to any operation in which the fuel line will be disconnected, the high pressure in the system must first be eliminated. This procedure is described in Chapter 1. Disconnect the negative battery cable to eliminate the possibility of sparks occurring while fuel is present.

c) Prior to removing any EFI component, be sure the ignition switch is Off and the negative battery cable is disconnected.

d) The EFI wiring harness should be kept at least four inches (10 mm) away from adjacent harnesses. This includes a CB antenna feeder cable. This is to prevent electrical pulses in other systems from interfering with EFI operation.

e) Be sure all EFI wiring connections are tight, clean and secure, as a poor connection can cause extremely high voltage surges in the ignition coil which could drain the IC circuit.

f) The accelerator should *Not* be depressed prior to starting the engine. Immediately after starting, do not rev the engine unnecessarily.

The electric fuel pump uses relays, located in the engine compartment, that are designed so that should the engine stop (causing the alternator to turn off and the oil pressure to drop), the fuel pump will cease to operate.

Some basic checks of the EFI components are included in this Chapter. However, the complexity of the system prevents many problems from being accurately diagnosed by the home mechanic. If a problem develops in the system which cannot be pinpointed by the checks listed here, it is best to take the vehicle to a Nissan dealer to locate the fault.

2 ECCS components — general description

Refer to illustrations 2.1, 2.2a, 2.2b, 2.2c, 2.3, 2.4, 2.5, 2.6, 2.8, 2.9, 2.10, 2.11, 2.12, 2.14, 2.15, 2.17, 2.18 and 2.24.

ECCS control unit

1   The Electronic Concentrated Control System (ECCS) control unit is a microcomputer with electrical connectors for receiving input/output signals and for power supply, inspection lamps and a diagnostic mode selector. The control unit regulates the amount of fuel that is injected, as well as the ignition timing, idle speed, fuel pump operation and the feedback of the mixture ratio (see illustration).

Crank angle sensor

2   The crank angle sensor could be regarded as the right hand to the ECCS control unit, as it is the basic signal sensor for the entire ECCS system. It monitors the engine speed, piston position and it sends signals to the ECCS control unit for control of the fuel injection, ignition timing, idle speed, fuel pump operation and the EGR function (see illustrations).

The crank angle sensor has a rotor plate and a wave forming circuit. The assembly consists of a rotor plate with 360 slits representing 1° signals (engine speed signal) and six slits for 120° signals (crank angle

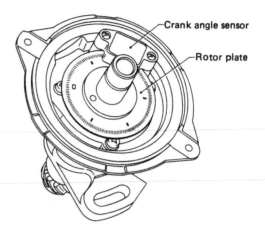

2.2a The crank angle sensor is the main signal sensor for the ECCS

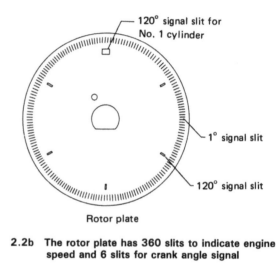

2.2b The rotor plate has 360 slits to indicate engine speed and 6 slits for crank angle signal

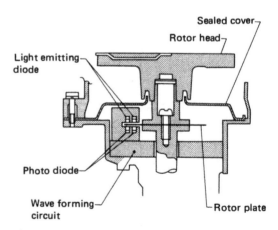

2.2c Light passes through the slits in the rotor plate, sending signals in the form of on-off pulses to the control unit

signal). Light emitting diodes (LED) and photo diodes are built into the wave forming circuit.

In operation, the signal rotor plate passes through the space between the LED and photo diode and the slits in the signal rotor plate intermittantly cut off the light sent to the photo diode from the LED. This causes an alternating voltage and the voltage is converted into an on/off pulse by the wave forming circuit. The on/off signal is sent to the control unit for processing.

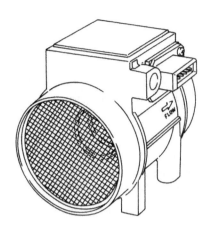

2.3   The air flow meter measures the mass flow rate of intake air

2.5   The exhaust gas sensor monitors the quantity of oxygen in the exhaust gas

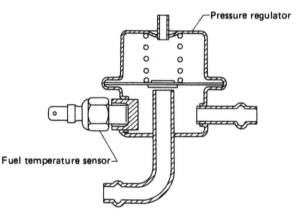

2.8   The fuel temperature sensor lets the E.C.C.S. control module know to enrich the mixture when the fuel temperature is too high

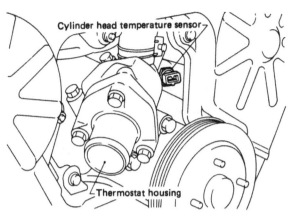

2.4   The cylinder head temperature sensor is located behind the thermostat housing and timing belt cover

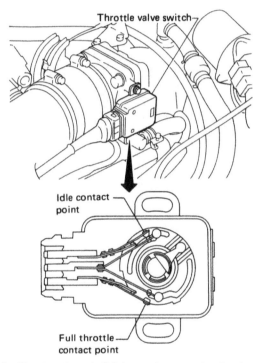

2.6   The throttle switch informs the control unit when the accelerator is at idle and on electronically controlled automatic transmission equipped models it informs the control unit when the throttle is wide open

Air Flow meter

3   The air flow meter measures the mass flow rate of intake air. The control circuit emits an electrical output signal which varies in relation to the amount of heat dissipated from a hot wire placed in the stream of intake air (see illustration).

Cylinder head temperature sensor

4   The cylinder head temperature sensor is located in the left cylinder head and monitors changes in the cylinder head temperature and transmits a signal to the ECCS control unit (see illustration).

Exhaust gas sensor

5   Mounted in the exhaust manifold, the exhaust gas sensor monitors the quantity of oxygen in the exhaust gases (see illustration).

Throttle valve switch

6   The throttle valve switch is attached to outside of the throttle chamber and actuates in response to accelerator pedal movement. The switch is equipped with two contacts, one for idle and the other for

full throttle contact. The idle contact closes when the throttle valve is positioned at idle and opens when it is at any other position. The full throttle contact is used for electronic controlled automatic transmissions only (see illustration).

Vehicle speed sensor

7   The vehicle speed sensor provides a vehicle speed signal to the ECCS control unit. Two types of speed sensor are employed, depending upon the type of speedometer installed. Needle type speedometer models utilize a reed switch, which is installed in the speedometer unit and transforms vehicle speed into a pulse signal which is sent to the control unit. The digital type speedometer consists of an LED, photo diode, shutter and wave forming circuit.

Fuel temperature sensor

8   The fuel temperature sensor is built into the pressure regulator and senses the fuel temperature. When the fuel temperature is higher than the specified level, the ECCS control unit enriches the mixture injected to compensate. The fuel temperature sensor should not be removed from the regulator. Always replace as an assembly (see illustration).

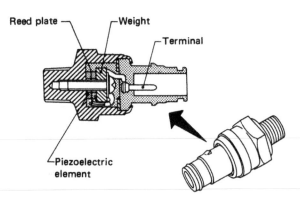

2.9   A detonation sensor is installed on the engine block of VG30ET models to sense engine knock conditions

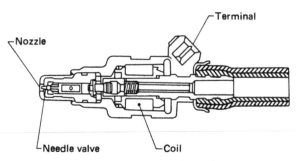

2.10   The fuel injector recieves a signal from the control unit and the needle valve opens to inject fuel into the intake manifold. The amount of fuel injected is determined by the pulse duration

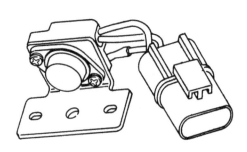

2.11   The ignition signal is amplified by the power transistor, which triggers the proper high voltage in the secondary circuit

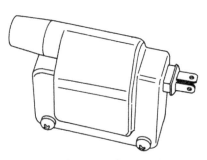

2.12   The ignition coil

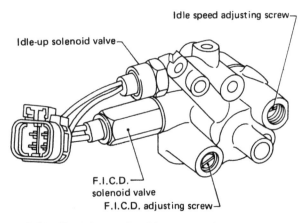

2.14   The idle-up solenoid acts to stabilize idle speed when engine load is heavy

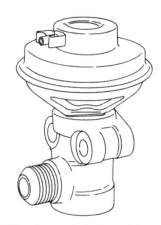

2.15   The EGR valve controls the amount of exhaust gas to circulated into the intake manifold

Detonation sensor (turbo models only)

9   Attached to the cylinder block, the detonation sensor is capable of sensing engine knock conditions. Any knocking vibration from the cylinder block is applied as pressure to the piezoelectric element. This pressure is then converted into a voltage signal which is delivered as output to the control unit (see illustration).

Fuel injector

10   The fuel injector supplies each cylinder with fuel. The injector is a small, precision solenoid valve. As the ECCS control unit outputs an injection signal to each fuel injector, the coil built into the injector pulls the needle valve back and fuel is sprayed through the nozzle into the intake manifold. The amount of fuel injected is controlled by the ECCS control unit by injection pulse duration (see illustration).

Power transistor

11   The ignition signal from the ECCS control unit is amplified by the power transistor, which connects and disconnects the coil primary circuit to induce the proper high voltage in the secondary circuit (see illustration).

Ignition coil

12   The molded type ignition coil provides spark for the combustion process (see illustration).

Auxiliary air control (AAC) valve (turbo models only)

13   The AAC valve is attached to the intake collector. The ECCS control unit actuates the AAC valve by an on/off pulse. The longer the on-duty signal is left on, the larger the amount of air that will be allowed to flow through the AAC valve.

Idle-up solenoid valve (non-turbo models only)

14   The idle-up solenoid valve is attached to the intake collector. The solenoid is actuated to stabilize idle speed when the engine load is heavy

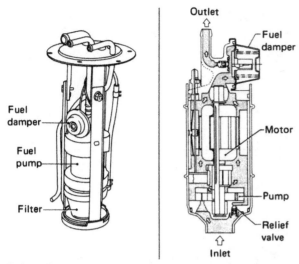

2.17   The fuel pump is a wet type pump, where the vane rollers are directly coupled to a electric motor which is filled with fuel

because of electric load, such as power steering pump, air conditioning, etc. (see illustration).

EGR contol valve

15   The EGR control valve controls the quantity of exhaust gas circulated to the intake manifold through movement of the taper valve connected to the diaphragm, to which vacuum is applied in response to the opening of the throttle valve. **Note:** *When installing the EGR guide tube, be careful of its direction. The outlet faces the rear of the engine. Otherwise the distribution efficiency of the exhaust gas will be reduced (see illustration).*

EGR control solenoid valve

16   The EGR control solenoid valve cuts the intake manifold vacuum signal for EGR control. The solenoid valve actuates in response to the on/off signal from the ECCS control unit. When the solenoid is off, a vacuum signal from the intake manifold is fed into the EGR control valve. As the control unit outputs an on signal, then the coil pulls the plunger downward, and cuts the vacuum signal (see illustration).

Fuel pump

17   The fuel pump, which is located in the fuel tank, is a wet type pump where the vane rollers are directly coupled to a motor which is filled with fuel (see illustration).

Air regulator

18   The air regulator gives an air bypass when the engine is cold to allow fast idle during warm-up. A bimetal heater and rotary shutter are built into the air regulator. When the bimetal temperature is low, the air bypass post is open. As the engine starts and electric current flows through a heater, the bimetal begins to rotate the shutter to close off the bypass port. The air passage remains closed until the engine is stopped and the bimetal temperature drops (see illustration).

Air injection valve (AIV)

19   The air injection valve sends secondary air to the exhaust manifold by means of a vacuum caused by exhaust pulsation in the exhaust manifold. When the exhaust pressure is below atmospheric pressure (vacuum), secondary air is sent to the exhaust manifold. When the exhaust pressure is above the atmospheric pressure the reed valve will prevent the secondary air from being sent back to the air cleaner.

AIV control solenoid valve

20   The AIV control solenoid valve controls an intake manifold vacuum signal for the AIV control. The solenoid valve actuates in response to the on/off signal from the ECCS control unit. When the solenoid is Off the vacuum signal from the intake manifold is cut. As this happens the control unit outputs an On signal and the coil pulls the plunger downward, feeding the vacuum signal to the AIV control valve.

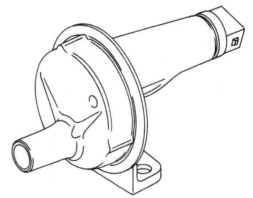

2.18   The air regulator gives an air bypass when the engine is cold to provide fast idle during warm-up

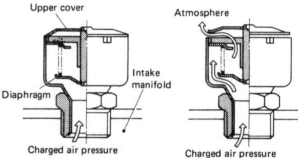

2.24   In case the bypass valve fails, the emergency relief valve will open to release pressure in the intake manifold

Pressure regulator control solenoid valve

21   The pressure regulator control solenoid valve cuts the intake manifold vacuum signal for pressure regulator control. The solenoid valve actuates in response to the on/off signal from the ECCS control unit. When it is Off, a vacuum signal from the intake manifold is fed into the pressure regulator. As the control unit outputs an On signal, the coil pulls the plunger downward, cutting the vacuum signal.

Vacuum control valve

22   The vacuum control valve is provided to reduce the engine lubricating oil consumption when the intake manifold vacuum increases to a very high level during deceleration. As the manifold vacuum increases beyond the specified value, the valve will open, allowing air to be drawn into the intake manifold.

Turbocharger

23   The turbocharger is mounted to the left exhaust manifold and is designed to utilize the exhaust gas to spin the turbine wheel, which is connected to the compressor wheel. Pressurized air is supplied by the compressor through the throttle chamber into the intake manifold.

   To prevent the turbocharger from an excessive rise in supercharging pressure, the turbine speed is maintained within a safe range by controlling the amount of exhaust gas that passes through the turbine. The system consists of a bypass valve, which allows some of the exhaust gas to bypass the turbine and to flow directly into the exhaust pipe.

Emergency relief valve

24   To prevent an abnormal rise in supercharging pressure, and possible engine damage, in case the bypass valve should fail to open properly, an emergency valve is installed as a safety device in the intake manifold. The valve opens when the pressure in the intake manifold is above 63.3 kPa (18.70 in Hg) (see illustration).

3.1a   On non-turbo models remove the three air hoses then loosen the large clamps at each end to remove the upper inlet air tube

3.1b   On turbo models, remove the two air hoses (one not shown, underneath) then loosen the end clamps. It will be necessary to remove the retaining bolt securing the rocker arm hose to the inlet tube (arrows)

3.3   Remove the two bolts attaching the lower inlet tube to the fender well (arrows)

3.4   Disconnect the lower inlet tube from the air flow meter by loosening the hose clamps (arrows)

3   Air inlet tubes — removal and installation

Refer to illustrations 3.1a, 3.1b, 3.3 and 3.4

Upper tube

1   Remove the three air hoses from the tube (two hoses on turbo models) and loosen the two large hose clamps at each end of the tube (see illustrations). On turbo models disconnect the retaining bolt attaching the air hose running into the rocker arm cover.

Lower tube

2   If equipped with air conditioning, remove the compressor.
3   Remove the two bolts securing the air inlet tube to the fender well on non-turbo models (see illustration).
4   Loosen the hose clamps at each end of the air inlet tube on turbo models and at the air flow meter on non-turbo models (see illustration).
5   Pull the air inlet tube from the air flow meter (and turbo, if equipped)

and lift the tube out of the engine compartment.
6   Installation of both tubes is the reverse of the removal procedures.

4   Throttle chamber — removal and installation

Refer to illustrations 4.1, 4.2, 4.3 and 4.5

1   Disconnect the throttle cable by raising the throttle chamber actuating valve and sliding the cable from the actuator (see illustration).
2   Disconnect the two electrical connectors from the throttle valve switch sensor (see illustration).
3   Disconnect the vacuum lines from each side of the throttle chamber (see illustration).
4   Use a Phillips screwdriver to loosen the coolant hoses from each side of the throttle chamber.
5   Remove the four Allen head bolts and remove the throttle chamber from the intake collector (see illustration).
6   Reverse the removal procedures for installation.

4.1   Raise the actuator and slide the accelerator cable end
out of the retaining slot

4.2   After pulling the white connector apart, pull the TVS
connector from the throttle valve switch

4.3   Disconnect the two vacuum lines from the lower part
of the chamber (arrows)

4.5   Remove the four Allen head bolts retaining the
chamber to the intake collector (arrows)

5   Air flow meter — removal and installation

Refer to illustration 5.3

1   Loosen the bolts securing the air filter housing to the body.
2   Disconnect the air flow meter electrical connector by pulling the
connector apart.
3   Loosen the large hose clamps at each end of the air flow meter
(see illustration).
4   Push the air filter housing forward and work the air flow meter
from the air tube.
5   To install, work the air flow meter into the air filter housing tube
then work the other end onto the air flow meter and tighten all retaining
bolts and hose clamps.

6   Cylinder head temperature sensor — removal and installation

1   Refer to timing belt removal in Chapter 2a and remove the timing
belt and timing belt rear cover.

5.3   Disconnect the air flow meter by removing the
electrical connector and loosening the large hose clamps at
each end of the meter (arrows)

7.1   Remove the heat shield retaining bolts (arrows)

7.5   Coat the threads with anit-sieze compound before installing the sensor

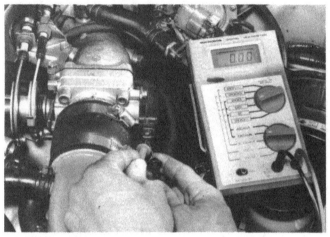

8.6   With an ohmmeter connected to the top two terminals, rotate the TVS counterclockwise and the meter should show no continuity

8.2   Remove the two throttle valve switch bolts (arrows)

2   Disconnect the temperature sensor electrical connector located on the front of the left cylinder head.
3   Remove the sensor from the head.
4   To install, wrap the new sensor threads with teflon tape then secure the sensor to the block.
5   Reverse the remaining removal procedures for installation.

---

7   Exhaust gas sensor — removal and installation

Refer to illustrations 7.1 and 7.5

1   Remove the two master cylinder heat shield retaining bolts attached to the master cylinder and the left fender well and remove the heat shield (see illustration).
2   Remove the left exhaust manifold heat shield retaining bolts and remove the heat shield.
3   Disconnect the exhaust gas sensor electrical connector.
4   Remove the exhaust gas sensor from the exhaust manifold.
5   If a new sensor is to be installed the threads should come with an anti-seize compound on them. If installing the old sensor, apply anti-

seize compound to the threads before installation (see illustration).
6   Reverse the remaining removal procedures for installation.

---

8   Throttle valve switch — removal, installation and adjustment

Refer to illustrations 8.2 and 8.5

1   Unplug both electrical connectors (only one on manual transmission equipped models) attached to the throttle valve switch.
2   Remove the two bolts and lift the throttle valve switch from the throttle chamber (see illustration). **Note:** *Any time the throttle valve switch bolts are loosened it will be necessary to readjust the switch.*
3   Reverse the removal procedures for installation but do not tighten down the retaining bolts.
4   With the engine idling at normal operating temperature, confirm that engine idle speed is within specifications. If not, correct idle speed, referring to Chapter 1 if necessary.
5   With the throttle valve switch electrical connector disconnected, connect an ohmmeter to the top two terminals.
6   Rotate the throttle valve switch counterclockwise and the meter should show no continuity (see illustration).
7   Rotate the throttle valve switch clockwise until continuity is established, then tighten the retaining bolts.
8   Replace the throttle valve switch electrical connector.

---

9   Fuel temperature sensor — removal and installation

Never remove the fuel temperature sensor from the pressure regulator. Always replace as an assembly. See the fuel pressure regulator procedure for removal and installation.

11.5 Cover the intake ports with a rag

11.8 After removing the fuel lines and injector retaining bolts the injectors will come off as an assembly

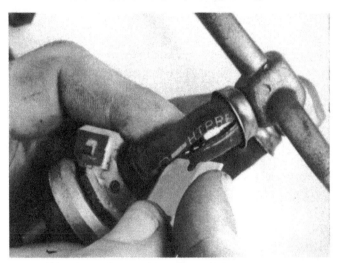

11.9 To replace an injector, carefully cut through the hose

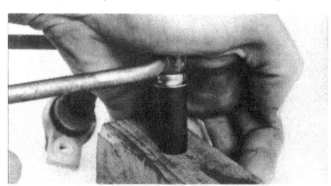

11.10a Be sure the hose seats against the umbrella

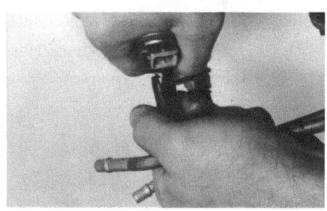

11.10b Even with the hose wet it will take considerable force to seat the injector

---

10 Detonation sensor (turbo models) — removal and installation

1 Locate the detonation sensor on the right side of the cylinder block.
2 Disconnect the electrical connector.
3 Remove the sensor from the cylinder block.
4 Reverse the removal procedure for installation.

---

11 Fuel injector — removal and installation

Refer to illustrations 11.5, 11.8, 11.9, 11.10a, 11.10b, 11.12a, 11.12b, 11.12c, 11.14 and 11.16

1 Refer to Chapter 1 and relieve the fuel pressure.
2 Disconnect the negative cable at the battery. Place the cable out of the way so it cannot accidentally come in contact with the negative terminal of the battery, as this would once again allow power into the electrical system of the vehicle.
3 Remove the upper air intake tube.
4 Remove the top part of the intake collector with the throttle chamber attached.
5 Cover the intake ports with a rag (see illustration).
6 Disconnect the spark plug wires at the plugs and remove the

distributor cap and move the wire assembly out of the work area.
7 Disconnect the fuel rails at the fuel regulator and fuel inlet tubes by loosening the hose clamps and pulling the hose from the tubing. **Note:** *When disconnecting the fuel lines place a rag around the connection to catch any leaking fuel.*
8 Remove each of the injector retaining bolts and lift off the injector assembly (see illustration).

Injector replacement

9 If any of the individual injectors are to be replaced, determine the injector to be replaced and carefully cut through the hose to free the injector from the rail assembly (see illustration).
10 To install the new injector wet the new hose with gasoline, place the umbrella onto the fuel rail and press the rubber hose onto the fuel rail. Be sure the hose seats against the umbrella (see illustrations).

11.12a After lubrictaing the fuel line hoses with gas, slip
the injector rail assembly into place

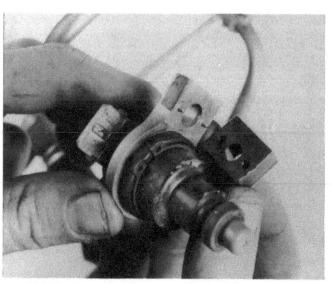

11.12b Make sure the flat spacer is installed properly

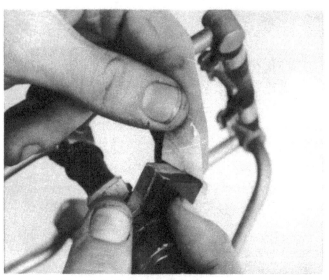

11.12c To faciltate installation tape the spacer to
each injector

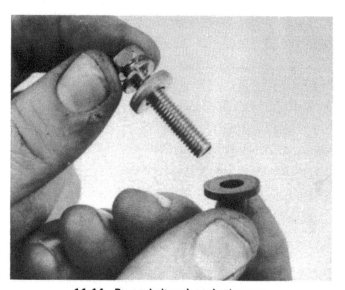

11.14 Proper bolt and washer/spacer
installation sequence

11.16 Be sure to install a new gasket between the intake
manifold and the collector

12.3 Disconnect the electrical connector and remove the
hold-down bolt to remove the power transistor (arrow)

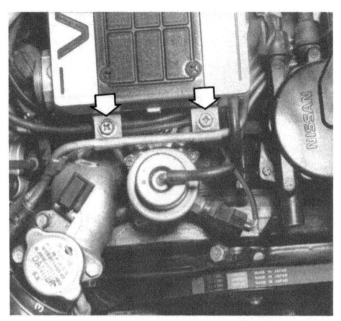

13.3   Remove the two injector assembly fuel lines (arrows)

Injector assembly installation

11   Install the fuel line hose clamps about a half of an inch up the hose.
12   Lubricate the fuel line hoses with gas and install the injector assembly into the fuel line hoses (see illustration). **Note:** *Make sure the injector flat side contacts the flat spacer (see illustrations).*
13   Once the hoses are attached insert each injector into its proper port.
14   Install and tighten the hold-down bolt for each injector (see illustration).
15   Working on the right side of the engine, install the metal spring clips onto the injector electrical connector and slip the connector onto each of the injectors on the right side.
16   Install a new gasket (see illustration) and set the intake collector in place, routing the right side wiring harness up on top of the collector.
17   Route the left side wiring harness up over the collector.
18   Install the fuel injector electrical connectors onto the injectors on the left side and secure with the metal clips.
19   Install the Allen head bolts that secure the collector to the intake manifold.
20   Connect the EGR tube to the collector.
21   Connect all electrical connectors and vacuum/fuel hoses that were disconnected during removal.
22   Connect all wiring harness hold-down straps to the collector.
23   Replace the air regulator to the collector and secure it with the two retaining bolts.
24   Install the electrical connector onto thé air regulator.
25   Reverse the remaining removal procedures for installation.

12   Power transistor − removal and installation

Refer to illustration 12.3

1   Disconnect the negative cable at the battery. Place the cable out of the way so it cannot accidentally come in contact with the negative terminal of the battery, as this would once again allow power into the electrical system of the vehicle.
2   Remove the power transistor electrical connector.
3   Remove the hold-down bolt and lift off the power transistor (see illustration).
4   Reverse the removal procedures for installation.

13   Fuel pressure regulator − removal and installation

Refer to illustrations 13.3 and 13.5

1   Relieve the fuel pressure.

13.5   Disconnect the fuel return line attached to the bottom of the regulator

14.5   Three bolts (arrows) hold the FICD to the intake collector (non-turbo model shown)

2   Disconnect the negative cable at the battery. Place the cable out of the way so it cannot accidentally come in contact with the negative terminal of the battery, as this would once again allow power into the electrical system of the vehicle.
3   From the top of the regulator remove the two injector assembly fuel lines (see illustration).
4   Disconnect the vacuum line and fuel temperature sensor electrical connector.
5   Disconnect the return fuel line located under the regulator (see illustration).
6   Remove the two bolts and remove the regulator.
7   Installation is the reverse of the removal procedures. **Note:** *Wet the inside of the fuel lines with gasoline to facilitate hose installation.*

14   FICD (Fast Idle Control Device) valve − removal and installation

Refer to illustration 14.5

Note: *On non-turbo models the idle-up solenoid valve is attached to the FICD. On turbo models the auxiliary air control valve (AAC) is attached to the FICD.*

1   Disconnect the negative cable at the battery. Place the cable out of the way so it cannot accidentally come in contact with the negative terminal of the battery, as this would once again allow power into the electrical system of the vehicle.
2   To gain access to the FICD remove the upper air inlet tube.
3   Disconnect the electrical connector.
4   Remove the wiring harness retaining bolt attaching the harness to the FICD valve.
5   Remove the three bolts retaining the FICD valve to the intake collector (see illustration).
6   Installation is the reverse of the removal procedure.

15.2   Disconnect the air regulator electrical connector

16.3   Disconnect the electrical connectors attached to the AAC and the FICD assembly (arrows)

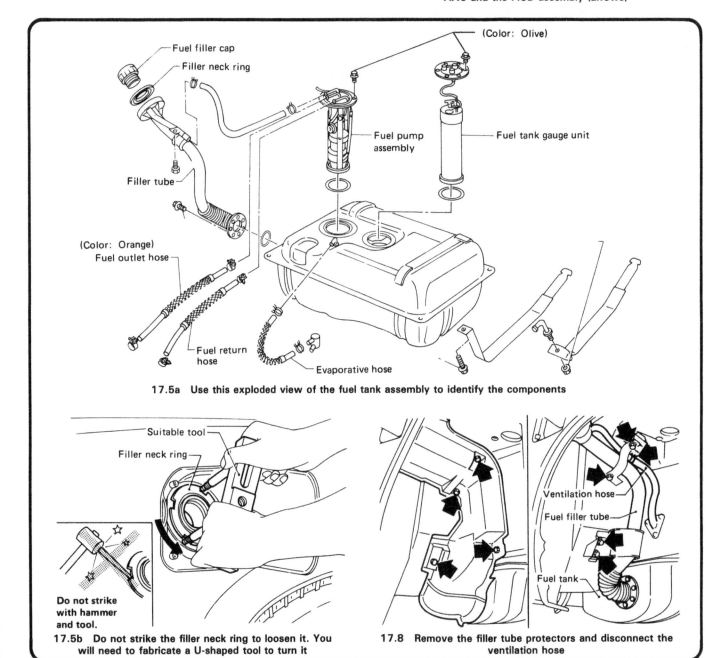

17.5a   Use this exploded view of the fuel tank assembly to identify the components

17.5b   Do not strike the filler neck ring to loosen it. You will need to fabricate a U-shaped tool to turn it

17.8   Remove the filler tube protectors and disconnect the ventilation hose

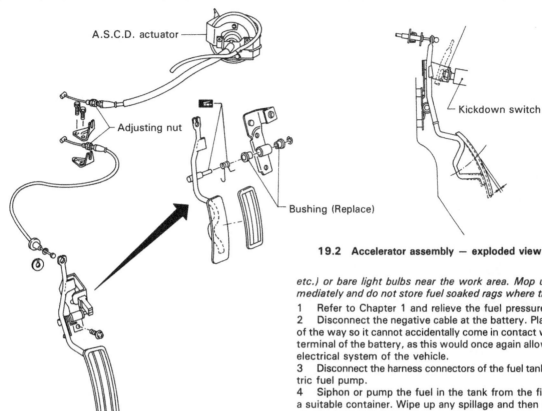

A.S.C.D. actuator

Adjusting nut

Bushing (Replace)

Kickdown switch (A/T)

19.2   Accelerator assembly — exploded view

etc.) or bare light bulbs near the work area. Mop up any spills immediately and do not store fuel soaked rags where they could ignite.
1   Refer to Chapter 1 and relieve the fuel pressure.
2   Disconnect the negative cable at the battery. Place the cable out of the way so it cannot accidentally come in contact with the negative terminal of the battery, as this would once again allow power into the electrical system of the vehicle.
3   Disconnect the harness connectors of the fuel tank gauge and electric fuel pump.
4   Siphon or pump the fuel in the tank from the filler opening into a suitable container. Wipe up any spillage and then use compressed air, if available, to blow away any fuel from around the filler opening.
5   Remove the fuel filler neck ring (see illustration). **Warning:** *Do not strike the ring with a hammer or tool which could create a spark. Fabricate a U-shaped removal tool to turn the filler neck ring (see illustration).*
6   Raise the vehicle and support it securely on jackstands.
7   Remove the right rear wheel.
8   Remove the fuel filler tube protectors and disconnect the ventilation hose (see illustration).
9   Disconnect the fuel hoses and evaporative hose on the fuel tube side.
10   Loosen the bolt and nut securing mounting bands and then remove the fuel tank with the filler tube attached.
11   Reverse the removal procedure for installation.

15   Air regulator — removal and installation

Refer to illustration 15.2
1   Disconnect the negative cable at the battery. Place the cable out of the way so it cannot accidentally come in contact with the negative terminal of the battery, as this would once again allow power into the electrical system of the vehicle.
2   Disconnect the air regulator electrical connector (see illustration).
3   Disconnect the air regulator air tube.
4   Remove the two bolts and remove the air regulator.
5   Reverse the removal procedures for installation. **Note:** *Use a new O-ring when installing the air regulator.*

16   Auxiliary air control (AAC) valve (turbo only) — removal and installation

Refer to illustration 16.3
1   Disconnect the negative cable at the battery. Place the cable out of the way so it cannot accidentally come in contact with the negative terminal of the battery, as this would once again allow power into the electrical system of the vehicle.
2   To gain access to the AAC remove the upper air inlet tube.
3   Disconnect the electrical connectors by pulling the connectors apart (see illustration).
4   Remove the wiring harness bolt attaching the harness to the AAC valve.
5   Remove the three bolts retaining the AAC (and FICD component assembly) valve to the intake collector.
6   Installation is the reverse of the removal procedure.

17   Fuel tank — removal and installation

Refer to illustrations 17.5a, 17.5b and 17.8
Warning: *There are certain precautions which must be taken when inspecting or servicing fuel system components. Work in a well ventilated area and do not allow open flames (cigarettes, appliance pilot lights,*

18   Fuel pump — removal and installation

Warning: *There are certain precautions which must be taken when inspecting or servicing fuel system components. Work in a well ventilated area and do not allow open flames (cigarettes, appliance pilot lights, etc.) or bare light bulbs near the work area. Mop up any spills immediately and do not store fuel soaked rags where they could ignite.*
1   Refer to Chapter 1 and relieve the fuel pressure.
2   Refer to the previous Section and remove the fuel tank.
3   Disconnect the fuel pump electrical connector.
4   Loosen the hose clamps and pull the three hoses from the top of the fuel pump assembly.
5   Remove the retaining bolts securing the pump to the tank and lift the pump assembly from the tank.
6   Installation is the reverse of the removal procedures.

19   Accelerator cable — removal, installation and adjustment

Refer to illustrations 19.2 and 19.5
1   Disconnect the throttle cable at the throttle chamber by raising the actuator and sliding the cable end out of the retaining slot.
2   Disconnect the cable from the hold down/adjusting bracket by backing off on the throttle cable adjusting nut then lifting the cable from the hold down/adjusting bracket (see illustration).

3   Pull the cable from the two retaining clips located inside the engine compartment.
4   From inside the vehicle disconnect the cable from the accelerator pedal by pulling the white plastic grommet forward (toward you), releasing the cable from the accelerator pedal.
5   Lift the retaining slotted washer from the rubber firewall retainer (see illustration).
6   From the engine compartment, pull the cable out of the firewall.
7   Installation is the reverse of the removal procedures.
8   Adjust slack from the cable by turning the front adjusting nut. **Note:** *A properly adjusted cable should go from idle to full open with no slack in the cable when it returns to idle.*

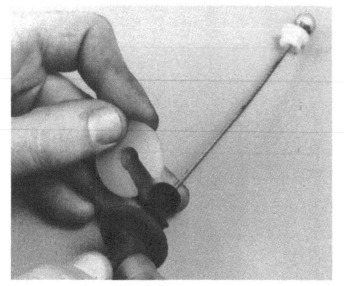

19.5   The slotted cable retaining washer is shown removed from the vehicle for clarity

20   Turbocharger — general information

Refer to illustrations 20.1 and 20.2

1   The turbocharger, installed on the left exhaust manifold, is designed to compress the intake air flowing from the air cleaner (see illustration). The turbocharger uses two turbines. One turbine, situated in the exhaust gas flow, is turned by the gases leaving the exhaust manifold. It then turns the other turbine, which compresses the intake air and sends it into the intake manifold.
2   To prevent the supercharging pressure from becoming excessive, the system is equipped with an exhaust gas bypass valve, which opens at a predetermined pressure to allow exhaust gases to bypass the turbocharger and go straight into the exhaust pipe. An emergency relief valve is also incorporated into the intake collector to relieve excessive pressure at that point (see illustration).
3   Due to the failure of turbine bearings caused by exposure to extreme heat, water jackets were added in 1985 to the turbocharger and connected to the engine cooling system.

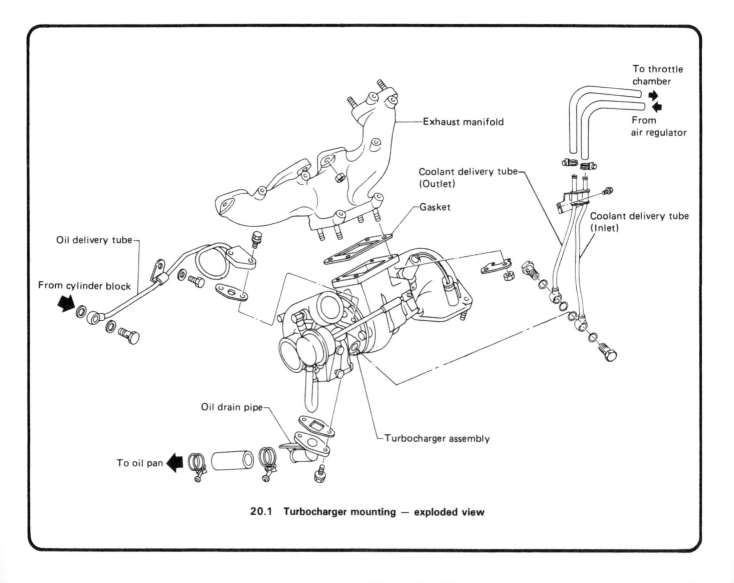

20.1   Turbocharger mounting — exploded view

21 Turbocharger — precautions

1   The turbocharger is driven by exhaust gases and routinely operates at extremely high temperatures. The turbocharger castings retain heat for a very long time and must not be touched for a period of at least three hours after the engine was last run. Even at this time, it is advisable to wear heavy gloves to prevent burns.

2   High rotation speed in the turbocharger means that bearing life and operation is dependent upon a constant flow of engine oil. Careful attention should be paid to the condition of the oil lines and the tightness of their fittings. Never overtighten the hollow bolts, as this will deform the unions and cause leakage.

3   Because the turbocharger is dependent upon clean oil, drain the engine oil and change the oil filter any time the turbocharger is removed. If a bearing fails in the turbocharger or in the engine, flush all engine and turbocharger oil passages completely before reassembly.

4   A turbocharger is a ducted fan, and, like any turbine, its greatest enemy is dirt and foreign objects. A stray nut, metal chip or rock passing through a turbine rotating at 120,000 rpm can cause extreme damage. The best defense against foreign object damage is to work on your engine and turbocharger unit only after the engine has been cleaned. Cover all inlets and pipes. Account for every nut, bolt, screw and washer before starting the engine after assembly is complete.

22 Turbocharger — removal and installation

1   If the turbo is equipped with coolant lines it will be necessary to drain the cooling system.

2   Lift the vehicle and support it securely on jackstands.

3   Refer to the exhaust Section and remove the front exhaust tube from the turbo.

4   Refer to Chapter 10 and remove the steering column lower joint.

5   Remove the retaining bolts from the heat shields and remove the shields.

6   Disconnect the exhaust gas sensor electrical connector.

7   Disconnect the oil delivery tube at the turbo.

8   Disconnect the oil drain pipe by loosening the rubber hose clamps and pulling the hose from the nipple.

9   Disconnect the inlet and outlet coolant delivery tubes from the turbo.

10   Disconnect the four turbo retaining bolts securing the turbo to the exhaust manifold.

11   Remove the turbo out the bottom of the engine compartment.

12   Clean all traces of gasket and sealant material from the gasket mounting surfaces.

13   Installation is the reverse of the removal procedure. **Note:** *Install a new gasket and replace the metal washers on the hollow bolts connecting the coolant and oil delivery lines.*

23 Turbocharger — inspection

Refer to illustration 23.7

Note: *If any of the internal components, such as the waste gate, bearing(s) or oil seals are defective it will be necessary to replace the turbocharger as an assembly. New and factory rebuilt turbochargers are available from your Nissan dealer or local parts store.*

1   If the engine is lacking power, check for air leaks around the compressor, inlet/outlet tubes or the intake manifold and correct any faulty connection.

2   If the engine shows excessively high engine power check for a disconnected or cracked rubber hose on the bypass valve. Replace the hose if a defect is found.

3   If the engine exhibits excessively high oil consumption or the exhaust shows pale blue smoke, check for oil leaks at the lubricating oil passage connection.

4   Further checks will require removal of the turbocharger. Refer to the appropriate Section in this Chapter and remove the turbocharger.

5   With the turbocharger on the work bench, inspect the turbine and compressor wheel by visually checking for cracks, clogging, deformity or other damage.

6   Spin the wheels and check for free movement and abnormal noise or friction.

Checking axial play

7   Check axial play using a dial indicator positioned through the inlet opening on the compressor shaft (see illustration).

8   Using hand pressure only, move the compressor fan as far away from the dial indicator as possible and zero the indicator. Push the compressor wheel toward the dial indicator and compare clearance to the Specifications. If the reading shows too much free play the turbocharger will have to be replaced.

24 Injector cooling fan — removal and installation

Refer to illustrations 24.2 and 24.4

1   Disconnect the negative cable at the battery. Place the cable out of the way so it cannot accidentally come in contact with the negative terminal of the battery, as this would once again allow power into the electrical system of the vehicle.

2   Remove the air blower tube retaining nut and bolt from the radiator

20.2   The turbocharger emergency relief valve is attached to the intake collector

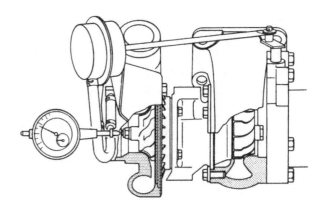

23.7   Check axial free play with a dial indicator positioned in the inlet opening on the compressor shaft

24.2   Remove the retaining nut and slide the bolt
out of the fan shroud to remove the injector
cooling fan blower tube

24.4   Disconnect the fan electrical connector and remove
the three retaining bolts to remove the fan

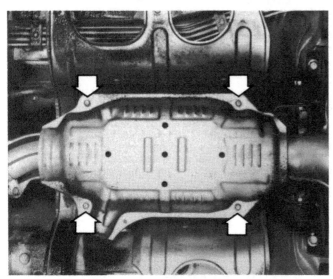

25.2   Remove the four bolts securing the heat shield to
the catalytic converter (arrows)

25.5   Loosen but do not remove the two bolts securing
the front tube to the catalytic converter (arrows)

25.16   Remove the retaining bolts (arrows) and remove
the crossover bracket

25.18   Remove the U-bolt and clamp attaching the
extension tube to the center tube

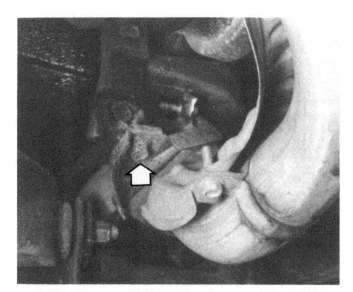

25.25 Remove the bolt securing the front of the muffler assembly to the body (arrow)

25.26 Remove the bolt securing the tail pipe and muffler assembly to the body

fan shroud and remove the tube from the fan (see illustration).
3   Disconnect the injector cooling fan electrical connector.
4   Remove the three bolts securing the fan to the fender well and remove the fan (see illustration).
5   Installation is the reverse of the removal procedure.

---

25 Exhaust components — removal and installation

Refer to illustrations 25.2, 25.5, 25.16, 25.18, 25.25 and 25.26

Front exhaust tube

1   Lift the vehicle and support it securely on jackstands.
2   Remove the lower catalytic converter shield by removing the four bolts (see illustration).
3   Use a piece of wire to secure the converter to the underside of the car.
4   Disconnect any brackets that hold the front exhaust tube to the transmission or body.
5   On non-turbo models, remove heat shield plates and the front exhaust tube as an assembly. On turbo models remove the heat shields. Loosen but do not remove the two bolts attaching the front tube to the catalytic converter (see illustration).
6   Remove the three bolts attaching the front tube to the exhaust manifold.
7   While supporting the front exhaust pipe, remove the two bolts holding the tube to the converter and lift out the tube.
8   Installation is the reverse of the removal procedure. **Note:** *Be sure to use new gaskets and before installing the attaching bolts apply anti-seize compound to the threads.*

Catalytic converter

9   Lift the vehicle and support it securely on jackstands.
10 Remove the lower catalytic converter heat shield.
11 Using a piece of strong wire, secure the exhaust center tube to the driveshaft.
12 While supporting the catalytic converter, break loose, but do not yet remove the four converter mounting bolts attaching the converter to the front and center exhaust tubes.

13 While supporting the converter, remove the four retaining bolts and remove the converter.
14 Installation is the reverse of the removal procedure. **Note:** *Be sure to use new gaskets and before installing the attaching bolts apply anti-seize compound to the threads.*

Exhaust center tube

15 Lift the vehicle and support it securely on jackstands.
16 Remove the rear crossover bracket by removing the four retaining bolts (see illustration).
17 Remove the heat shields from around the exhaust center tube.
18 Remove the U-bolt and clamp attaching the extension tube to the exhaust center tube (see illustration).
19 Lightly tap around the connection with a soft face hammer to break the connection loose.
20 Disconnect the exhaust center tube from the catalytic converter by removing the two retaining bolts.
21 The exhaust center tube can be removed by twisting and pulling it over the rear end housing.
22 Installation is the reverse of the removal procedure. **Note:** *Be sure to use new gaskets and before installing the attaching bolts apply anti-seize compound to the threads. Also use Nissan exhaust sealant kit 20720-N2225 or an equivalent to eliminate gas leakage from the center tube connection.*

Muffler assembly

23 Lift the vehicle and support it securely on jackstands.
24 Disconnect the U-bolt securing the exhaust center tube to the muffler assembly.
25 Remove the bracket retaining bolt securing the front of the muffler assembly to the body (see illustration).
26 Remove the bracket retaining bolt securing the tail pipe and muffler assembly to the body (see illustration).
27 Lower the muffler assembly from the vehicle.
28 Installation is the reverse of the removal procedure. **Note:** *Be sure to use new gaskets and before installing the attaching bolts apply anti-seize compound to the threads. Also, use Nissan exhaust sealant kit 20720-N2225 or an equivalent to eliminate gas leakage from the center tube connection.*

Chapter 5   Engine electrical systems

Contents

Specifications

Ignition system

Firing order	1-2-3-4-5-6
Rotation	counterclockwise
Ignition coil primary voltage	12 volts

Alternator

Hitachi type (turbo)	LR170-701B
normal rating	12V-70A
regulated output voltage	14.4 to 15.0 volts
Mitsubishi type (turbo)	A2T48195
normal rating	12V-70A
regulated output voltage	14.1 to 14.7 volts

Torque specifications	Kg-m	Ft-lbs
Distributor hold-down bolt	0.5 to 0.63	3.6 to 4.6
Alternator adjusting bolt	1.4 to 1.7	10 to 12
Alternator (Mitsubishi)		
through bolt	0.40 to 0.55	2.9 to 4.0
pulley nut	6.0 to 7.5	43 to 54
Alternator (Hitachi)		
through bolt	0.32 to 0.40	2.3 to 2.9
pulley nut	4.0 to 6.0	29 to 43
Solenoid to starter bolts	0.65 to 0.80	4.7 to 5.8
Starter housing through bolts	0.50 to 0.65	3.6 to 4.7

1   Ignition system — general information

In order for the engine to run correctly it is necessary for an electrical spark to ignite the fuel/air mixture in the combustion chamber at exactly the right moment in relation to engine speed and load. The ignition system is based on feeding low tension voltage from the battery to the coil, where it is converted to high tension voltage. The high tension voltage is powerful enough to jump the spark plug gap in the cylinders many times a second under high compression pressures, pro-

viding that the system is in good condition and that all adjustments are correct.

The low tension (primary) circuit consists of the ignition switch, ignition and accessory relay, the primary windings of the ignition coil, the transistorized IC (intregrated circuit) ignition unit, the pick-up assembly in the distributor and all connecting wires.

The high tension circuit consists of the secondary windings of the ignition coil, the heavy ignition lead from the center of the coil to the center of the distributor cap, the rotor, the spark plug leads and spark plugs.

2.3 Mark the relationship of the rotor to a fixed point on the engine

2.4 Disconnect the distributor electrical connector located on the right side of the engine

2.6 To keep the timing as close as possible mark the location of the distributor before removing the distributor

2.8 Replace the rubber O-ring if necessary before installing the distributor

The primary circuit of the ignition system is intregrated with the ECCS fuel injection/turbocharging system, described in detail in Chapter 4.

The engine speed and position of the pistons are monitored by a crank angle sensor, which, in this system, combines with the control unit to replace the distributor pick-up assembly, mechanical and vacuum advance. The necessary information is passed from the crank angle sensor to the control unit, which electronically determines the optimum spark timing and triggers the spark.

Because of the complexity of the ECCS system, if an ignition problem develops which cannot be isolated to the high tension circuit, the vehicle should be taken to a Nissan dealer or other specialist for proper diagnosis and correction. **Warning:** *Because of the high voltage generated by the electronic ignition system, extreme caution should be taken whenever an operation is performed involving ignition components. This not only includes the distributor, coil, control module and ignition wires, but related items which are connected to the system as well, such as the plug connections, tachometer and any testing equipment.*

2 Distributor — removal and installation

Refer to illustrations 2.3, 2.4, 2.6, 2.8, 2.9a, 2.9b, 2.12 and 2.13

1 Disconnect the negative battery cable at the battery. Place the cable out of the way so it cannot accidentally come in contact with the negative terminal of the battery, as this would once again allow power into the electrical system of the vehicle.

2 Remove the two retaining screws, remove the distributor cap and position the cap out of the way.

3 Mark the relationship of the rotor to a fixed point on the engine (see illustration).

4 Disconnect the distributor electrical connector on the right side of the engine (see illustration) and work the wire from under the intake collector.

5 Disconnect the ground wire on the side of the distributor.

6 Mark the relationship of the distributor housing to the engine block (see illustration) and remove the distributor hold-down bolt.

7 Lift straight up on the distributor to remove it from the vehicle.

8 Before installing the distributor check the rubber O-ring on the shaft and replace it if it is not soft and pliable (see illustration).

2.9a   Insert the distributor with the rotor about 45° from the alignment marks

2.9b   When the distributor gear engages fully with the drive gear the marks should line up

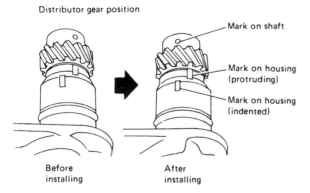

2.12   Before installing the distributor align the bottom mark on the shaft with the protruding mark just above the drive gear

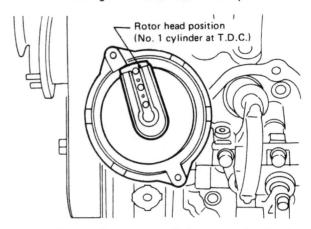

2.13   With the distributor installed the rotor should point towards the number one cylinder

3.3   Disconnect the electrical connector by removing the retaining screw (arrow) and unplugging the connector from the distributor

9   Insert the distributor with the rotor about 45° from the alignment marks (see illustration) so the distributor will line up with the mark when the distributor gear meshes into the drive gear (see illustration).

10   Reverse the remaining removal procedures for installation.

Installation if the engine was rotated

11   Set the engine at TDC for the number one piston. Refer to Chapter 2 if necessary.

12   Locate the alignment marks on the rotor shaft above and below the drive gear. Align the bottom mark with the protruding mark just above the drive gear and insert the distributor (see illustration).

13   Once the distributor is installed and the rotor is in place the rotor should be pointing toward the number one cylinder (see illustration). Confirm this and tighten the distributor hold-down bolt.

---

3   Crank angle sensor — removal and installation

Refer to illustrations 3.3, 3.4, 3.7, 3.8 and 3.9

Note: *A general description of the crank angle sensor can be found in Chapter 4.*

1   Remove the distributor.

2   Remove the rotor retaining screw and remove the rotor.

3   Disconnect the electrical connector retaining screw and disconnect the connector from the distributor (see illustration).

3.4   Remove the screws retaining the crank angle sensor
dust cover (arrows)

3.7   Carefully lift the rotor plate from the distributor

3.8   Lift the rotor plate positioner and note the sequence
of the wavy washer and base plate, then remove all three
as an assembly

3.9   Remove the crank angle sensor by removing the three
retaining screws (arrows)

4   Remove the crank angle sensor dust cover retaining screws (see
illustration) and lift off the cover.
5   Wrap the distributor gear with a rag and place the distributor in
a vise. **Caution:** *Apply just enough vise pressure to hold the distributor.*
6   Insert a screwdriver into the rotor shaft and remove the retaining
screw and shaft.
7   Carefully lift the rotor plate from the distributor (see illustration).
8   Lift the rotor plate positioner and note the wavy spring located
between the positioner and the rotor plate base plate (see illustration).
Set the rotor plate positioner back onto the assembly and lift the complete assembly off the distributor.
9   Remove the three retaining screws and remove the crank angle
sensor from the distributor (see illustration).
10   Reverse the removal procedure for installation.

---

4   Ignition coil — removal and installation

1   Disconnect the negative battery cable at the battery. Place the cable
out of the way so it cannot accidentally come in contact with the
negative terminal of the battery, as this would once again allow power

into the electrical system of the vehicle.
2   Unplug the high tension lead from the coil.
3   Disconnect the ignition coil electrical connector.
4   Remove the coil retaining screws and remove the coil.
5   Reverse the removal procedure for installation.

---

5   Starting system — general information

The starting system is composed of a starting motor, solenoid and
battery. The battery supplies the electrical energy to the solenoid, which
then completes the circuit to the starting motor.
All ZX models use a reduction gear type starter motor. This starter
operates in a similar manner to a conventional type except that increased torque is provided through a reduction gear located between
the armature and pinion gear.
An overrunning clutch is attached to the starter motor to transmit
the driving torque and to prevent the armature from overrunning when
the engine starts.

The solenoid and starting motor are mounted together at the right rear of the engine. No periodic lubrication or maintenance is required for the starting system components.

The electrical circuitry of the vehicle is arranged so that the starter motor can only be operated on automatic transmission models when the lever is in Park or Neutral.

Never operate the starter motor for more than 30 seconds at a time without pausing to allow it to cool for at least two minutes. Excessive cranking can cause overheating, which can seriously damage the starter.

6   Starting system — testing

Refer to illustration 6.10

1   If the starter motor does not rotate when the switch is operated on an automatic transmission model, check that the shifter is in the Neutral or Park position.

2   Check that the battery is well charged and all cables, both at the battery and starter solenoid terminals, are secure.

3   If the motor can be heard spinning but the engine is not being cranked, then the overrunning clutch in the starter motor is slipping and the assembly must be removed from the engine and replaced.

4   Often when the starter fails to operate, a click can be heard coming from the starter solenoid when the ignition switch is turned to the Start position. If this is heard, proceed to Step 12. If this click is not heard, disconnect the ignition wire (usually black and yellow) from the starter solenoid.

5   Connect a test light between this lead and the negative battery terminal.

6   Have an assistant turn the key, and note whether the test light comes on or not. If the light comes on, proceed to Step 12.

7   If the test light does not come on, connect a voltmeter between the positive battery terminal and the bayonet connector at the solenoid. The black and yellow wire should be connected.

8   Disconnect the ignition coil wire from the distributor cap and ground it to the engine.

9   Have an assistant attempt to crank the engine and note the reading on the voltmeter. If less than 1.5 volts is shown, there is a open circuit in the starter, and it should be replaced with a new or rebuilt unit.

10   If the voltmeter shows more than 1.5 volts, connect a jumper wire between the positive battery terminal and the S terminal on the solenoid (see illustration).

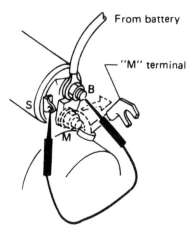

6.10   Connect a jumper wire between the battery positive terminal and the S terminal on the solenoid

11   Turn the ignition switch to the Start position and listen for a click from the solenoid. If there still is no click and the starter does not turn, both the solenoid and starter are defective and should be replaced. If a click is heard and the starter turns, then there is an open circuit in the ignition switch, inhibitor switch or relay (on automatic transmission models) or in the wires or connectors.

12   If a click was heard when the ignition switch was turned to the Start position in Step 4, or if the starter turns then the starter current should be tested. Disconnect the positive battery cable at the starter.

13   Connect an ammeter (set in the 500A range) between this cable and its terminal on the starter.

14   Remove the ignition coil wire from the distributor cap to keep the engine from starting.

15   Have an assistant crank the engine and note the reading on the ammeter and the speed at which the starter turns.

16   If the ammeter reading is less than 150 amps and the starter speed is normal, the starter is okay and the problem lies elsewhere.

17   If the ammeter reading is less than 150 amps but the starter turns slowly, test for a voltage drop in the starter positive circuit.

18   Connect a voltmeter so the positive lead is on the positive battery post and the negative lead is connected to the *M* solenoid terminal (this is the terminal with the braided copper strap leading to the starter motor).

19   Remove the ignition coil wire from the distributor cap and ground it on the engine.

20   Have an assistant crank the engine and note the reading on the voltmeter. If less than 1 volt is shown, proceed to Step 23.

21   If more than 1 volt is shown, connect the negative voltmeter lead to the solenoid terminal (this is the terminal which connects to the battery).

22   Have an assistant crank the engine and note the reading on the voltmeter. If more than 1 volt is shown, then the problem is a bad connection between the battery and the starter solenoid. Check the positive battery cable for looseness or corrosion. If the reading indicated less than 1 volt, then the solenoid is defective and should be replaced.

23   If the voltmeter test in Step 20 shows less than 1 volt, test for a voltage drop in the starter ground circuit.

24   Connect the negative lead of the voltmeter (set on the low scale) to the negative battery terminal and hold the positive voltmeter lead to the starter housing. Be sure to make a good connection.

25   Have an assistant crank the engine and note the voltmeter reading. If the reading shows more than 0.5 volts, then there is a bad ground connection. Check the negative battery cable for looseness or corrosion. Also check the starter motor ground connections and the tightness of the starter motor mounting bolts.

7   Starter motor — removal and installation

Refer to illustrations 7.3 and 7.4

1   Disconnect the negative battery cable from the battery. Place the cable out of the way so it cannot accidentally come in contact with the negative terminal of the battery, as this would once again allow power into the electrical system of the vehicle.

2   Lift the vehicle and support it securely on jackstands.

3   Unplug the ignition switch wire connecter to the solenoid and remove the battery cable by removing the retaining nut (see illustration).

7.3   Unplug the ignition switch wire connector to the solenoid and remove the battery cable wire by removing the retaining nut

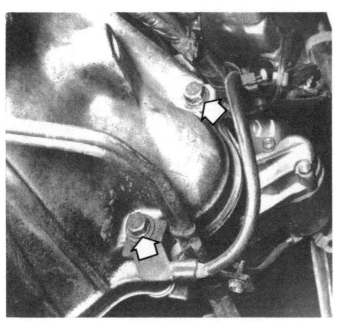

7.4 Remove the the two starter retaining bolts (arrows)

8.2 Remove the nut securing the braided copper wire to the solenoid

4 Remove the two starter retaining bolts (see illustration).
5 Disengage the starter/solenoid assembly from the flywheel/driveplate and remove the assembly from the vehicle.
6 Installation is the reverse of the removal procedure.

8 Starter solenoid — removal and installation

Refer to illustrations 8.2 and 8.3
1 Remove the starter motor as described in Section 12.
2 Remove the nut holding the braided copper strap to the solenoid (see illustration).
3 Remove the two solenoid mounting bolts (see illustration) and lift the solenoid from the starter motor.
4 Installation is the reverse of the removal procedure. **Note:** *During installation be sure to engage the solenoid plunger with the starter motor shift lever.*

9 Charging system — general information

The charging system is made up of the alternator, voltage regulator and the battery. These components work together to supply electrical power for the engine ignition, lights, radio, etc.
The alternator is turned by a drivebelt at the front of the engine. When the engine is operating, voltage is generated by the internal components of the alternator to be sent to the battery for storage.
The alternator uses a solid state regulator that is mounted inside the alternator housing. The purpose of this voltage regulator is to limit the alternator voltage to a preset value. This prevents power surges, circuit overloads, etc., during peak voltage output. The regulator voltage setting cannot be adjusted.
The charging system does not ordinarily require periodic maintenance. The drivebelts, electrical wiring and connections should, however, be inspected during normal routine maintenance.

10 Charging system — testing

1 Where a faulty alternator is suspected, first check to be sure that the battery is fully charged.
2 Visually inspect all wires and connections to make sure they are clean, tight and in good condition.
3 Turn the ignition switch to the On position and check that the alternator warning light in the instrument cluster illuminates. If it does, pro-

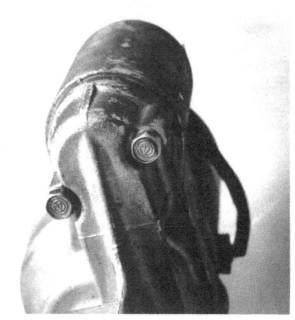

8.3 After removing the two retaining bolts the solenoid can be separated from the starter

ceed to Step 6.
4 If the warning light does not come on, check that the warning light bulb is not burned out. To do this, disconnect the S/L connector at the rear of the alternator and ground the L lead wire (brown/white) with a jumper wire. Again, turn the ignition switch to the On position and check the warning light. If the warning light is still off, a burned bulb or loose connection between the alternator and the warning light is indicated.
5 If the warning light did come on in the previous test, the bulb is in good condition and a faulty alternator or regulator is indicated. To determine which, reconnect the S/L connector and use a jumper wire to ground the F terminal. Turn the ignition switch On and check the warning light. If the light comes on, the voltage regulator is faulty. If the light remains off, the alternator is defective. Replace either component by referring to the appropriate Section of this Chapter.
6 If the warning light does come on with the ignition switch in the On position, start the engine and allow it to idle. The light should go out. If it remains on, even dimly or as a flicker, a faulty alternator is indicated. Replace it with a new or rebuilt unit.

11.7   Remove the two front alternator retaining bolts
(arrows)

11.8   Turn the alternator around and unplug the electrical
connectors and remove the two retaining bolts securing the
attached wires

7   Slowly increase the engine speed to 1500 rpm.
8   Turn the headlights on and again check the alternator warning light.
If it remains off, proceed to Step 10.
9   If the warning light comes on dimly, lower the engine speed to
normal idle. Connect the voltmeter between the B (white) and L terminals and measure the voltage. If more than 0.5 volts is shown, the
alternator should be replaced. If less than 0.5 volts is shown, the unit
is okay.
10 If the warning light remains off when the lights are turned on, maintain the engine speed at 1500 rpm and measure the voltage at the B
terminal. Be sure the S terminal (white) is correctly connected.
11 If more than 15.5 volts is shown, the IC regulator is defective.
Replace the alternator.
12 If the voltage reading is between 13 and 15 volts, lower the engine
speed to idle.
13 Turn the headlights on and again check the alternator warning light
in the dash. It should remain off. If the light comes on, a defective alternator is indicated.

11   Alternator — removal and installation

Refer to illustrations 11.7 and 11.8

1   Disconnect the negative battery cable. Place the cable out of the
way so it cannot accidentally come in contact with the negative terminal
of the battery, as this would once again allow power into the electrical
system of the vehicle.
2   Remove alternator drivebelt.
3   Lift the vehicle and support it securely on jackstands.
4   Remove sway bar-to-body brackets and let the sway bar hang
down.
5   Remove the splash shield retaining bolts and remove the shield.
6   From the rear of the alternator remove the one retaining bolt securing the alternator to the mounting bracket.
7   Remove the two front alternator retaining bolts (see illustration).
8   Turn the alternator around and squeeze the electrical connector
and pull it from the alternator. Unplug the capacitor by pulling the connector apart. Remove the two retaining bolts and remove the attached
wires (see illustration).
9   Installation is the reverse of the removal procedure.

Chapter 6 Emissions control systems

Contents

Specifications

Related specifications can be found in Chapter 4

Torque specifications	Kg-m	Ft-lbs
EGR control valve	1.8 to 2.3	13 to 17
EGR tube	3.5 to 4.5	25 to 33

1 General information

As pollution control standards have become more stringent, the emissions control systems developed to meet these requirements have not only become increasingly more diverse and complex, but are now designed as integral parts of the operation of the engine. Where once the anti-pollution devices used were installed as peripheral "add-on" components, the present systems work closely with such other systems as the fuel, ignition and exhaust systems, and in some cases even control vital engine operations. **Note:** *Refer to Chapter 4 for fuel/emissions related components.*

Because of this close integration of systems, disconnecting or not maintaining the emissions control systems, besides being illegal, can adversely affect engine performance and life, as well as fuel economy. The emissions systems are not particularly difficult for the home mechanic to maintain and service. You can perform general operational checks, and do most (if not all) of the regular maintenance with common tune-up and hand tools.

2 Emissions control systems – general component description

Refer to illustrations 2.1a, 2.1b and 2.6

Positive Crankcase Ventilation (PCV) system

The positive crankcase ventilation system (see illustrations) reduces hydrocarbon emissions by circulating fresh air through the crankcase. This air combines with blow-by gases, and the combination is then sucked into the intake manifold to be burned by the engine.

This process is achieved by using one air pipe running from the air duct (suction hose on Turbo models) to the rocker arm cover, a one-way PCV valve located on the left underside of the intake manifold and a second air pipe running from the crankcase to the PCV valve.

During partial throttle operation the vacuum created in the intake manifold is great enough to suck the gases from the crankcase, through the PCV valve and into the manifold. The PCV valve allows

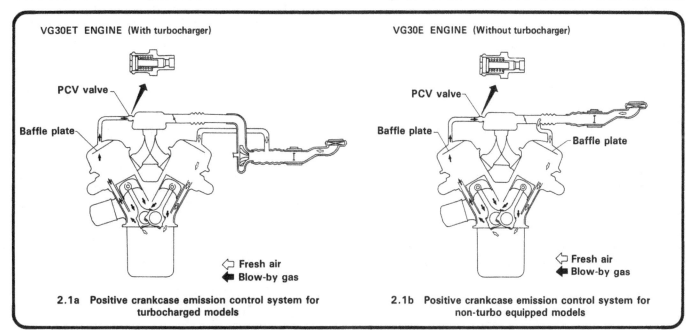

2.1a **Positive crankcase emission control system for turbocharged models**

2.1b **Positive crankcase emission control system for non-turbo equipped models**

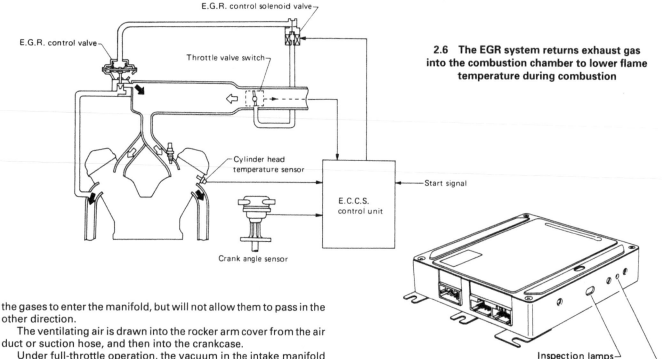

2.6   The EGR system returns exhaust gas into the combustion chamber to lower flame temperature during combustion

3.3   Verify the diagnosis mode selector is turned fully counterclockwise using a small screwdriver

the gases to enter the manifold, but will not allow them to pass in the other direction.

The ventilating air is drawn into the rocker arm cover from the air duct or suction hose, and then into the crankcase.

Under full-throttle operation, the vacuum in the intake manifold is not great enough to draw the gases in. Under this condition the blow-by gases flow backwards into the rocker arm cover, through the air tube and into the air duct or suction hose, where it is carried into the intake manifold in the normal air intake flow.

Exhaust Gas Recirculation (EGR) system

The EGR system (see illustration), returns exhaust gas into the combustion chamber to lower the flame temperature during combustion. This results in a reduction of the nitrogen oxide density in the exhaust gas. To check the EGR valve, refer to Chapter 1.

When the EGR control valve is open some of the exhaust gas is led from the exhaust manifold to the EGR tube. The exhaust gas is then regulated by the EGR valve and introduced into the intake manifold. A signal from the ECCS, sent to the EGR control solenoid valve, cuts the vacuum line for the EGR control valve on demand.

Catalytic converter

Once the exhaust gas, which contains unburned, harmful gases, leaves the combustion chamber it is passed through the catalytic converter. The three-way catalytic converter activates to oxidize and reduce harmful gases (HC, CO and NO_x) into harmless gases (CO_2, H_2O and N_2).

Evaporative emission control system

The fuel vapor from the sealed fuel tank is led into a canister which contains activated carbon and the vapor is stored while the engine is not running. Once the engine is started the gas vapor is drawn through the purge line into the intake manifold. The purge line control valve is closed during idle and opens as engine speed increases.

---

3   ECCS system – self diagnosis

1984 thru 1986

Refer to illustration 3.3

Self-diagnostic procedure

1   Locate the ECCS control module under the passenger side kick panel.

2   Remove the module retaining bolts and pull the module out so you can handle it. **Caution:** *The self-diagnosis results are retained in the memory by a small current flow from the battery. Disconnecting the battery cable or the ECCS's 15-pin connector erases the memory*

stored. *Always perform the self-diagnosis regarding intermittent checks before disconnecting.*

3   Verify the diagnosis mode selector (see illustration) is turned fully counterclockwise using a small screwdriver.

4   Turn the ignition switch On.

5   Check that the inspection lamps stay on to check the bulbs.

6   Turn the mode selector fully clockwise.

7   Confirm that the lamps are displaying codes 23, 24 and 31 on turbo models. If not, write down the malfunction code.

8   Depress the accelerator pedal one time and release.

9   On VG30ET only, shift the transmission through the gears, ending in neutral.

10   On 1984-85 models make sure the inspection lamps are displaying codes 24 (VG30ET) and 31. If not, write down the malfunction code.

11   Start the engine, and on 1986 VG30ET models with automatic transmission, apply the brake and move the selector to "D".

12   Confirm the codes displayed: They should be 14 for VG30ET engines and 31 for others.

13   If a VG30ET model, drive the vehicle at more than 6 mph.

14   Make sure the lights are flashing code 31. If not, write down the malfunction code.

15   Add a load to the system by turning the air conditioning switch On then Off.

16   The lights should display code 44.

17   Turn the diagnosis mode selector fully counterclockwise.

18   Turn the engine off.

19   See the decoding chart for trouble code identification.

20   Check the malfunctioning area, then erase the memory. **Caution:** *The crank angle sensor plays an important part in the electronic computer control system and a malfunctioning sensor is sometimes accompanied by a display which shows malfunctions in other signal systems. If this happens, start with the crank angle sensor.*

Memory erasure

21   Turn the switch to the On position.

22   Turn the diagnosis mode selector fully clockwise and hold it there for more than two seconds.

23   Turn the ignition switch to Off.

24 After correcting a malfunctioning system be sure to erase the memory.

25 Reverse the removal procedures to install the computer control module.

1987 on

26 Turn the ignition switch to the On position. Turn the diagnostic mode selector on the ECU fully clockwise and wait until the inspection lamps flash. Count the number of flashes to find which mode you are in, then turn the diagnostic mode selector fully counterclockwise.

27 When the ignition switch is turned off during diagnosis, in each mode, and then turned on again after the power to the ECU has dropped off completely, the diagnosis will automatically return to Mode I.

28 The check engine light on the instrument panel (California models only) comes on when the ignition switch is turned on or in Mode I when the emission system malfunctions (with the engine running).

29 Malfunctions related to fuel and emission control systems can be diagnosed using the self diagnostic codes of Mode III.

30 To start the diagnostic procedure, remove the dash side finish panel so you can watch the inspection lamps on the ECU. Start the engine and warm it up to normal operating temperature. Turn the diagnostic mode selector on the ECU fully clockwise. After the inspection lamps have flashed 3 times, turn the diagnostic mode selector fully counterclockwise. The ECU is now in Mode III. Check the trouble code chart for the particular malfunction.

31 After the tests have been performed and the repairs completed, erase the memory by turning the diagnostic mode selector on the ECU fully clockwise. After the inspection lamps have flashed 4 times, turn the mode selector fully counterclockwise. This will erase any signals the ECU has stored concerning a particular component.

Trouble codes*	Circuit or system	Probable cause
Code 11 (1 red flash, 1 green flash)	Crank angle sensor/circuit	Faulty crank angle sensor or circuit.
Code 12 (1 red flash, 2 green flashes)	Air flow meter/circuit	The air flow meter source or ground circuit(s) may be shorted or open. Check the air flow meter.
Code 13 (1 red flash, 3 green flashes)	Cylinder head temperature sensor	The sensor source or ground circuit(s) may be shorted or open. Check the temperature sensor/circuit(s).
Code 14 (1 red flash, 4 green flashes)	Vehicle speed sensor	The vehicle speed sensor signal circuit is open. This repair must be performed by a dealer service department.
Code 21 (2 red flashes, 1 green flash)	Ignition signal	The ignition signal in the primary circuit is not entered during engine cranking or running. This repair must be performed by a dealer service department.
Code 22 (2 red flashes, 2 green flashes)	Fuel pump circuit	The fuel circuit is open or shorted.
Code 23 (2 red flashes, 3 green flashes)	Idle switch circuit	The idle switch signal circuit is open.
Code 24 (2 red flashes, four green flashes)	Park/neutral switch	The park/neutral switch circuit is open or shorted.
Code 31 (3 red flashes, 1 green flash)	Load signal circuit or ECU	One of the load signal (air conditioning, power steering or radiator fan) circiuits is open or shorted or the ECU is malfunctioning.
Code 32 (1986 and earlier) (3 red flashes, 2 green flashes)	Starter signal circuit	The starter signal circuit is open or shorted.
Code 32 (1987 and later) (3 red flashes, 2 green flashes)	EGR Function	The EGR control valve does not operate.
Code 33 (3 red flashes, 3 green flashes)	Exhaust gas sensor	The exhaust gas signal circuit is open.
Code 34 (3 red flashes, 4 green flashes)	Detonation sensor	The detonation sensor circuit is open or shorted. This repair must be performed by a dealer service department.
Code 35 (3 red flashes, 5 green flashes)	Exhaust gas temperature circuit	The temperature sensor signal circuit is open or shorted.
Code 41 (1984 only) (4 red flashes, 1 green flash)	Air temperature sensor or circuit	Faulty air temperature sensor or circuit.
Code 41 (1985 and later) (4 red flashes, 1 green flash)	Fuel temperature sensor or circuit	Faulty fuel temperature sensor or circuit.
Code 42 (4 red flashes, 2 green flashes)	Fuel temperature sensor	The fuel temperature sensor circuit is open or shorted.
Code 43 (4 red flashes, 3 green flashes)	Throttle sensor	The throttle sensor circuit is open or shorted. This repair must be performed by a dealer service department.
Code 45 (4 red flashes, 5 green flashes)	Injector leak	The injector(s) are leaking.
Code 44 or 55 (5 red flashes, 5 green flashes)	No malfunction	Normal operation

* Not all codes apply to all models

4.1   Allow time for the EGR valve and surrounding components to cool before beginning work

4.3   Disconnect the EGR inlet pipe (arrow)

5.4   Remove the two remaining screws (arrows) to remove the EGR control solenoid valve

4   EGR control valve – removal and installation

Refer to illustrations 4.1 and 4.3
Note: *EGR valve testing is found in Chapter 1.*
1   Allow time for the EGR control valve (see illustration) and adjacent accessories to cool.
2   Disconnect the EGR vacuum hose.
3   Remove the EGR inlet pipe from the EGR control valve (see illustration).
4   Remove the two retaining nuts and lift the control valve off the studs.
5   After cleaning the gasket mounting surfaces and installing a new gasket, reverse the removal procedures for installation. **Caution:** *When installing the EGR guide tube, be careful of its direction. The outlet faces the rear of the engine. Otherwise the distribution efficiency of the exhaust gas will be reduced.*

5   EGR control solenoid valve – removal and installation

Refer to illustration 5.4
1   Disconnect the negative cable at the battery. Place the cable out of the way so it cannot accidentally come in contact with the negative terminal of the battery, as this would once again allow power into the electrical system of the vehicle.
2   Disconnect the electrical connector by pulling it apart.
3   Remove the upper and lower vacuum hoses from the valve.
4   Remove the two EGR control solenoid valve retaining screws (see illustration).
5   Installation is the reverse of the removal procedures.

6   Pressure regulator control solenoid valve – description

Refer to illustration 6.1
   The pressure regulator control solenoid valve (see illustration) cuts the intake manifold vacuum signal for pressure regulator control. The solenoid valve is actuated by an on/off signal from the ECCS control unit.
   Once the valve is turned off, a vacuum signal from the intake manifold is fed into the pressure regulator. When the control unit outputs an on signal, the coil pulls the plunger downward and cuts the vacuum signal.

7   Throttle valve switch – description

Refer to illustration 7.1
1   Attached to the throttle chamber, the throttle valve switch (see

6.1   The pressure regulator control solenoid cuts the intake manifold vacuum signal

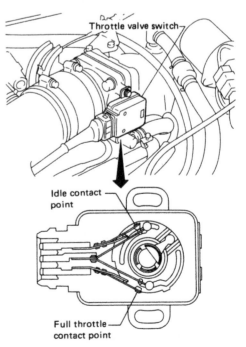

7.1   The throttle valve switch has two throttle contacts. At idle the contact is closed, and it opens when the throttle valve is at any other position

When the pressure in the intake manifold is below 55.3 kPa (415 mmHg, 16.34 inHg)

When the pressure in the intake manifold is above 63.3 kPa (475 mmHg, 18.70 inHg)

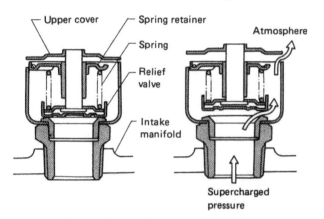

8.1   The emergency relief valve is a pop-off valve to prevent an abnormal rise in supercharging pressure

9.1   Remove the retaining bolt (arrow) and move the climate control vacuum reservoir out of your work area

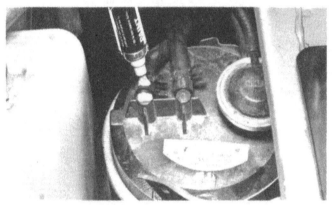

9.2   Make installation of hoses easier by marking them with paint or colored tape before removing

9.5   It's a tight squeeze but the canister can be pulled straight up to remove it from the vehicle

illustration) actuates in response to accelerator pedal movement. The switch has idle contact and full throttle contact. When the throttle valve is positioned at idle, the idle contact is closed, and it opens when it is at any other position.

2   The full throttle contact is used only for electronically controlled automatic transmission models.

8   Emergency relief valve — description, removal and installation

Refer to illustration 8.1

1   The emergency relief valve (see illustration) is installed on the intake collector of the turbocharged models as a safety device to prevent an abnormal rise in supercharging pressure, and possible engine damage, in case the bypass valve should fail to open properly.

2   The valve is screwed into the intake collector and simply requires the proper size wrench to remove it from the collector.

3   Installation is the reverse of the removal procedure.

9   Evaporative emissions control canister — removal and installation

Refer to illustrations 9.1, 9.2 and 9.5

1   Remove the retaining bolt from the climate control vacuum reservoir and move the reservoir to gain access to the evaporative emissions canister (see illustration).

2   To avoid confusion, mark each vacuum hose with different colored tape or paint before removing the hoses from the canister (see illustration).

3   Squeeze the hose retaining clamps and pull the hoses off the canister.

4   Lift the retaining spring up and over the canister.

5   Pull the canister up and out of the vehicle. It will be necessary to pull the upper radiator hose over, and even then it is a tight squeeze (see illustration).

6   Reverse the removal procedures for installation.

Chapter 7 Part A Manual transmission

Contents

Specifications

FS5W71C 5 speed manual transmission

Gear ratios	
1st	3.321:1
2nd	1.902:1
3rd	1.308:1
4th	1.000:1
overdrive	0.759:1
reverse	3.382:1
Oil capacity	See Chapter 1
Gear backlash	
main drive gear	0.0020 to 0.0039 in
1st gear	0.0020 to 0.0079 in
2nd gear	0.0020 to 0.0079 in
3rd gear	0.0020 to 0.0079 in
overdrive	0.0020 to 0.0079 in
reverse idler gear	0.0020 to 0.0079 in
Gear end play	
1st gear	
1986	0.0122 to 0.0161 in
1984-1985	0.0122 to 0.0146 in
2nd gear	
1886	0.0043 to 0.0083 in
1984-1985	0.0043 to 0.0067 in
3rd gear	
1986	0.0043 to 0.0083 in
1984-1985	0.0043 to 0.0067 in
overdrive gear	
1986	0.0126 to 0.0197 in
1984-1985	0.0126 to 0.0138 in
reverse idler gear	
all	0.0020 to 0.0197 in
Baulk ring to gear clearance	
1st and 2nd gears	
1986	0.0394 to 0.0630 in
1984-1985	0.0394 to 0.0551 in
3rd and main drive gears	
1986	0.0472 to 0.0630 in
1984-1985	0.0472 to 0.0591 in
overdrive gear	
all	0.0394 to 0.0551 in

Wear limit
  1st and 2nd gears
    1986 . . . .   0.031 in
    1984-1985 . . . .   0.020 in
  3rd and main drive gears . . . .   0.031 in
  overdrive gear . . . .   0.020 in
Available snap rings
  main drive gear bearing . . . .   0.0681 in
    0.0709 in
    0.0736 in
    0.0764 in
    0.0791 in
    0.0819 in
  mainshaft front . . . .   0.094 in
    0.098 in
    0.102 in
  mainshaft rear end bearing . . . .   0.043 in
    0.047 in
    0.051 in
    0.055 in
  counter drive gear . . . .   0.055 in
    0.059 in
    0.063 in
Available shims for countershaft front bearing
  Distance from bearing surface to transmission case
    1986
      0.1780 to 0.1854 in . . . .   not necessary
      0.1740 to 0.1776 in . . . .   0.004 in
      0.1701 to 0.1736 in . . . .   0.008 in
      0.1661 to 0.1697 in . . . .   0.012 in
      0.1622 to 0.1657 in . . . .   0.016 in
      0.1583 to 0.1618 in . . . .   0.020 in
      0.1543 to 0.1579 in . . . .   0.024 in
    1984-1985
      0.1346 to 0.1382 in . . . .   0.004 in
      0.1307 to 0.1343 in . . . .   0.008 in
      0.1268 to 0.1303 in . . . .   0.012 in
      0.1228 to 0.1264 in . . . .   0.016 in
      0.1189 to 0.1224 in . . . .   0.020 in
      0.1150 to 0.1185 in . . . .   0.024 in

Model BW T-5 (FS5R90A) 5 speed manual transmission
Gear backlash
  main drive gear . . . .   0.0020 to 0.0059 in
  1st gear . . . .   0.0020 to 0.0059 in
  2nd gear . . . .   0.0020 to 0.0059 in
  3rd gear . . . .   0.0020 to 0.0059 in
  overdrive gear . . . .   0.0020 to 0.0059 in
  reverse idler gear . . . .   0.0020 to 0.0079 in
Gear end play
  1st gear . . . .   0.0039 to 0.0098 in
  2nd gear . . . .   0.0039 to 0.0098 in
  3rd gear . . . .   0.0059 to 0.0157 in
  5th counter gear . . . .   0.0039 to 0.0189 in
  counter gear . . . .   0.0098 to 0.0197 in
Preload of main drive gear bearing . . . .   0.0051 to 0.0098 in
Main drive gear bearing adjusting shims . . . .   0.0122 in
  0.0142 in
  0.0161 in
  0.0181 in
  0.0201 in
  0.0220 in
  0.0362 in
  0.0441 in

Torque specifications   **Ft-lbs**
Clutch operating cylinder . . . .   22 to 30
Transmission to engine . . . .   29 to 36
Engine gusset to transmission . . . .   22 to 29
Crossmember to body . . . .   23 to 31
Rear mounting insulator to rear extension . . . .   23 to 31
Starter motor to transmission . . . .   22 to 39

Torque specifications (continued)

	Ft-lbs
Model FS5W71C transmission gear assembly	
rear extension to transmission case	12 to 14
front cover to transmission case	12 to 15
control housing to rear extension	10 to 13
ball pin	14 to 25
filler plug	18 to 25
drain plug	18 to 25
speedometer sleeve installation	2.9 to 3.6
return spring plug	
1985-1986	14 to 22
1984	5.8 to 7.2
reverse check sleeve to transmission case	2.9 to 3.6
reverse lamp switch	14 to 22
check ball plug	14 to 18
mainshaft lock nut	101 to 123
countershaft lock nut	72 to 94
striking lever lock nut	6.5 to 8.7
bearing retainer to adapter plate	14 to 18
Model BW T-5 FS5R90A transmission gear assembly	
clutch housing to case	30 to 51
cover to case	6.0 to 10.8
front retainer to case	11 to 20
filler plug	14 to 25
drain plug	14 to 25
speedometer sleeve	2.9 to 3.6
reverse pivot pin	15 to 25
neutral switch	14 to 22
extension to case	20 to 46
control lever housing to extension	11 to 20

1 General information

The manual transmissions installed in the ZX models are conventional five-speeds. The non-turbo models are equipped with either an FS5W71C (all years) or an FS5R30A (1987 on) transmission. The turbo models are equipped with a BW T-5 FS5R90A type. All transmissions have synchromesh on all forward gears. **Note:** *Overhaul procedures and specifications for the model FS5R30A could not be included at the time this manual was published.*

The main driveshaft gear is meshed with the counter drive gear. The forward speed gears on the countershaft are in constant mesh with the main gears.

When the shift lever is operated, the relevant coupling sleeve is caused to slide on the synchronizer hub and engages its inner teeth with the outer teeth on the mainshaft gear. Moving the shift lever to the Reverse gear position moves the mainshaft reverse gear into engagement with the reverse idler gear.

2 Transmission mount – check and replacement

Refer to illustration 2.8 and 2.9

Check

1   Raise the vehicle and support it securely on jackstands. Make sure the vehicle is stable as it will be necessary to jostle it during the course of checking and replacing the mounts.

2   Push up on the transmission extension housing and note the amount the housing moves.

3   Pull down on the extension housing and note the amount of movement available.

4   If the extension housing can be pushed upwards a greater distance than it can be pulled down, it is an indication that the rubber is worn and the mount has bottomed out.

5   If the rubber portion of the mount separates from the metal plate when you push upwards, the mount should be replaced.

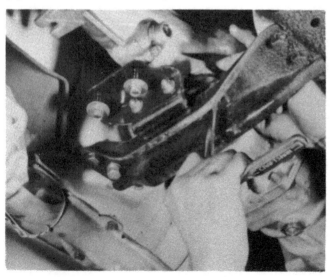

2.8 Remove the two through bolts securing the mount to the crossmember

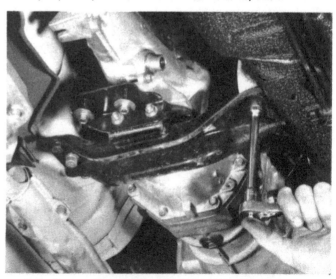

2.9 Remove the four bolts securing the crossmember to the body and remove the crossmember

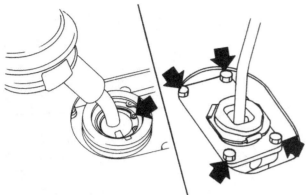

3.4 Remove the FS5W71C shifter by removing the snap ring (left) and remove the FS5R90A shifter by removing the four retaining bolts (right)

Replacement

6 Remove the front exhaust tube (Chapter 4). **Note:** *Wire the catalytic converter to the body.*

7 Use a jack positioned under the transmission to support the weight of the transmission. **Caution:** *Place a block of wood between the jack and the transmission to protect the transmission case.*

8 Remove the two through bolts securing the mount to the crossmember (see illustration).

9 Remove the four retaining bolts securing the crossmember to the body and remove the crossmember (see illustration).

10 Remove the two bolts retaining the mount to the transmission and remove the mount.

11 Reverse the removal procedure for installation.

---

3 Transmission — removal and installation

Refer to illustration 3.4

1 Disconnect the negative cable at the battery. Place the cable out of the way so it cannot accidentally come in contact with the negative terminal of the battery.

2 Raise the vehicle and support it at the four corners with jackstands. Be sure to raise the vehicle enough so the transmission can be slid from underneath when removed.

3 From inside the vehicle remove the console. Refer to Chapter 11 if necessary.

4 If equipped with a FS5W71C transmission (non-turbo), remove the shift lever by removing the large snap-ring retaining the shifter ball to the transmission. On FS5R90A models, remove the shifter by removing the four retaining bolts and lift off the shifter (see illustration).

5 From underneath the vehicle, remove the exhaust front tube, catalytic converter and crossover connecting pipe.

6 Disconnect the six retaining bolts and remove the catalytic converter heat shield.

7 Disconnect the wires from the back-up light switch, the top gear switch, the neutral switch and/or the overdrive switch as equipped, at the transmission. Mark the connectors with tape or paint for identification.

8 Disconnect the speedometer cable at the transmission (Section 5).

9 Drain the oil from the transmission.

10 Remove the driveshaft as described in Chapter 8.

11 Remove the exhaust pipe support bracket.

12 Unbolt and remove the clutch operating cylinder and tie it back out of the way.

13 Support the engine under the oil pan, using a jack and a block of wood as an insulator. **Caution:** *Do not support engine at the drain plug.*

14 Place a jack under the transmission.

15 Remove the transmission mounting insulator nuts and the mounting-to-frame bolts (Section 2).

16 Remove the starter motor from the bellhousing (Chapter 5).

17 Remove the bolts which secure the bellhousing to the engine. These

4.6a Removing the transmission rear extension seal with a screwdriver

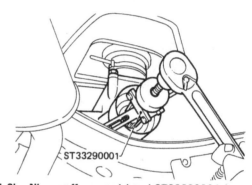

4.6b Nissan offers special tool ST33290001 for pulling the rear extension seal

bolts are of unequal length and should be layed out in order so they can be returned to their original locations.

18 Lower each of the two jacks simultaneously, until the transmission can be withdrawn to the rear and removed from beneath the vehicle. **Caution:** *Do not allow the weight of the transmission to hang on the input shaft while it is still engaged with the clutch plate splines.*

19 Installation is the reverse of the removal. Smear a small amount of grease on the input shaft splines and check clutch mechanism alignment as described in Chapter 8.

20 Check the clutch pedal free travel and adjust if necessary (Chapter 8). Remember to fill the transmission with the correct grade and quantity of lubricant.

---

4 Extension housing oil seal — replacement

Refer to illustration 4.6a, 4.6b and 4.7

1 Disconnect the negative cable at the battery. Place the cable out of the way so it cannot accidentally come in contact with the negative terminal of the battery.

2 Raise the vehicle and support it at four corners with jackstands.

3 From underneath the vehicle, remove the exhaust front tube and catalytic converter (Chapter 4).

4 Disconnect the retaining bolts and remove the catalytic converter heat shield.

5 Remove the driveshaft (Chapter 8).

6 Use a long screwdriver to pry out the old seal (see illustration). Be careful not to damage the output shaft of the transmission housing. An alternative method of removal is to sink a sheet metal screw into the metal face of the seal and use the screw (or screws) to work the seal out of its recess. Nissan tool ST33290001 is used by factory mechanics to pull the seal from the transmission (see illustration).

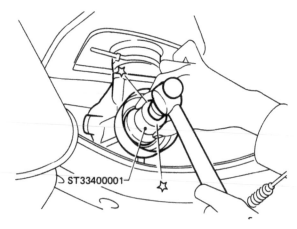

4.7   Carefully drive the seal into its recess using either the special tool shown or use a socket or piece of pipe

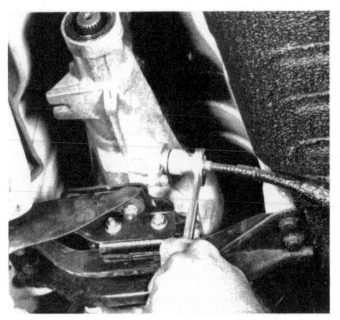

5.3   Unscrew the speedometer cable from the transmission

5.4   Remove the speedometer lock plate hold-down bolt

7   Apply a coat of gear oil to the lips of the new seal and apply sealant to the outside diameter of the seal, then carefully drive it into place using a socket or piece of pipe. Special tools are available from Nissan for this procedure. Nissan tool ST33400001 is used to drive the seal into position (see illustration).
8   The remainder of the installation is the reverse of the removal procedures.

---

5   Speedometer pinion gear and seal — replacement

Refer to illustrations 5.3, 5.4 and 5.5
1   The speedometer pinion gear and seal can be removed without removing the transmission from the vehicle. **Warning:** *Exhaust components are extremely hot after the engine has been running, allow the exhaust system cool completely before starting this project.*
2   Raise the vehicle and support it securely on jackstands.
3   Unscrew the speedometer cable from the transmission (see illustration).
4   Disconnect the speedometer lock plate hold-down bolt and remove the lock plate (see illustration). **Note:** *If equipped, make note of the ground wire connected to the hold-down bolt and be sure to install it on installation.*
5   It may be necessary to gently pry the gear from the extension hous-

5.5   Remove the speedometer pinion gear from the transmission

ing with a screwdriver. Remove the pinion gear (see illustration).
6   Pry off the O-ring and replace with a new one.
7   Installation is the reverse of the removal, but lubricate the new seal with transmission lubricant and hold the assembly so that the slot in the fitting is toward the lock plate boss on the extension.

---

6   5-speed transmission (FS5W71C) — overhaul

Refer to illustrations 6.4, 6.6a, 6.6b, 6.7, 6.8, 6.9, 6.10, 6.12, 6.16, 6.18, 6.20, 6.22a, 6.22b, 6.34, 6.38, 6.64a, 6.64b, 6.72a, 6.72b, 6.73, 6.75, 6.76, 6.84a, 6.84b, 6.94, 6.96, 6.97 and 6.105

Initial disassembly

1   With the transmission removed, thoroughly clean the external surfaces.

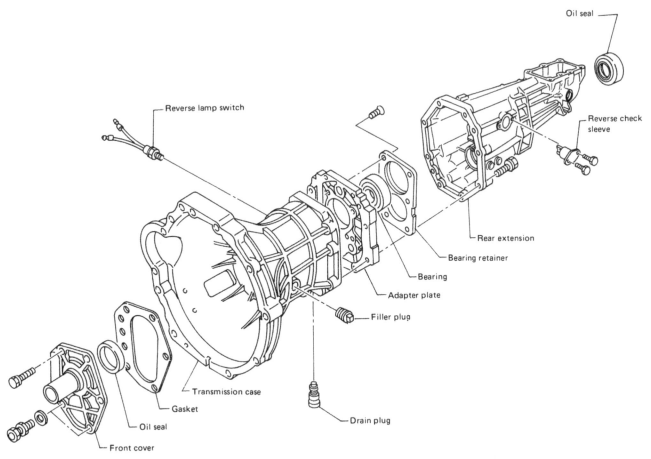

6.4 Case components of the FS5W71C 5-speed manual transmission

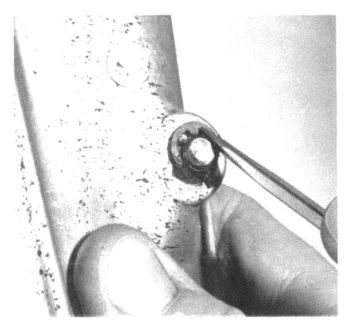

6.6a Pry off the E-ring from the stopper guide pin

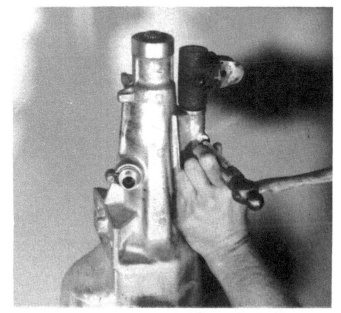

6.6b Use a small punch to drive out the stopper guide pin

2   Remove the rubber dust boot from the withdrawal lever opening in the clutch bellhousing.
3   Remove the release bearing and hub, together with the withdrawal lever (Chapter 8).
4   Remove the reverse lamp switch, neutral switch, top gear switch, and/or overdrive gear switch, as equipped (see illustration).
5   Remove the screw retaining the speedometer gear to the rear extension housing and lift out the gear (Section 3).
6   Pry off the E-ring from the stopper guide pin (see illustration) and drive out the pin (see illustration).

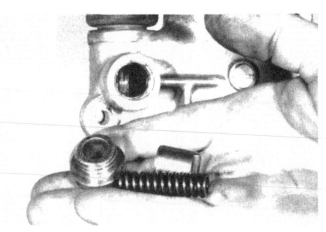

6.7   Remove the plug and lift out the return spring and plunger

6.8   Remove the cover and the reverse check sleeve

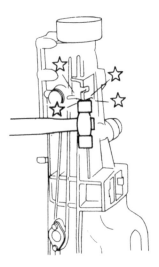

6.9   Remove the bolts and use a soft face hammer to drive the rear extension housing from the main casing

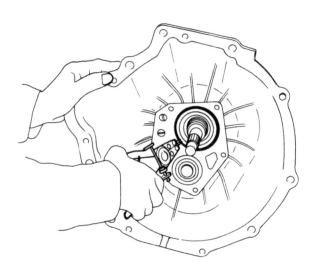

6.10   Extract the countershaft bearing shim and input shaft bearing snap ring

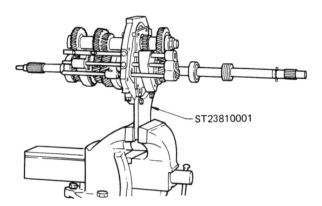

ST23810001

6.12   A suitable plate can be fabricated to support the adapter and gear assembly while working on it or you can purchase one from Nissan

7   Remove the return spring plug and lift out the return spring and plunger (see illustration).
8   Remove the two screws retaining the reverse check cover to the housing and lift out the reverse check sleeve (see illustration).
9   Remove the bolts and drive the rear extension housing from the main transmission casing, using a soft mallet (see illustration).

10   Remove the front cover retaining bolts. Remove the front cover with gasket and extract the countershaft bearing shim and the input shaft bearing snap-ring (see illustration).
11   Drive off the one-piece bellhousing/transmission casing from the adapter plate, using a soft face hammer and holding the exposed rear shaft.
12   Make up a suitable plate and bolt it to the transmission adapter, or purchase tool ST23810001 from Nissan, and support the adapter in a vise (see illustration).
13   Drive out the securing pins from each of the shift forks, using a suitable thin drift.
14   Remove the three check ball plugs.
15   Withdraw the selector rods from the adapter plate.
16   Catch the shift forks and extract the balls and springs as the selector rods are withdrawn. The four smaller balls are the interlock balls (see illustration).
17   Inspect the gears and shafts for any wear, chipping or cracking.
18   Use a feeler gauge between each mainshaft gear to determine the amount of gear end play that exists. Compare the results to the Specifications (see illustration).
19   If the gear end play is not within specifications, or if the gears or shafts show signs of wear or damage, the gear assemblies should be disassembled and the defective parts replaced.
20   Lock the gears and, using a two leg puller, draw the front bearing from the countershaft (see illustration).

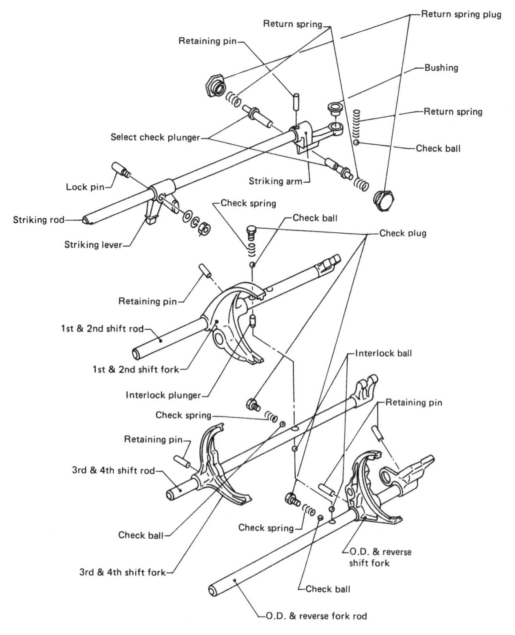

Return spring

Return spring plug

Retaining pin

Bushing

Return spring

Check ball

Select check plunger

Striking arm

Lock pin

Check spring

Check ball

Check plug

Striking rod

Striking lever

Retaining pin

1st & 2nd shift rod

1st & 2nd shift fork

Interlock plunger

Interlock ball

Retaining pin

Check spring

Retaining pin

3rd & 4th shift rod

Check ball

Check spring

O.D. & reverse shift fork

3rd & 4th shift fork

Check ball

O.D. & reverse fork rod

6.16   Shift forks — exploded view

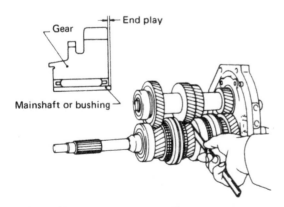

End play

Gear

Mainshaft or bushing

6.18   Check gear end play with a feeler gauge and compare the results to the Specifications

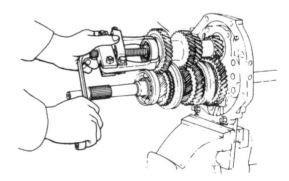

6.20   Use a puller to remove the countershaft front bearing

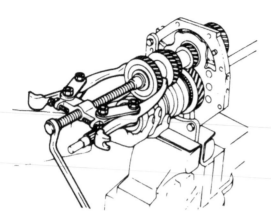

6.22a   Extract the counter drive gear with the main drive gear assembly

21 Extract the now exposed snap-ring from the countershaft.
22 Withdraw the countershaft gear together with the input shaft (see illustration). Take care not to drop the needle roller bearing which is located on the front of the mainshaft (see illustration).
23 Extract the snap-ring from the front of the mainshaft, followed by the thrust washer.
24 Withdraw third and fourth synchronizer unit, followed by third gear.
25 Both the mainshaft nut and the countershaft nut are staked to prevent them from loosening. Use a hammer and punch to drive out the staking.
26 Remove the countershaft nut. Once removed, this nut should not be reused.
27 Use a gear puller to remove the countershaft overdrive gear and bearing.
28 Remove the reverse counter gear and spacer.
29 Remove the snap-ring from the reverse idler shaft and remove the reverse idler gear.
30 Remove the snap-ring retaining the speedometer gear to the mainshaft, then remove the speedometer gear and steel ball.
31 Remove the two snap-rings from behind the speedometer gear and withdraw the overdrive mainshaft bearing.

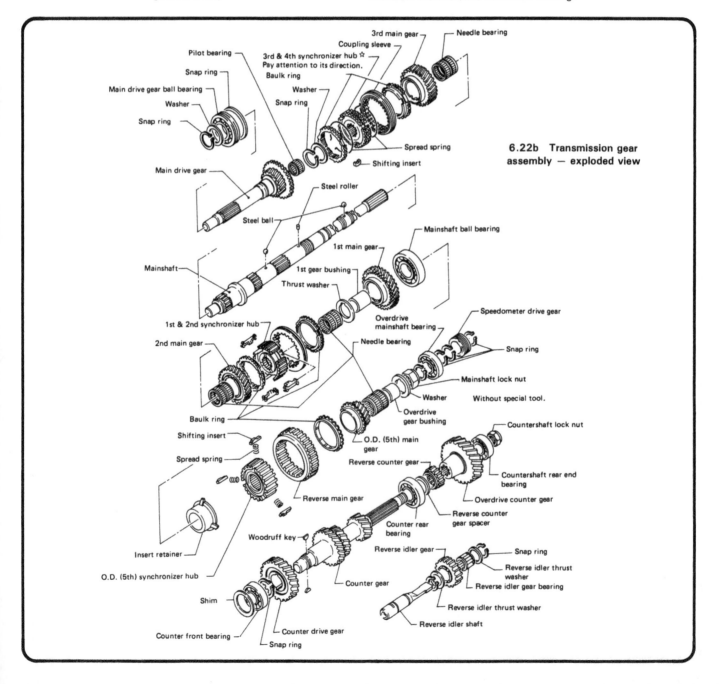

6.22b   Transmission gear assembly — exploded view

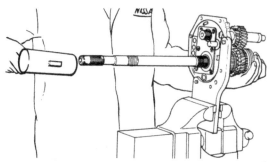

6.34 Remove the mainshaft and countershaft as an assembly, using a rubber mallet to drive the assembly from the adapter

32 Drive out the staking on the mainshaft nut and remove it. Once removed, this nut should not be reused.

33 Remove the thrust washer, overdrive gear bushing, needle bearing, overdrive gear, reverse main gear, overdrive synchronizer assembly and insert retainer.

34 Drive the mainshaft and countershaft assemblies simultaneously from the adapter plate, using a soft-faced hammer (see illustration).

Mainshaft

35 Carefully examine the gears and shaft splines for chipping of the teeth or wear, then dismantle the gear train into its component parts, replacing any worn or damaged items.

36 Examine the shaft itself for scoring or grooving and the splines for twist, taper or general wear.

37 Examine the synchromesh units for cracks or wear and replace them if evident, particularly if there has been a history of noisy gear changes or where the syncromesh can be easily "beaten".

38 Press the baulk ring tight against the synchromesh cone and measure the gap between the two components. If it is less than specified, replace the components (see illustration).

39 When reassembling the synchromesh unit ensure that the ends of the snap-rings on opposite sides of the units do not engage in the same slot.

40 Begin assembly of the mainshaft by installing the second gear needle bearing, second gear, the baulk ring followed by the first/second synchromesh unit, noting carefully the direction of installing the latter.

41 Install the first gear baulk ring, needle bearing, steel ball, thrust washer, bushing and first gear. Be sure the steel ball is well greased when installed.

Countershaft

42 The countershaft front bearing was removed at the time of dismantling the transmission into major units.

43 The countershaft rear bearing was left in position in the adapter plate.

44 Withdraw the countershaft drive gear and extract the two Woodruff keys.

45 Check all components for wear, especially the gear teeth and shaft splines for chipping. Reinstall the Woodruff keys and the snap-ring.

46 Reassembly is a reversal of dismantling.

Input shaft (main drive gear)

47 Remove the snap-ring and spacer.

48 Withdraw the bearing using a two-leg puller or a press. Once removed (by means of its outer race), discard the bearing.

49 Press the new bearing onto the shaft, applying pressure to the center race only.

50 Reinstall the washer.

51 Several thicknesses of snap-rings are available for the main input shaft bearing, as listed in the Specifications. Choose a size that will eliminate bearing end play.

Oil seals

52 Pry out the oil seal from the rear extension and drive in a new one, with the seal lips facing inwards.

53 Reinstall the speedometer pinion sleeve O-ring seal.

54 Reinstall the oil seal in the front cover by prying out the old one and driving in a new one with an appropriate sized socket.

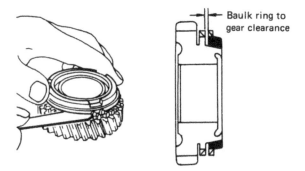

6.38 The baulk ring should be checked for wear by mating it with its gear and measuring the gap with a feeler gauge

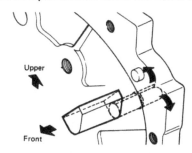

6.64a Install the dowel pin in the adapter plate so that it extends an equal distance above and below the plate

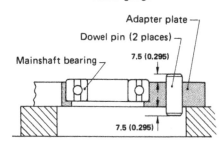

6.64b Install the oil gutter in the adapter plate and expand the rear side

Rear extension housing

55 Loosen the nut on the end of the striking rod lock pin, until it is half off the threads (see illustration 6.16).

56 Using the nut as a guide, drive the lock pin from the striking rod with a punch.

57 Slide the striking lever from the striking rod and withdraw the rod from the rear of the housing.

58 Check the rod and lever for wear or damage and replace if necessary. Replace the O-ring on the striking rod.

59 Inspect the bushing in the rear of the extension housing. If this bushing is worn or cracked, the entire rear extension housing must be replaced.

60 Reinstall the striking rod assembly by reversing the removal procedure.

Reassembly

61 Before beginning to reassemble the transmission, the mainshaft and countershaft adapter plate bearings should be removed, examined and replaced if worn. To do this, unscrew the six screws retaining the bearing retainer plate to the adapter plate. The use of an impact driver will probably be required for this operation.

62 With the bearing retainer plate removed, press the mainshaft and countershaft bearings from the adapter plate. Apply pressure only to the outer races of the bearings.

63 Check the bearings for wear by first washing them in clean solvent and them drying in air. Spin them with your fingers, and if they are noisy or loose in operation, replace them.

64 Check that the dowel pin and oil gutter are correctly positioned on the adapter plate (see illustrations).

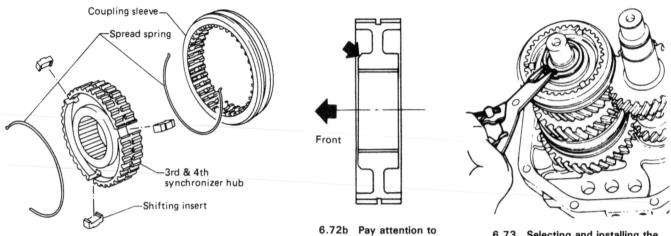

6.72a   Third and fourth gear synchronizer assembly

6.72b   Pay attention to the installed direction of the synchronizer

6.73   Selecting and installing the proper snap ring will minimize end play in the mainshaft

65 Tap the mainshaft bearing lightly and squarely into position in the adapter plate.
66 Drive the reverse idler shaft into the adapter plate so that two-thirds of its length is projecting rearwards. Ensure that the cutout in the shaft is positioned to receive the edge of the bearing retainer plate.
67 Install the bearing retainer plate and tighten the screws to specifications.
68 Stake each screw in two places to keep them from loosening.
69 Tap the countershaft rear bearing into position in the adapter plate.
70 Press the mainshaft assembly into position in the bearing in the adapter plate. Support the rear of the bearing center race during this operation.
71 Press the countershaft assembly into position in the bearing in the adapter plate. Again, support the rear of the bearing center race during this operation.
72 Install the needle bearing, third gear, baulk ring and the third/fourth synchromesh unit onto the front of the mainshaft (see illustrations).
73 Install the thrust washer, then choose a snap-ring from the sizes listed in the Specifications that will minimize end play (see illustration).
74 Insert the needle pilot bearing in its recess at the end of the input shaft.
75 Mesh the countershaft drive gear with the fourth gear on the input shaft. Push the drive gear and input shaft onto the countershaft and mainshaft simultaneously, using Nissan tools ST2386000 and KV31100401 and a press, or use a piece of tubing to drive the countershaft gear into position while supporting the rear of the countershaft (see illustration).
76 Select a countershaft drive gear snap-ring from the sizes listed in the Specifications, so that the gear end play will be minimized (see illustration).
77 Using a socket, drive the front bearing onto the countershaft.
78 Install the reverse counter gear spacer onto the rear of the countershaft.
79 Install the snap-ring, thrust washer, needle bearing, reverse idler gear, reverse idler thrust washer and rear snap-ring onto the reverse idler shaft.
80 Onto the rear side of the mainshaft, install the synchronizer assembly, reverse gear, overdrive gear bushing, needle bearing and baulk ring.
81 Install the reverse counter gear on the countershaft.
82 Mesh the overdrive gear with the overdrive counter gear and install them on their respective shafts, with the overdrive gear on the mainshaft and the overdrive counter gear on the countershaft.
83 Apply grease to the steel ball and install it and the thrust washer onto the rear of the mainshaft.
84 Install a new locknut onto the rear of the mainshaft and torque it to specification. **Note:** *In order to accurately tighten the nut to its torque specification, a wrench adapter should be used (see illustration). Used with the adapter, the torque reading on the wrench will not be accurate and should be converted to the correct torque by referring to the chart (see illustration).*
85 Install the countershaft rear end bearing onto the countershaft.

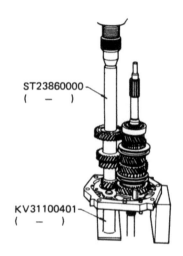

ST23860000
(—   )

KV31100401
(—   )

6.75 A press and special tools are called for to press the counter shaft gear into position, but a piece of tubing can be used with a hammer to drive the gear into position

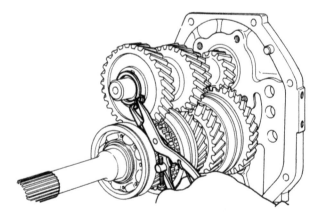

6.76   Select the correct snap ring to minimize end play

86 Install the countershaft locknut and torque it to specification.
87 Use a hammer and punch to stake both the mainshaft and countershaft locknuts so they engage the grooves in their respective shafts.
88 Once again measure the gear end play, as described in Step 18.
89 Fit a snap-ring onto the mainshaft and then install the overdrive mainshaft bearing.

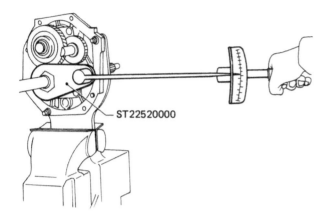

6.84a The mainshaft locknut should be tightened using a torque wrench and adapter

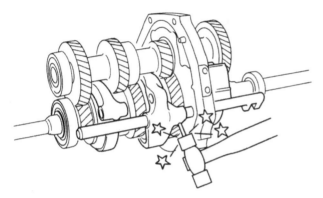

6.94 Drive in a new retaining pin

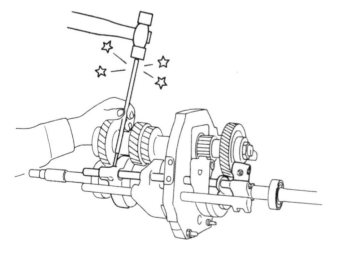

6.96 Drive in the retaining pin for the third and fourth gear selector rod

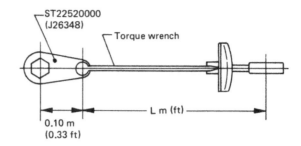

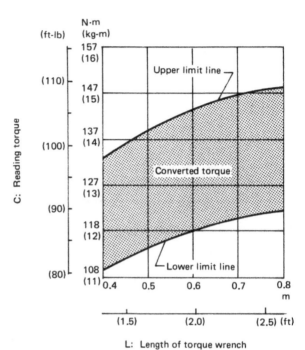

L: Length of torque wrench

6.84b When tightening the mainshaft locknut, the true torque can be found by matching the torque reading on the wrench with the chart above

90 Choose a snap-ring from the sizes listed in the Specifications to eliminate end play of the mainshaft rear bearing.
91 Install the next snap-ring, then grease the steel ball and install the ball and the speedometer drive gear onto the mainshaft. Install the last snap-ring.
92 Locate the first/second shift fork onto the first/second synchronizer unit. The long end of the shift fork must be towards the countershaft. Locate the third/fourth shift fork onto the third/fourth synchronizer unit. The long end of the shift fork must be the opposite side to the first/second shift fork.
93 Locate the overdrive reverse shift fork onto the overdrive synchronizer so that the upper rod hole is in line with the third/fourth shift fork.

94 Slide the first/second selector rod through the adapter plate and into the first/second shift fork. Align the hole in the rod with the hole in the fork and drive in a new retaining pin (see illustration).
95 Align the notch in the first/second selector rod with the check ball bore, then install the check ball and spring and screw in the check ball plug. Apply thread sealant to the check ball plug.
96 Invert the adapter plate assembly (hold the third/fourth and OD/reverse shift forks in position) so that the check ball plug installed in Step 95 is lowermost. Drop two interlock balls into the third/fourth detent ball plug hole and, using a suitable thin probe, push them up against the first/second selector rod. If the adapter plate is correctly positioned, the interlock balls will drop into position. Slide the third/fourth selector rod through the upper hole of the OD/reverse shift fork and the adapter plate, ensuring that the interlock balls are held between this selector rod and the first/second selector rod, and into the third/fourth shift fork. Align the holes in the shift fork and selector rod, and drive in a new retaining pin (see illustration). Install a check ball, spring and check ball plug (with thread sealant applied) to the third/fourth check ball plug bore. Ensure that the notch in the third/fourth selector rod is aligned with the check ball plug bore before assembling the check ball.
97 Drop two interlock balls into the remaining ball plug bore, ensuring that they locate against the third/fourth selector rod. Slide the overdrive/reverse selector rod through the overdrive reverse shift fork and into the adapter plate. Ensure that the two interlock balls are held in position between the third/fourth selector rod and the overdrive/reverse selector rod, sliding the overdrive/reverse selector rod into the adapter plate until the notch in the selector rod aligns with the check ball plug bore. Insert the check ball, spring and check ball plug as before. Drive

in a new retaining pin to retain the overdrive/reverse shift fork to the overdrive/reverse selector rod (see illustration).

98 Tighten the three detent ball plugs to the specified torque.

99 Thoroughly oil the entire assembly and check to see that the selector rods operate correctly and smoothly.

100 Clean the mating faces of the adapter plate and the transmission

casing and apply gasket sealant to both surfaces.

101 Tap the transmission casing into position on the adapter plate, using a soft face hammer, taking particular care that it engages correctly with the input shaft bearing and countershaft front bearing.

102 Fit the outer snap-ring to the input shaft bearing.

103 Clean the mating faces of the adapter plate and rear extension housing and apply gasket sealant.

104 Arrange the shift forks in their neutral mode and lower the rear extension housing onto the adapter plate so that the striking lever engages correctly with the selector rods.

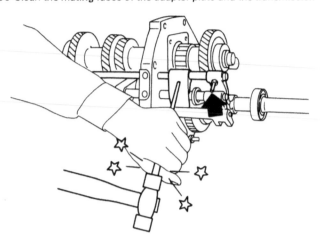

6.97   Install the retaining pin for the overdrive/reverse shift fork to the overdrive/reverse selector rod

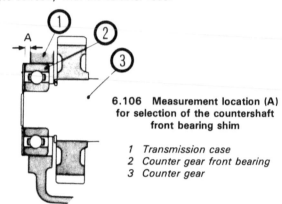

6.106   Measurement location (A) for selection of the countershaft front bearing shim

1   Transmission case
2   Counter gear front bearing
3   Counter gear

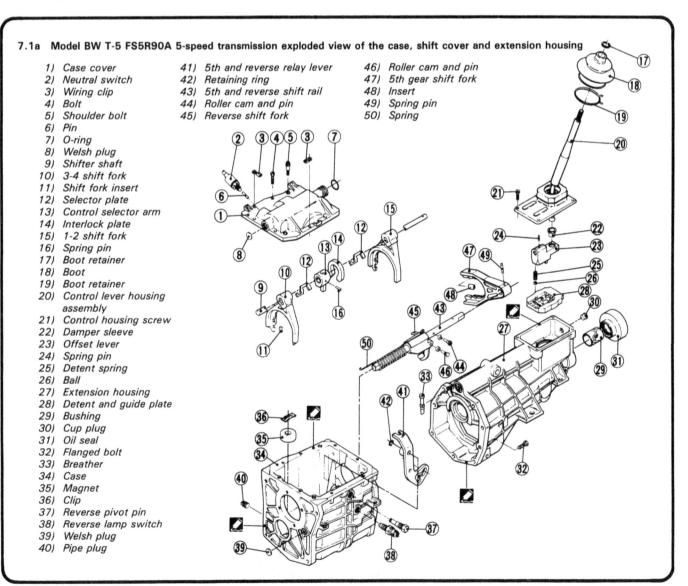

7.1a   Model BW T-5 FS5R90A 5-speed transmission exploded view of the case, shift cover and extension housing

1) Case cover
2) Neutral switch
3) Wiring clip
4) Bolt
5) Shoulder bolt
6) Pin
7) O-ring
8) Welsh plug
9) Shifter shaft
10) 3-4 shift fork
11) Shift fork insert
12) Selector plate
13) Control selector arm
14) Interlock plate
15) 1-2 shift fork
16) Spring pin
17) Boot retainer
18) Boot
19) Boot retainer
20) Control lever housing assembly
21) Control housing screw
22) Damper sleeve
23) Offset lever
24) Spring pin
25) Detent spring
26) Ball
27) Extension housing
28) Detent and guide plate
29) Bushing
30) Cup plug
31) Oil seal
32) Flanged bolt
33) Breather
34) Case
35) Magnet
36) Clip
37) Reverse pivot pin
38) Reverse lamp switch
39) Welsh plug
40) Pipe plug
41) 5th and reverse relay lever
42) Retaining ring
43) 5th and reverse shift rail
44) Roller cam and pin
45) Reverse shift fork
46) Roller cam and pin
47) 5th gear shift fork
48) Insert
49) Spring pin
50) Spring

105 Fit the bolts which secure the sections of the transmission together and tighten them to the specified torque.

106 Measure the amount by which the countershaft front bearing protrudes from the transmission casing front face. Use feeler gauge blades for this and then select the appropriate shims after reference to the table found in Specifications (see illustration):

107 Install the shim using a dab of thick grease, then fit the front cover to the transmission casing (within the clutch bellhousing) complete with a new gasket and taking care not to damage the oil seal as it passes over the input shaft splines.

108 Tighten the securing bolts to the specified torque, making sure that the bolt threads are coated with gasket sealant to prevent oil seepage.

109 Complete the reassembly by reversing the procedures described in Steps 1 through 8 of this Section.

7 5-speed transmission (BW T-5 FS5R90A) — overhaul

Refer to illustrations 7.1a, 7.1b, 7.1c, 7.1d, 7.1e, 7.3, 7.6, 7.7, 7.8, 7.14, 7.19, 7.21 and 7.35

Disassembly

1   Study the exploded view drawings of the transmission.
2   Remove the offset lever roll pin using a pin punch and hammer.
3   Remove the extension housing-to-transmission case bolts. Remove

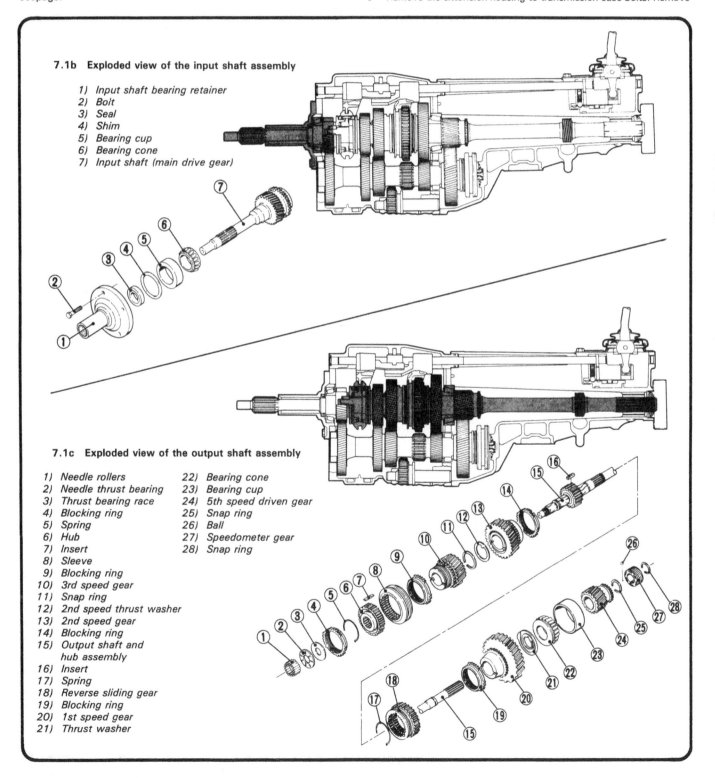

7.1b Exploded view of the input shaft assembly

1) Input shaft bearing retainer
2) Bolt
3) Seal
4) Shim
5) Bearing cup
6) Bearing cone
7) Input shaft (main drive gear)

7.1c Exploded view of the output shaft assembly

1) Needle rollers
2) Needle thrust bearing
3) Thrust bearing race
4) Blocking ring
5) Spring
6) Hub
7) Insert
8) Sleeve
9) Blocking ring
10) 3rd speed gear
11) Snap ring
12) 2nd speed thrust washer
13) 2nd speed gear
14) Blocking ring
15) Output shaft and hub assembly
16) Insert
17) Spring
18) Reverse sliding gear
19) Blocking ring
20) 1st speed gear
21) Thrust washer
22) Bearing cone
23) Bearing cup
24) 5th speed driven gear
25) Snap ring
26) Ball
27) Speedometer gear
28) Snap ring

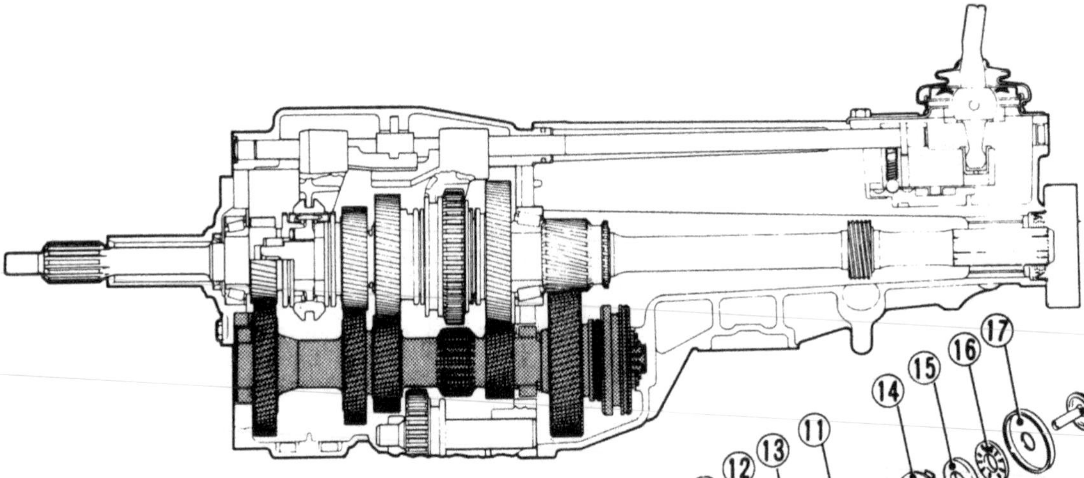

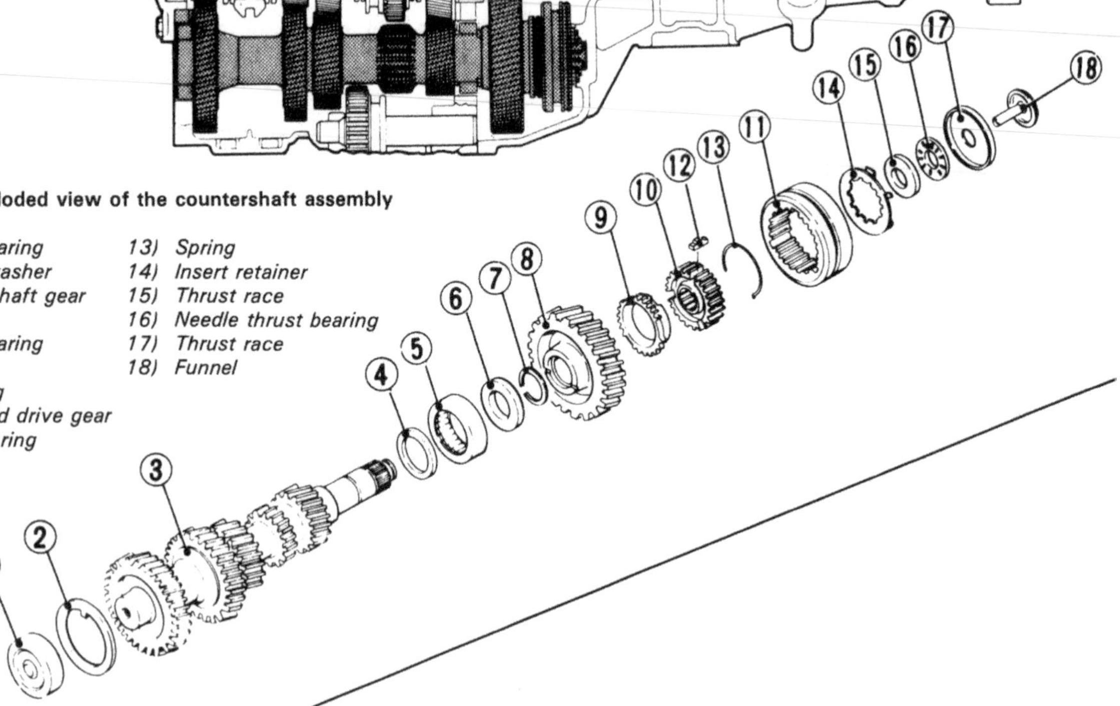

7.1d Exploded view of the countershaft assembly

1) Roller bearing
2) Thrust washer
3) Countershaft gear
4) Spacer
5) Roller bearing
6) Spacer
7) Snap ring
8) 5th speed drive gear
9) Blocking ring
10) Hub
11) Sleeve
12) Insert

13) Spring
14) Insert retainer
15) Thrust race
16) Needle thrust bearing
17) Thrust race
18) Funnel

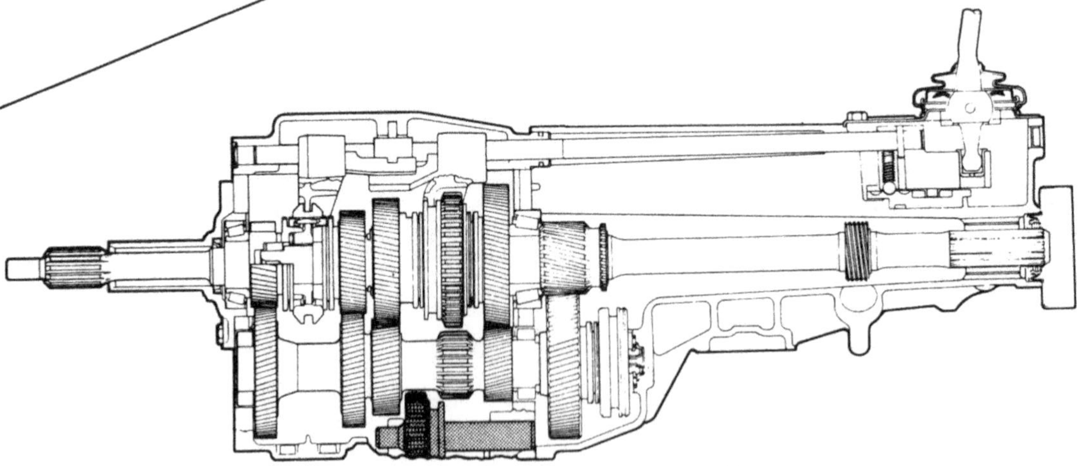

7.1e Exploded view of the reverse idler assembly

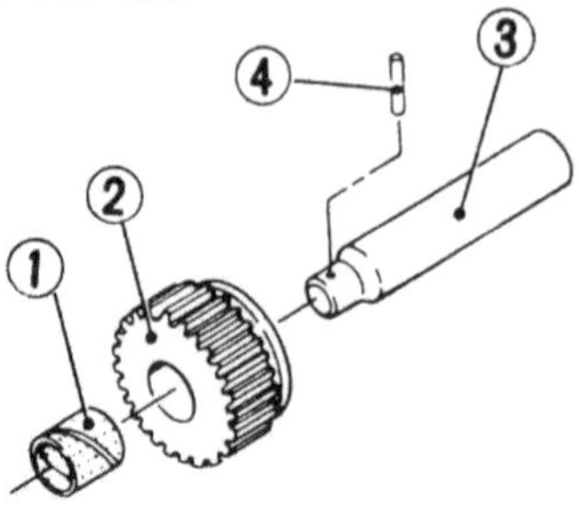

1) Reverse idler bushing
2) Reverse idler gear
3) Reverse idler shaft
4) Spring pin

7.3 Remove the retaining bolts and separate the extension housing from the case

7.6 Remove the cover bolts and slide the cover to the right side of the case, then lift it off

7.7 Drive out the roll pin to remove the overdrive (5th gear) shift fork

7.8 From the rear of the case, use snap ring pliers and remove the snap ring and thrust race from the overdrive (5th gear) synchronizer

the housing and offset lever as an assembly (see illustration).

4 Remove the detent ball, spring and roll pin from the offset lever.

5 Remove the plastic funnel, thrust bearing race and thrust bearing from the end of the countershaft or the inside of the extension housing.

6 Remove the transmission cover bolts and remove the cover by sliding the cover to the right side of the case and lifting it off (see illustration). Note the location of the dowel-type alignment bolts for assembly reference.

7 Remove the roll pin from the fifth gear shift fork using a hammer and punch (see illustration). **Note:** *To prevent damage to the reverse shift rail, place a wood block under the fifth gear shift fork during roll pin removal.*

8 Remove the fifth gear synchronizer snap-ring and shift fork (see illustration). Remove the fifth gear synchronizer assembly and detach fifth gear from the rear of the countershaft.

9 Remove the insert retainer, insert springs and inserts from the fifth gear synchronizer sleeve. Mark the position of the sleeve and hub for assembly reference.

10 Remove the snap-ring and separate the fifth speed driven gear from the rear of the output shaft.

11 Punch alignment marks on the front bearing cap and the transmission case for assembly reference.

12 Remove the bearing cap mounting bolts and detach the bearing cap.

13 Remove the bearing race and end play shims from the cap. Pry out the bearing cap oil seal with a screwdriver.

14 Rotate the clutch shaft until the gear flat faces the countershaft, then remove the clutch shaft from the transmission case (see illustration). Remove the clutch shaft needle bearings, thrust bearing and race.

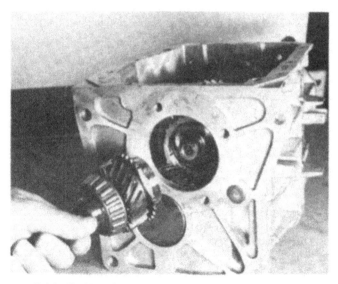

7.14 Position the flat area of the main drive gear toward the counter gear and pull the input shaft and main drive gear assembly through the front of the case

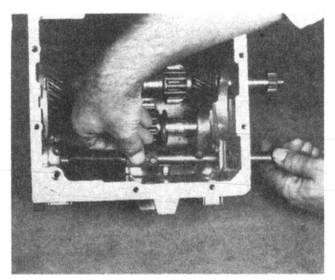

7.19   Remove the overdrive and reverse shift rail, reverse shift fork and the overdrive and reverse relay lever and spring as an assembly

7.21   Use a press to remove the countershaft bearing

7.35   Use diagonal cutters to remove the first speed gear assembly roll pin

25 Remove the idler shaft and gear from the transmission case.
26 Using a press, remove the countershaft front bearing from the case if replacement is necessary.
27 Remove the clutch shaft front bearing.
28 Remove the rear extension housing seal using a flat drift and a hammer.
29 Remove the thrust bearing washer from the output shaft.
30 Remove the third/fourth synchronizer assembly and blocking ring. Mark the sleeve and hub for assembly reference.
31 Remove the third/fourth synchronizer insert springs and inserts and remove the sleeve from the hub.
32 Remove third gear from the output shaft.
33 Remove the second gear snap-ring, tabbed thrust washer and second gear from the shaft.
34 Remove the output shaft bearing.
35 Remove the first gear thrust washer, roll pin (using diagonal cutters) (see illustration), first gear and blocking ring.
36 Scribe alignment marks on the first/second gear synchronizer sleeve and the output shaft hub for assembly reference.
37 Remove the insert spring and inserts from the first/reverse sliding gear and remove the gear from the output shaft hub. **Note:** *The first/second/reverse hub is part of the output shaft and is not removable.*
38 Place the selector arm plates and shift rail in the Neutral (centered) position.
39 Rotate the shift rail counterclockwise until the selector arm disengages from the selector arm plates and the selector arm roll pin is accessible.
40 Pull back on the shift rail until the selector arm contacts the first/second shift fork.
41 Remove the roll pin using a 3/16-inch pin punch and remove the shift rail.
42 Remove the shift forks, selector arm plates, selector arm and interlock plate.
43 Pry out the shift rail oil seal and O-ring using a screwdriver.

15 Remove the output shaft rear bearing race and lift the shaft assembly out of the case.
16 Unhook the overcenter link spring from the rear of the case. A home made spring removal tool made of welding rod or wire would be helpful here.
17 Remove the reverse lever C-clip.
18 Rotate the reverse shift rail until it disengages from the reverse lever and remove the rail from the rear of the case.
19 Remove the reverse lever pivot pin, disengage the reverse lever from the idler gear and remove the reverse lever and fork assembly from the transmission case (see illustration).
20 Remove the rear countershaft snap-ring and spacer.
21 Using a press and a brass drift inserted through the clutch shaft opening in the front of the case, carefully press the countershaft assembly to the rear to remove the countershaft bearing (see illustration).
22 Move the countershaft assembly to the rear and lift it out of the transmission case.
23 Remove the countershaft rear bearing spacer.
24 Remove the idler shaft roll pin.

44 Remove the shift rail plug using a hammer and punch.

45 Remove the nylon inserts and selector arm plates from the shift forks, noting their position for assembly reference.

Inspection

46 Wash the transmission components thoroughly with solvent. Inspect the transmission case for cracks in the bores, sides, bosses and bolt holes and for stripped threads in the bolt holes. Check the gear train and shift mechanism for broken, chipped or worn gear teeth, bent or broken inserts, weak or broken insert springs, damaged roller or needle bearings and bearing bores in the countershaft and hub, clutch shaft or reverse idler gear shaft. Check the snap-rings for distortion and lack of tension. Inspect the front and rear bearings for galling, damage and roughness. Inspect the shift mechanism for worn, damaged or bent inserts, forks, rails, arms, plates, interlocks and levers.

Reassembly

47 Begin assembling the cover by first installing the nylon inserts and selector arm plates in the shift forks.

48 Coat the edges of the shift rail plug with sealer and install the plug.

49 Coat the shift rail and shift rail bores with petroleum jelly and insert the rail into the cover until the end is flush with the inside edge of the cover.

50 Position the first/second shift fork in the cover and push the shift rail through the fork. **Note:** *The first/second shift fork is the larger of the two shift forks.*

51 Position the selector arm and interlock plate and insert the shift rail. The widest part of the plate faces away from the cover and the roll pin faces down and toward the rear of the cover.

52 Position the third/fourth shift fork with the selector arm plate under the first/second shift fork selector arm plate.

53 Insert the shift rail through the third/fourth shift fork and into the shift rail bore in the cover.

54 Rotate the shift rail until the selector arm plate at the forward end faces away from, but parallel to, the cover.

55 Align the roll pin holes and install the roll pin. Be sure it is installed flush with the selector arm surface.

56 Install the O-ring in the shift rail oil seal groove.

57 Install the shift rail seal by first placing an oil seal protector over the threaded end of the shift rail.

58 Lubricate the lip of the oil seal and slide it over the protector and onto the shift rail.

59 Seat the oil seal in the transmission cover using an oil seal installer tool or an appropriate sized socket as an equivalent.

60 Coat the output shaft and gear bores with transmission lubricant.

61 Install the first/second synchronizer sleeve on the output shaft using the reference marks for alignment.

62 Install the three synchronizer inserts and two insert springs in the first/second synchronizer sleeve. The insert spring tangs engage in the same synchronizer insert, but the open ends of the springs face away from each other.

63 Install the blocking ring and second gear on the output shaft.

64 Install the thrust washer and second gear snap-ring. Be sure the washer tab is seated in the output shaft notch.

65 Install the blocking ring and first gear on the output shaft.

66 Install the first gear roll pin.

67 Install the rear bearing on the output shaft using a bearing installer and a press.

68 Install the first gear thrust washer.

69 Install third gear, the third and fourth gear synchronizer hub inserts and the sleeve on the output shaft. Be sure that the portion of the hub with the extended hose is toward the front.

70 Install the thrust bearing washer on the output shaft.

71 Coat the countershaft front bearing outer cage with locking compound and install the countershaft front bearing flush with the case using an arbor press.

72 Coat the countershaft thrust washer with petroleum jelly. Install the washer with the tab corresponding to the depression in the case.

73 Stand the case on end and install the countershaft in the front bearing bore.

74 Install the countershaft rear bearing spacer.

75 Coat the rear bearing with petroleum jelly and install it with a bearing installer and a sleeve tool. The installed bearing should extend beyond the case surface. See the Specifications for the installed height.

76 Position the reverse idler gear and install the idler shaft from the rear of the case. Install the roll pin.

77 Install the output shaft.

78 Install the front clutch shaft bearing on the clutch shaft using a bearing installer and a press.

79 Coat the roller bearings with petroleum jelly and install them in the clutch shaft.

80 Install the clutch bearing and race in the clutch shaft.

81 Install the rear output shaft bearing race cap.

82 Install the fourth gear blocking ring.

83 Install the clutch shaft, engaging it in the third/fourth synchronizer sleeve and blocking ring.

84 Install the front bearing cap oil seal.

85 Install the front bearing race in the front bearing cap, but do not yet install the preload shims.

86 Temporarily install the front bearing cap.

87 Install the reverse lever, pivot bolt and C-clip. Coat the pivot bolt threads with non-hardening gasket sealer. Be sure the reverse lever fork engages the idler gear.

88 Install the fifth speed driven gear and snap-ring.

89 Install fifth gear on the countershaft.

90 Insert the reverse rail from the rear of the case and rotate it until it engages in the fifth speed reverse lever.

91 Install the reverse lever overcenter link spring.

92 Using reference marks for alignment, assemble the fifth gear synchronizer sleeve, insert springs and insert retainer.

93 Install the plastic inserts on each side of the fifth speed shift fork.

94 Place the fifth gear synchronizer assembly on the fifth speed shift fork and slide them onto the countershaft and reverse rail.

95 Place a wood block under the rail and fork assembly and install the roll pin.

96 Install the thrust race against the fifth speed synchronizer hub and install the snap-ring.

97 Coat the thrust race and bearing with petroleum jelly and install the bearing against the thrust race.

98 Install the lipped thrust race over the thrust bearing and install the plastic funnel in the end of the countershaft gear.

99 Temporarily install the extension housing.

100 Turn the transmission case on end and mount a dial indicator on the extension housing with the indicator stylus on the end of the output shaft.

101 Rotate the output shaft and zero the dial indicator.

102 Pull up on the output shaft until the end play is removed. Read the dial indicator and use the dimension to determine the thickness of the bearing preload shim.

103 Select a shim pack to remove the end play. See the Specifications for shim sizes.

104 Set the transmission down on its side and remove the bearing cap and race.

105 Add the necessary shims to the bearing cap and install the clutch shaft bearing race in the cap.

106 Apply some non-hardening gasket sealer to the case mating surface of the front bearing cap. Install the cap using the alignment marks and tighten the attaching bolts to specifications.

107 Recheck the end play. There must be no end play.

108 Remove the extension housing.

109 Move the shift forks and synchronizer rings to the Neutral position.

110 Apply non-hardening gasket sealer to the cover mating surfaces of the transmission case. Lower the cover assembly while aligning the shift forks and synchronizer sleeves. Center the cover to engage the reverse lever and install the two dowel bolts.

111 Install the remaining cover bolts and tighten them to meet specifications. Apply non-hardening gasket sealer to the extension housing mating surface of the transmission case and install the extension housing to a position where the shift rail just enters the shift cover opening.

112 Install the offset lever and spring with the detent ball in the neutral guide plate detent.

113 Install the extension housing bolts and tighten them to meet specifications.

114 Install the roll pin and the damper sleeve in the offset lever.

115 Coat the backup light switch with non-hardening gasket sealer and install the switch in the case.

116 Refer to the Specifications and fill the case with the proper type and amount of gear lubricant.

117 Install the transmission (Section 3).

8   Back-up light switch — checking and replacement

1   Disconnect the negative cable at the battery. Place the cable out of the way so it cannot accidentally come in contact with the negative terminal of the battery.
2   Place the transmission in the Reverse position.
3   Raise the vehicle and support it securely on jackstands.
4   Locate the back up wire on the side of the transmission.
5   Disconnect the electrical contact leads from the switch.
6   Place the leads of an ohm meter onto the terminals of the switch and the meter should show continuity. Have an assistant shift the shifter through the gears and check for continuity. There should be no continuity except in Reverse.
7   If the switch fails the test, replace it by unscrewing it from the transmission case and replacing it with a new one.

Chapter 7 Part B Automatic transmission

Contents

Specifications

Transmission model number		
all except 1984 turbo	E4N71B	
1984 turbo	4N71B	
Transmission model code number		
1985-86		
non-turbo	X8084	
turbo	X8179	
1984		
non-turbo	X8075	
turbo	X8006	
Torque converter assembly (stamped on converter)		
1986		
non-turbo	GXA	
turbo	G	
1985		
non-turbo	GK	
turbo	NGD	
1984		
non-turbo	GK	
turbo	GC	
Stall speed		
non-turbo	2150 to 2450 rpm	
turbo	2500 to 2800 rpm	
Stall torque ratio	2.0:1	
Transmission gear ratios		
1st gear	2.458	
2nd gear....................................	1.458	
3rd gear	1.000	
overdrive	0.686	
reverse	2.182	
Fluid type....................................	Dexron	
Oil capacity	7.0 liters (7-3/8 US qts)	

Torque specifications	Kg-m	Ft-lbs
Converter housing to engine	4.0 to 5.0	29 to 36
Driveplate to crankshaft	14 to 16	101 to 116
Driveplate to torque converter	4.0 to 5.0	29 to 36
Inhibitor switch to transmission case.................	0.5 to 0.7	3.6 to 5.1
Oil pan to transmission...........................	0.5 to 0.7	3.6 to 5.1
Rear mounting bracket to transmission	3.2 to 4.3	23 to 31
Rear mounting bracket to rear insulator	3.2 to 4.3	23 to 31
Rear mounting member to body	6.0 to 8.0	43 to 58
Transmission case to converter housing	4.5 to 5.5	33 to 40
Transmission case to rear extension	2.0 to 2.5	14 to 18

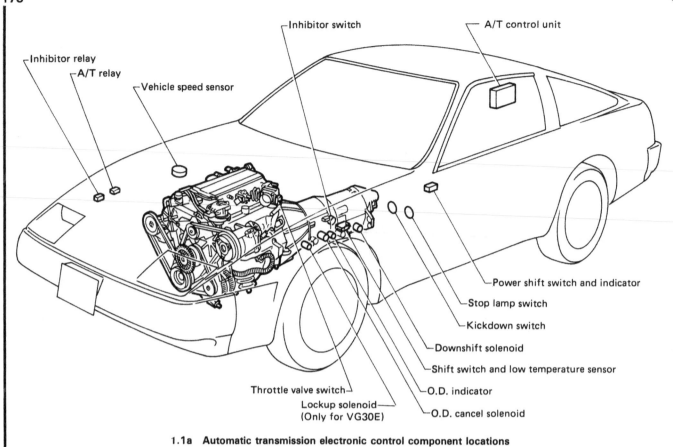

Inhibitor relay

A/T relay

Vehicle speed sensor

Inhibitor switch

A/T control unit

Power shift switch and indicator

Stop lamp switch

Kickdown switch

Downshift solenoid

Shift switch and low temperature sensor

O.D. indicator

O.D. cancel solenoid

Throttle valve switch

Lockup solenoid
(Only for VG30E)

1.1a Automatic transmission electronic control component locations

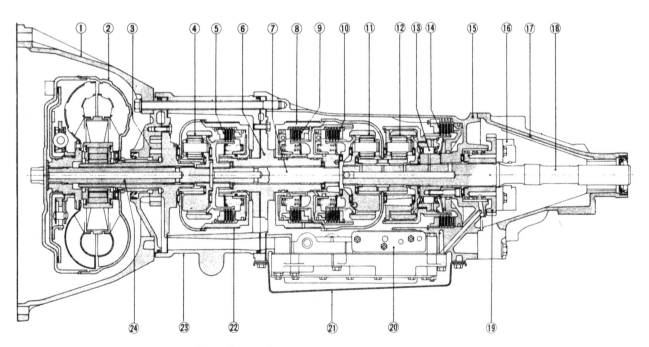

1.1b Automatic transmission internal components

1 Converter housing	7 Intermediate shaft	13 One-way clutch	19 Oil distributor
2 Torque converter	8 2nd band brake	14 Low and reverse clutch	20 Control valve assembly
3 Oil pump assembly	9 High-reverse clutch (Front)	15 Transmission case	21 Oil pan
4 O.D. planetary gear	10 Forward clutch (Rear)	16 Governor valve assembly	22 O.D. band brake
5 Direct clutch	11 Front planetary gear	17 Rear extension	23 O.D. case
6 Drum support	12 Rear planetary gear	18 Output shaft	24 Input shaft

2.4 Location of the selector rod locknuts

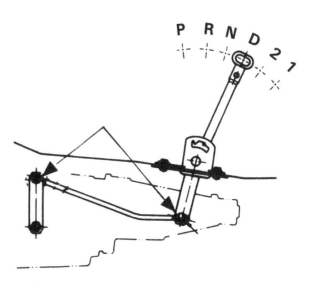

2.8 Arrows indicate the shifter linkage grommets

1 General information

Refer to illustrations 1.1a and 1.1b

The transmission provides four forward gears and one reverse (see illustrations). Changing of the forward gears is by microcomputer control. The computer receives information from four sets of sensors controlling the upshift and downshift in relation to the vehicle speed and engine torque output, and is dependent upon the vacuum in the intake manifold to actuate the gear change mechanism at the precise time. A manual override switch is provided next to the shifter if the driver desires not to use the overdrive function.

By use of a microcomputer attached to the right rear inside fender panel, the electronically controlled lockup system permits lockup of all forward speeds on 1984–85 models (except 85 turbo models). The 1986 models provide a lockup function of 3rd and overdrive by electronic control.

The transmission has six selector positions:

P — Parking position, which locks the output shaft to the interior wall of the transmission housing. The engine may be started with Park selected and this position should always be selected when adjusting the engine while it is running. Never attempt to select Park when the vehicle is in motion

R — Reverse gear.

N — Neutral. Select this position to start the engine or when idling in traffic for long periods.

D — Drive, for all normal motoring conditions.

2 — Locks the transmission in second gear for wet road conditions or steep hill climbing or descents. **Caution:** *The engine can be over-revved in this position.*

1 — The selection of this gear above speeds of approximately 25 mph (40 kph) will engage second gear and, as the speed drops below 25 mph (40 kph) the transmission will lock into first gear. Provides maximum retardation on steep descents.

Due to the complexity of the automatic transmission unit, any internal adjustment or servicing should be left to the Nissan dealer or other qualified transmission specialist. The information given in this Chapter is therefore confined to those operations which are considered within the scope of the home mechanic. An automatic transmission should give many miles of service provided normal maintenance and adjustment is carried out. When the unit requires major overhaul, consideration should be given to exchanging the old transmission for a reconditioned one, the removal and installation being well within the capabilities of the home mechanic as described later in this Chapter.

The routine maintenance schedule in Chapter 1 calls for an automatic transmission fluid change once every 30,000 miles. This interval should be shortened to every 15,000 miles if the car is normally driven under one or more of the following conditions: Heavy city traffic, where the outside temperature normally reaches 90° or higher; In very hilly or mountainous areas; If a trailer is frequently pulled. Refer to Chapter 1 for the proper procedures for checking and changing the automatic transmission fluid and filter.

The automatic transmission uses an oil cooler, located at the radiator, to prevent excessive temperatures from developing inside the transmission. Should the oil cooler need flushing or other servicing, take it to a dealer or other radiator specialist.

If rough shifting or other malfunctions occur in the automatic transmission, check the following items first before assuming the fault lies within the transmission itself: The fluid level, the kickdown switch adjustment, manual shift linkage adjustment and engine tune. All of these elements can adversely affect the performance of the transmission.

Periodically clean the outside of the transmission housing, as the accumulation of dirt and oil is liable to cause overheating of the unit under extreme conditions.

When towing a vehicle equipped with an automatic transmission restrict the towing speed to below 20 mph and towing distance to less than 20 miles. If speed or distance is greater it is advised to remove the driveshaft so not to damage the transmission.

2 Shift linkage — adjustment

Refer to illustrations 2.4 and 2.8

1   To check the manual shift linkage adjustment, move the shifter through the entire range of gears. You should be able to feel the detents in each gear. If these detents are not felt or if the pointer is not properly aligned with the correct gear selection, the shift linkage should be adjusted in the following manner.

2   With the engine off, place the shift lever in Neutral.

3   Raise vehicle and support it securely on jackstands.

4   Working underneath the car, loosen the shifter linkage locknuts (see illustration).

5   Move the shift lever so that it is correctly aligned with the *N* position.

6   Move the selector lever on the transmission so that it is also correctly aligned in the Neutral position.

7   Tighten the locknuts and recheck the levers. There should be no tendency for the selector rod to push or pull one rod against the other.

8   Again run the shifter through the entire range of gear positions. If there are still problems, the grommets that connect the selector rods with the levers may be worn or damaged and should be replaced (see illustration).

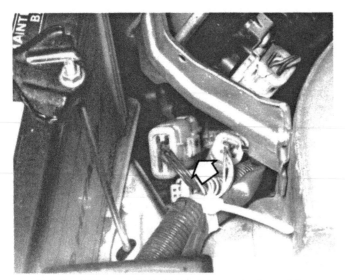

3.1 The inhibitor switch electrical connector (arrow) is located next to the battery so tests can be performed without raising the vehicle

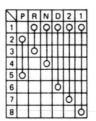

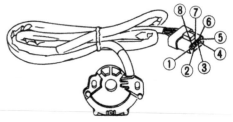

3.3   Inhibitor switch electrical pin numbering for all models except 1984 turbo

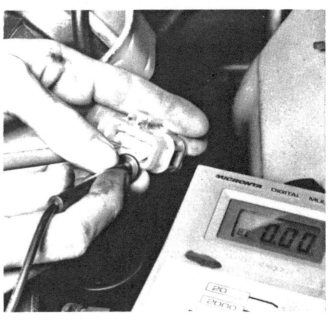

3.5   With the shifter in Neutral and the ohmmeter leads on terminals one and four there should be continuity

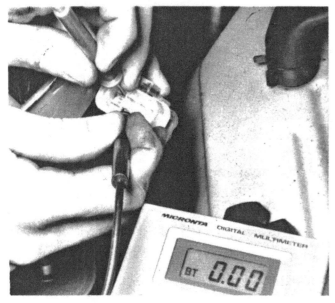

3.4   With the ohmmeter leads on the one and three terminals there should be continuity

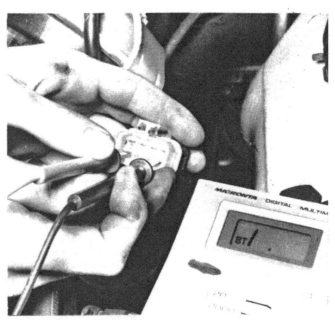

3.6   With the shifter in Drive there should be no continuity through the one and four terminals

3   Inhibitor switch — adjustment and testing

Refer to illustrations 3.1, 3.3, 3.4, 3.5, 3.6, 3.9, 3.15 and 3.16

Testing — all models except 1984 turbo

1   The inhibitor switch electrical connector is located next to the battery. Identify the connector by its eight attaching wires (all models except 1984 turbo) (see illustration).

2   Disconnect the connector.

3   Verify that the shift selector is in the Park position and place the leads of an ohm meter to the number two and five terminals (see illustration) and the meter should show continuity.

4   Place the shifter in Reverse and there should be no continuity between the two and five terminals. Continuity should now be between the one and three terminals (see illustration).

5   Place the shifter in Neutral and terminals one and three will no longer have continuity and terminals one and four will show continuity (see illustration).

6   Place the shifter lever in Drive and you will lose continuity in the one and four terminals (see illustration).

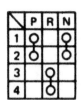
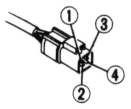

3.9 Inhibitor switch numbering for 1984 turbo only

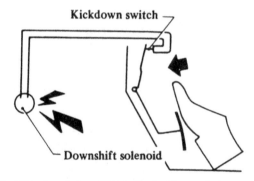

3.16 Pass a wire through the adjusting hole then tighten the adjusting bolts

4.2 With the engine off but the ignition switch on you should hear the solenoid click when the throttle is fully depressed

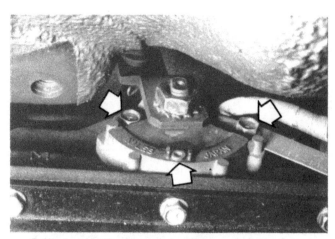

3.15 Loosen the two switch adjustment bolts and then remove the adjusting hole screw (arrows)

4.6 The meter should show no continuity with the pedal up

Testing — 1984 turbo model

7   The inhibitor switch electrical connector is located next to the battery. Identify the connector by four attaching wires for the 1984 turbo model. If identifying it is a problem locate the inhibitor switch on the transmission and follow the wire up to the connector.

8   Disconnect the connector.

9   Verify that the shift selector is in the Park position and place the leads of ohmmeter on the number one and two terminals (see illustration) and the meter should show continuity.

10   Place the shifter in Reverse and there should be no continuity at the one and two terminals. Continuity should now be between the three and four terminals.

11   Place the shifter in Neutral and terminals three and four will not have continuity and terminals one and two will show continuity.

12   Place the shifter lever in drive and you will lose continuity in all terminals.

Adjustment

13   Place the shifter in Neutral.

14   Raise the vehicle support it securely on jackstands.

15   Remove the inhibitor switch adjustment hold-down screw and loosen the two adjustment bolts (see illustration).

16   Rotate the inhibitor switch until the hole left by the removed screw will accept a 0.079 inch diameter wire (see illustration). Once the hole accepts the wire, tighten the two adjustment bolts and remove the wire.

17   Replace the adjustment hold-down screw.

18   Repeat ohmmeter tests. If any of the tests fail, replace the inhibitor switch.

4   Kickdown switch — adjustment

Refer to illustrations 4.2, 4.6 and 4.7

1   The kickdown switch, coupled with the downshift solenoid, causes the transmission to downshift when the accelerator pedal is fully depressed. This is to provide extra power when passing. If the transmission is not downshifting upon full throttle, the system should be inspected.

2   With the engine off, but the ignition on, depress the accelerator pedal all the way and listen for a click just before the pedal bottoms (see illustration).

3   If no click is heard, locate the kickdown switch at the upper post of the accelerator pedal. Loosen the locknut and, with the pedal still depressed, extend the switch until it makes contact with the post and clicks. The switch should click only as the pedal bottoms. If it clicks too soon, it will cause the transmission to downshift on part throttle.

4   Tighten the locknut and recheck the adjustment.

5   If the kickdown switch adjustment is correct but the transmission still will not downshift, check that there is continuity when the switch is activated by removing the connector from the switch and connecting an ohmmeter to the switch.

6   With the pedal not depressed, the meter should show no continuity (see illustration).

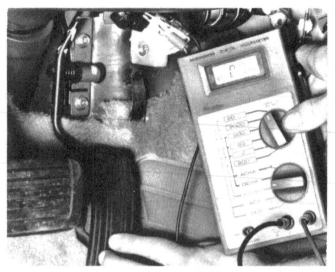

4.7   Depress the pedal all the way and the meter should
show continuity

5.2   Carefully pry the oil seal from the rear extension
housing of the transmission

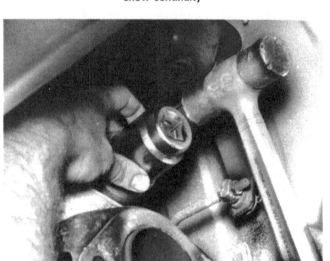

5.3   Use a socket to drive the new seal into the
extension housing

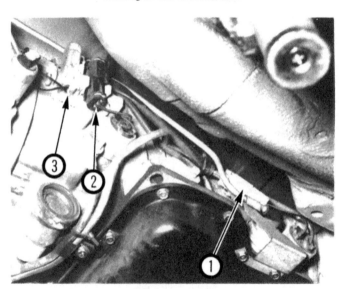

6.7   Locate the electrical connectors (arrows) and
disconnect them

7   Fully depress the pedal and the meter should show continuity (see
illustration).
8   If either continuity test fails, replace the kickdown switch.

---

5   Transmission rear oil seal — replacement

Refer to illustrations 5.2 and 5.3

1   Remove the driveshaft as described in Chapter 8.
2   Being careful not to damage the output shaft or transmission housing, use a screwdriver to pry out the old seal (see illustration).
3   Apply a coat of gear oil to the lips of the new seal and drive it into
place using a socket (see illustration).
4   Reinstall the driveshaft.

---

6   Transmission — removal and installation

Refer to illustrations 6.7, 6.9, 6.10, 6.11, 6.13, 6.17 and 6.19

1   Disconnect the negative cable at the battery. Place the cable out
of the way so it cannot accidentally come in contact with the negative
terminal of the battery.

2   Raise the vehicle and support it securely on jackstands.
3   Drain the transmission fluid and replace the pan to keep dirt from
entering the transmission (Chapter 1).
4   Remove the driveshaft (Chapter 8).
5   Remove the starter (Chapter 5).
6   Remove the exhaust front tube (Chapter 4).
7   On the left side of the transmission, locate the three electrical connectors and disconnect them (see illustration). Move the attached wires
out of the way, which will require cutting the tie straps gathering the
wires.
8   Trace the two remaining wires to the transmission to their connectors near the battery and disconnect them. These wires will remain
attached to the transmission.
9   Disconnect the vacuum line from the vacuum modulator (see illustration) and detach the metal line from its hold-down bracket.
10 Disconnect the two cooling lines at the transmission case and and
note number and location of the sealing rings on the cooling line retaining bolts (see illustration). Follow the lines and disconnect their holding
brackets. Disconnect the lines at the radiator and remove the lines.
Note: *Be prepared to catch fluid leaking from the lines.*
11 Unscrew the speedometer cable from the transmission (see
illustration).
12 Disconnect the ground wire attached to the body next to the
speedometer cable connector.

6.9 Disconnect the vacuum modulator hose and remove the metal line from its bracket (arrow)

6.10 Disconnect the coolant lines and note the number and location of the sealing rings on the banjo bolts (arrows)

6.11 Unscrew the speedometer cable at the transmission

6.13 Remove the cotter pin to disconnect the shifter linkage from the control linkage

6.17 Remove the starter support bracket by removing the two retaining bolts

13 Remove the cotter pin on the shift linkage and disconnect the shifter linkage from the transmission linkage (see illustration).

14 Place a jack under the engine oil pan. Use a piece of wood between the jack and pan so not to damage the pan.

15 Remove all but two of the transmission to engine bolts and remove the dust shield located between the two components.

16 Place a suitable transmission jack under the transmission pan.

Warning: *Transmissions are heavy so it is advisable to rent a suitable transmission jack to facilitate removal of the transmission.*

17 Remove the two bolts retaining the starter support plate to the engine block (see illustration) and remove the plate.

18 Disconnect the torque converter retaining bolts. Use a wrench on the crankshaft bolt to turn the crankshaft to gain access to each of the four retaining bolts through the starter opening.

19 Remove the four bolts securing the rear transmission mount crossmember to the body.

20 Remove the two remaining transmission retaining bolts and slide the transmission back and down to remove it from the vehicle.

21 Installation is the reverse of the removal. Be sure and fill the transmission with the proper amount of fluid and replace the pan gasket with a new one.

7   Overdrive/power pattern control switch — description

The automatic transmission comes equipped with a manual switch to operate the overdrive function of the transmission. The switch is located on the console on 1984–85 models and on the shifter handle on 1986 models. The vehicle shifts into overdrive when the switch is turned on and the indicator light comes on to indicate its function. When the vehicle cruise control *accel* button is pushed the transmission will shift into 3rd from overdrive even with the switch turned on. The transmission will automatically shift through its gears with the switch turned off but will not go into overdrive until the switch is turned on.

The 1984–85 models have as an option, in place of the overdrive switch, a power pattern change switch. The power pattern switch is located in the same place on the console and offers normal or power pattern driving functions. The normal function is available when the switch is turned off and is most effective for everyday driving, giving the driver maximum fuel efficiency. Even in the off position the switch can be electronically overidden when the accelerator is fully depressed for increased power. The power function comes when the switch is in the *On* position. The indicator light comes on to indicate this function. The transmission shifts down into 3rd gear if the switch is turned on while driving in overdrive.

The power pattern driving mode mode is most effective for quick acceleration for passing or entering freeways or uphill driving for better control of engine acceleration. The power pattern also offers better engine braking for long downhill slopes.

The vehicle should not be driven for long periods at highway speeds with the overdrive switch turned off. The overdrive function is advised for normal freeway driving but should be turned off in mountainous areas. **Caution:** *If any abnormality occurs in the automatic transmission the overdrive/power pattern indicator light will flicker. If this occurs, the transmission should be checked.*

Chapter 8   Clutch and drivetrain

Contents

Specifications

Clutch

Type	Single plate, diaphragm spring with hydraulic actuation
Fluid type	Dot 3 brake fluid
Pedal	
height	195 to 205 mm (7.68 to 8.07 in)
free play	1 to 3 mm (0.04 to 0.12 in)
Wear limit of clutch surface to rivet head	0.3 mm (0.012 in)

Driveaxles

Tripod-Tripod type	
boot length – wheel side	111.5 mm (4.39 in)
boot length – carrier side	92.5 mm (3.642 in)
grease capacity	
wheel side	185 to 195 gm (6.52 to 6.88 oz)
differential carrier side	155 to 165 gm (5.47 to 5.82 oz)
Double Offset-Birfield type	
boot length – wheel side	94 mm (3.70 in)
boot length – carrier side	91 mm (3.58 in)
grease capacity	115 to 155 gm (4.06 to 5.47 oz)

Torque specifications

	Kg-m	Ft-lbs
Pressure plate bolts		
1987 and later turbo models	3.5 to 4.5	2.5 to 3.3
all others	2.2 to 3.0	16 to 22
Clutch hose to operating cylinder and clutch		
damper securing nut	1.7 to 2.0	12 to 14
Clutch tube flare nut	1.5 to 1.8	11 to 13
Damper cover to cylinder		
1984 thru 1986	0.3 to 0.6	2.2 to 4.3
1987 on	0.4 to 0.5	2.7 to 3.7
Clutch damper bolt	0.8 to 1.2	5.8 to 8.7
Bleeder screw	0.7 to 0.9	5.1 to 6.5
Operating cylinder bolt	3.1 to 4.1	22 to 30
Supply valve stopper	0.15 to 0.3	1.1 to 2.2
Master cylinder nut	0.8 to 1.2	5.8 to 8.7
Master cylinder push rod lock nut	0.8 to 1.2	5.8 to 8.7
Fulcrum pin nut	1.6 to 2.2	12 to 16
Clutch switch lock nut	1.2 to 1.5	9 to 11
Pedal stopper lock nut	0.8 to 1.2	5.8 to 8.7
Differential carrier to mounting insulator	10 to 12	72 to 87
Carrier mounting bracket to body		
bolt	3 to 4	22 to 29
nut	6 to 8	43 to 58
Carrier to suspension member	6 to 8	43 to 58
Rear stabilizer bar to suspension arm	1.6 to 2.1	12 to 15
Stabilizer bar clamp to suspension member	3.2 to 4.3	23 to 31

Torque specifications (continued)

Driveshaft to companion flange

	Kg-m	Ft-lbs
turbo		
1984 thru 1987	6.0 to 7.0	43 to 51
1988 on	7.0 to 8.0	51 to 58
non-turbo	4.0 to 5.0	29 to 36
Wheel bearing lock nut	30 to 40	217 to 289

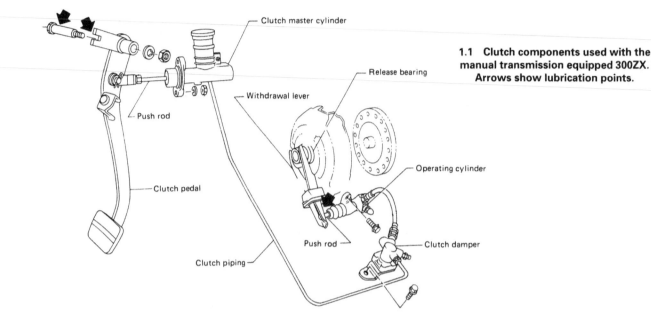

1.1 Clutch components used with the manual transmission equipped 300ZX. Arrows show lubrication points.

Labels: Clutch master cylinder, Release bearing, Withdrawal lever, Push rod, Clutch pedal, Operating cylinder, Push rod, Clutch damper, Clutch piping

1 Clutch — general information

Refer to illustration 1.1

The clutch is located between the engine and the transmission and its main components are the clutch disc, the pressure plate assembly and the release bearing. Other components which make up the hydraulically operated clutch system (see illustration) are the clutch pedal, clutch master cylinder, operating cylinder, and release lever.

The clutch pedal is connected to the clutch master cylinder by a short pushrod. The master cylinder and hydraulic reservoir are mounted on the engine side of the bulkhead in front of the driver.

Depressing the clutch pedal moves the piston in the master cylinder forward and forces hydraulic fluid through the clutch hydraulic pipe to the operating cylinder.

The piston in the operating cylinder actuates the clutch release lever by means of a short pushrod.

The other end of the release lever, located inside the clutch housing, is fork-shaped. This fork engages the clutch release bearing and forces the bearing against the pressure plate release fingers.   Because access to the clutch components is difficult, any time either the engine or the transmission is removed, the clutch disc pressure plate assembly and release bearing should be carefully inspected and, if necessary, replaced. Since the clutch disc is the highest wear item, it should be replaced as a matter of course if there is any question as to its condition.

2 Clutch assembly — diagnosis

1   Due to the slow-wearing qualities of the clutch, it is not easy to decide when remove the transmission in order to check the wear on the friction lining. The only positive indication that something needs attention is when it starts to slip or when squealing noises on engagement indicate that the friction lining has worn down to the rivets.

2   A clutch will wear according to the way in which it is used. Intentional slipping of the clutch while driving — rather than the correct selection of gears — will accelerate wear. It is best to assume, however, that the friction disc will need renewal at about 40,000 miles (64,000 km).

3   The clutch cannot be worked on without removing either the engine or transmission. If repairs which would require removal of the engine are not needed, the quickest way to gain access to the clutch is by removing the transmission (Chapter 7A).

3 Clutch assembly — removal and installation

Refer to illustrations 3.3a, 3.3b, 3.4, 3.9 and 3.10

1   With the transmission removed (Chapter 7), but before removing the pressure plate assembly from the flywheel, check that none of the metal fingers on the pressure plate are distorted or bent. If any damage is evident the pressure plate will need to be replaced.

2   The pressure plate need not be marked in relation to the flywheel as it can only be fitted one way due to the positioning dowels.

3   Loosen the attaching bolts in a diagonal pattern a little at a time until the spring pressure is relieved to keep from distorting the pressure plate (see illustration). If the flywheel begins to turn, insert a screwdriver through the starter motor opening (see illustration) and engage it in the teeth of the flywheel.

4   While supporting the pressure plate assembly, remove the bolts, then remove the pressure plate and clutch disc (see illustration).

5   Clean the pressure plate, flywheel mating surfaces and the bearing retainer outer surfaces of any oil and grease.

6   Inspect clutch assembly components (Section 4).

7   While servicing the clutch components, it is recommended that the release bearing be replaced.

8   Prior to clutch installation, apply a light coat of grease to the splines of the transmission input shaft.

9   The clutch disc must be installed with the damper springs offset toward the transmission (see illustration). The flywheel side should be identified as such by stamped letters in the disc.

10 To install, hold the clutch disc against the flywheel and insert centering tool ST20600000 or an equivalent through the center (see illustration).

11 Locate the clutch pressure plate so that the cover engages with the dowels and install the mounting bolts.

12 Tighten the the mounting bolts in steps and in a diagonal pattern until they are tightened to specifications.

13 Install the transmission (Chapter 7A).

3.3a   Loosen the clutch pressure plate attaching bolts a little at a time in a diagonal pattern so not to distort the pressure plate

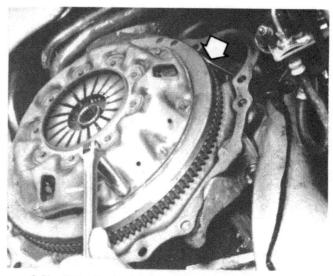

3.3b   If the flywheel turns while removing the pressure plate, insert a screwdriver through the starter opening to wedge against the flywheel teeth

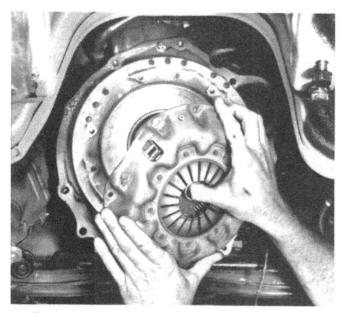

3.4   After removing the bolts, remove the pressure plate and clutch disc

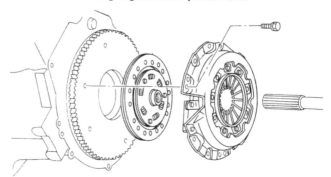

3.9   Install the clutch disc with the damper spring offset toward the transmission

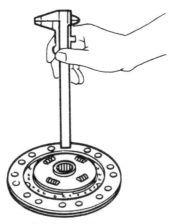

4.2   Measure the rivet to friction surface height and replace the disc if the lining is close to the wear limit

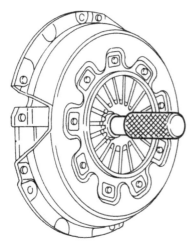

3.10   Align the clutch using an alignment tool

4   Clutch – inspection

Refer to illustration 4.2

1   Examine the pressure plate surface where it contacts the clutch disc. This surface should be smooth, with no scoring, gouging or warping. Check the pressure plate cover and fingers for damage. If any fault is found with the pressure plate assembly it must be replaced as a unit.

2   Inspect the clutch disc for lining wear. Check for loose or broken rivets or springs. Measure the lining wear limit above the rivets (see illustration) and compare to the specifications. If the lining material

5.3　Paint or scribe an alignment mark (arrow) to indicate the relationship of the flywheel and crankshaft

shows signs of breaking up or black areas where oil contamination has occurred, it should also be replaced.

3　Inspect the surface of the flywheel for rivet grooves, burned areas or scoring. If the damage is slight, the flywheel can be removed (Section 5) and reconditioned by a machine shop. If the damage is extensive, the flywheel should be replaced. Check that the ring gear teeth are not broken, cracked or seriously burned.

4　If any traces of oil are detected on the clutch components the source should be found and corrected. If oil is coming from the center of the flywheel, this indicates a failure of the rear oil seal (Chapter 2). Oil at the rear of the clutch assembly may indicate the need to replace the transmission input shaft seal (Chapter 7A).

5　Flywheel — removal, inspection and installation

Refer to illustrations 5.3 and 5.4

1　Raise the vehicle and support it securely on jackstands.
2　Remove the pressure plate assembly and clutch disc (Section 3).
3　Use paint to make an alignment mark from the flywheel to the end of the crankshaft for reinstallation purposes (see illustration).
4　Remove the bolts which secure the flywheel to the crankshaft rear flange. If some difficulty is experienced in removing the bolts due to the movement of the crankshaft, wedge a screwdriver through the starter assembly opening to keep the flywheel from turning (see illustration).
5　Remove the flywheel from the crankshaft flange.
6　Clean any grease or oil from the flywheel. Inspect the surface of the flywheel for rivet grooves, burned areas or scoring. Light scoring may be corrected using emery cloth. Check for any cracked or broken teeth. Lay the flywheel on a flat surface and use a straightedge to check for warpage.
7　Clean the mating surface surfaces of the flywheel and the crankshaft.
8　Position the flywheel against the crankshaft, matching the alignment marks made during removal. Before installing the retaining bolts, put a dab of sealing agent on the threads of the bolts and tighten the bolts finger tight.
9　Wedge a screwdriver through the starter motor opening to keep the flywheel from turning as you tighten the bolts.
10　The remainder of the installation is the reverse of the removal procedures.

6　Clutch throw out bearing — replacement

Refer to illustrations 6.3, 6.4, 6.5, 6.6, 6.7, 6.8 and 6.9

1　The sealed release bearing, although designed for long life, is worth

5.4　Wedge a screwdriver (arrow) into a tooth of the flywheel to keep it from turning

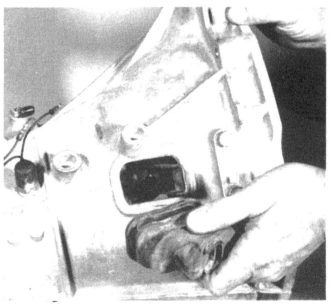

6.3　Remove the rubber dust boot from the transmission housing

replacing at the same time the other clutch components are being replaced or serviced.

2　Deterioration of the release bearing should be suspected when there are signs of grease leakage or if the unit is noisy when spun with the fingers.

3　With the transmission removed (Chapter 7) remove the rubber dust boot which surrounds the release lever at the bellhousing opening (see illustration).

4　Using a screwdriver, detach the retainer spring from the ball-pin in the front transmission cover (see illustration).

5　Remove the release lever (see illustration).

6　The clutch release bearing and sleeve assembly can now be removed (see illustration).

7　If necessary, remove the release bearing from its sleeve using a puller (see illustration).

8　Press on the new bearing, but apply pressure only to the center track (see illustration).

9　Apply mulit-purpose grease to the internal recess of the release bearing sleeve (see illustration).

6.4 Use a screwdriver to unhook the spring from the ball-pin in the front of the transmission

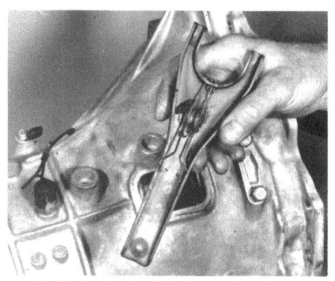

6.5 Lift the release lever from the transmission housing after unhooking the retaining spring

6.6 Remove the release bearing and sleeve by pulling it off the input shaft

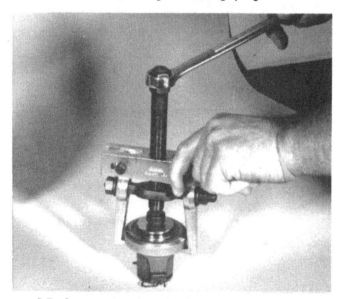

6.7 Separate the release bearing from the sleeve using a puller

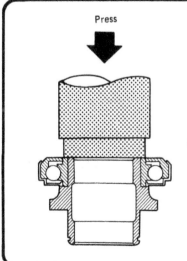

Press

6.8 Press on the new bearing, making sure you apply pressure only to the center track

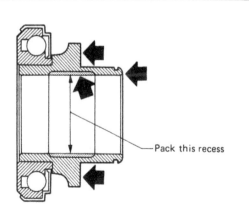

Pack this recess

6.9 Pack the release bearing with grease and apply a thin coat of grease to the points indicated by arrows

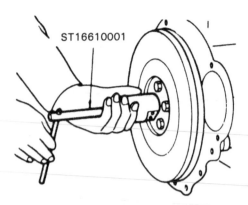

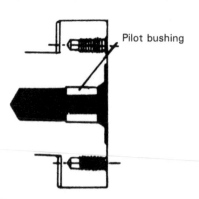

Pilot bushing

7.2a   A pilot bushing puller, such as the Nissan tool shown, is recommended to remove the pilot bushing

7.2b   Pack the area behind the bushing with grease to force the bushing out of the crankshaft

10 Apply grease to the pivot points of the clutch release lever, the sliding surface of the bearing sleeve and the splines on the transmission input shaft. **Note:** *Apply only a thin coat of grease to these points, as too much grease will run onto the friction plates when hot, causing damage to the clutch disc surfaces.*
11 Install the transmission.

the bushing inside diameter. Strike the rod with a hammer the grease will force the bushing from the recess.
3   Before installing the new bushing, thoroughly clean the bushing recess.
4   Using an appropriate sized socket or piece of tubing, insert the new bushing into the recess.
5   Reinstall the clutch assembly and transmission.

7   Pilot bearing — replacement

Refer to illustrations 7.2a and 7.2b

1   With the transmission removed (Chapter 7) remove the clutch disc and pressure plate assembly as described in Section 3.
2   A special tool is recommended to remove the pilot bushing (see illustration). This tool can be obtained from a Nissan dealer or from an auto parts store. If the tool is not available, pack the recess behind the bushing with grease (see illustration). Insert a rod the same diameter

8   Clutch master cylinder — removal and installation

Refer to illustration 8.1
Caution: *Hydraulic fluid can damage paint, so take care not to spill any on the painted surfaces of the car.*
1   From inside the vehicle disconnect the master cylinder pushrod from the pedal arm (see illustration).

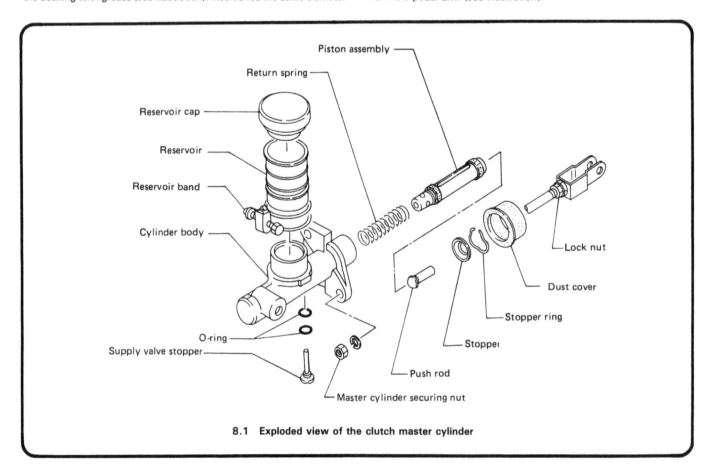

8.1   Exploded view of the clutch master cylinder

9.6 The clutch operating cylinder is attached to the transmission housing by two bolts

2 Place newspapers or rags under the master cylinder to catch any spilled hydraulic fluid.
3 Siphon the reservoir or disconnect the fluid line from the master cylinder and drain the fluid.
4 For clearance, remove the windshield washer tank by lifting it straight up from its retaining bracket, then position it so it will not interfere with the removal of the master cylinder.
5 Remove the master cylinder flange mounting bolts and withdraw the unit from the engine compartment.
6 To reinstall place the master cylinder flange onto the bulkhead mounting bolts and install the nuts.
7 From inside the vehicle attach the master cylinder pushrod to the pedal arm.
8 Fill the master cylinder reservoir with hydraulic fluid (Chapter 1).
9 Adjust the clutch pedal (Section 12).
10 Bleed the clutch hydraulic system (Section 11).
11 Remove the newspapers or rags from under the master cylinder. Be careful not to drip any fluid onto the paint.
12 Replace the windshield washer fluid reservoir by sliding it into its retaining bracket.

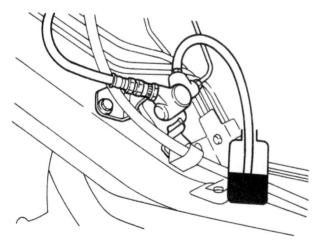

11.4 Bleed the clutch damper by attaching a hose to the damper bleed valve and inserting the hose into a bottle containing a few inches of clean brake fluid. The fluid will keep air from re-entering the system

9 Clutch operating cylinder — removal and installation

Refer to illustration 9.6

1 In order to prevent excessive loss of hydraulic fluid when the operating cylinder hose is disconnected, remove the clutch master cylinder reservoir cap and place a piece of polyethylene sheeting over the open reservoir. Screw on the cap and thus create a vacuum which will stop the fluid from running out of the open hose.
2 Raise the vehicle and support it securely on jackstands.
3 Loosen the clutch line flare nut at the bracket mounted on the body sidemember.
4 Remove the locking clip from the bracket and disengage the hose.
5 Remove the clutch hose from the operating cylinder.
6 Remove the two bolts which secure the operating cylinder to the clutch housing (see illustration). The operating cylinder can now be removed.
7 Lubricate the operating cylinder push rod at the point where it contacts the operating arm.
8 Install the operating cylinder onto the transmission housing and tighten the bolts.
9 Install the clutch hose on the operating cylinder.
10 Slide the hose into its retaining bracket and tighten the connection. Secure the hose to the bracket with the locking clip.
11 Remove the polyethylene sheeting from the clutch reservoir and replace the cap.
12 Bleed the hydraulic system (Section 11).
13 Remove the jackstands and lower the vehicle.

10 Clutch damper — removal and installation

1 Raise the vehicle and support it securely on jackstands.
2 In order to prevent excessive loss of hydraulic fluid when the clutch damper hose is disconnected, remove the clutch master cylinder reservoir cap and place a piece of polyethylene sheeting over the open reservoir. Screw on the cap and thus create a vacuum which will stop the fluid from running out of the open hose.
3 Using a flare nut wrench, disconnect the hydraulic hose from the reservoir at the clutch damper.
4 Disconnect the hydraulic hose to the clutch operating cylinder at the clutch damper.
5 Remove the bolts and lift the damper from the body.
6 Install the damper to the body and tighten the bolts.
7 Install the hoses to the damper and tighten with a flare nut wrench.
8 Remove the polyethylene sheeting from the clutch reservoir and replace the cap.
9 Bleed the hydraulic clutch system (Section 11).
10 Remove the jackstands and lower the vehicle.

11 Clutch hydraulic system — bleeding procedure

Refer to illustrations 11.4 and 11.7

1 Raise the vehicle and support it securely on jackstands.
2 Bleeding will be required whenever the hydraulic system has been opened allowing air to enter the system.
3 Fill the fluid reservoir with clean brake fluid. Never use fluid which has been drained from the system or has been bled out on a previous occasion.
4 Fit a rubber or plastic tube to the bleed screw on the clutch damper (see illustration) and immerse the open end of the tube in a glass jar containing an inch or two of fluid.
5 Open the bleed screw about half a turn and have an assistant depress the clutch pedal fully. Tighten the bleed screw and then have the clutch pedal slowly released. Repeat this sequence of operations until air bubbles are no longer ejected from the open end of the tube beneath the fluid in the jar.
6 After two or three strokes of the pedal, check that the fluid level in the reservoir has not fallen too low. Keep it topped-up with fresh fluid.

7   Repeat Steps 4 through 6 on the clutch operating cylinder (see illustration).
8   Remove the jackstands and lower the vehicle.

12 Clutch pedal — removal, installation and adjustment

Refer to illustration 12.10

1   Remove the lower instrument trim piece (Chapter 11), located directly below the steering column.
2   Disconnect the master cylinder pushrod from the pedal by prying off the snap-ring and withdrawing the clevis pin.
3   Pry off the E-ring from the adjusting rod and disengage the rod from the pedal.
4   Remove the fulcrum bolt.
5   Disengage and remove the return spring and assist return spring.
6   Remove the assist spring clutch lever.
7   Clean the parts in solvent and replace any that are damaged or excessively worn.
8   Installation is the reverse of the removal procedure. **Note:** *During installation, apply multi-purpose grease to all friction surfaces.*
9   Following installation, check the height and free play of the clutch pedal.

Adjustment

10 It is important to have the clutch pedal height and free play correctly adjusted (see illustration). The height of the pedal is the distance it stands away from the floor (distance H in illustration 12.10) and the free play is the pedal slack or the distance the pedal can be depressed before it begins to have any effect on the clutch (measurement A in illustration 12.10). Both of these measurements can be found in the Specifications of this Chapter. If these measurements are not as specified, the clutch pedal should be adjusted.
11 Ensure that the pedal height is correct. If this distance is not correct, loosen the locknut on the pedal stopper or clutch switch. Turn the pedal stopper or clutch switch in toward the pedal until the distance between the top center of the pedal pad and the floor panel is within specifications, then retighten the pedal stopper/clutch switch locknut. **Note:** *When making this adjustment, be sure that the pedal is not depressed and that the pushrod is not pushed beyond its free play.*
12 Check the pedal free play and correct if not within specifications. To correct free play, loosen the push rod lock nut and turn the push rod until free play of the pedal is within specifications. Tighten the push rod locknut.
13 Following all adjustments, fully depress and release the pedal several times to ensure that the clutch linkage is operating smoothly and that there is no binding or squeaks.
14 Install the lower trim panel.

13 Clutch safety switch — checking and replacement

1   Remove the dash lower trim panel (Chapter 11).
2   Locate the clutch switch on the top of the clutch pedal and disconnect the electrical connector.
3   Use a voltmeter to test for continuity. There should be no continuity if the pedal is up, but it should show continuity when the pedal is depressed.
4   Check clutch pedal height and free play (Section 12). If the height and free play are correct and the switch does not pass the continuity test, replace the switch.
5   Replace the switch by loosening the switch locking nut and turning the switch until the switch is released from its retaining bracket.
6   Install the switch and tighten the switch locking nut.
7   Connect the switch electrical connector.
8   Adjust pedal height and free play.
9   Replace the lower trim panel.

14 Drivetrain — general information

A single-piece driveshaft is installed on all models. The driveshaft is supported at the rear by the pinion flange of the differential and at

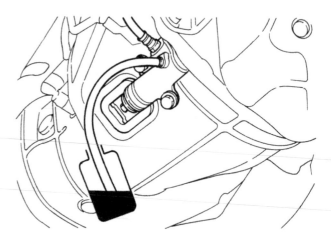

11.7   Bleed the clutch operating cylinder using the same method as described for the damper

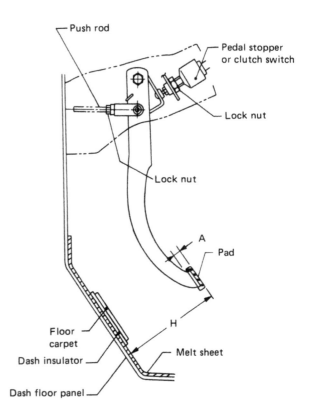

12.10   The adjustment heights of the clutch pedal components

A 0.04 to 0.12 in
H 7.68 to 8.07 in

the front by the transmission rear extension housing. It incorporates two universal joints, one at either end of the shaft.
The driveshaft is balanced during manufacture and, in the case of worn universal joints, a replacement driveshaft will have to be purchased. When removing the driveshaft for other than replacement reasons, the yoke-to-companion flange relationship must be marked so the driveshaft can be reinstalled in the exact position it was prior to removal.
The main rear axle component is the hypoid final drive and differential unit, which is fixed to the body at the rear using a cross-bracket located in rubber mountings. The front of the differential is mounted to the

15.3 Remove the heat shield and the rear cross bar

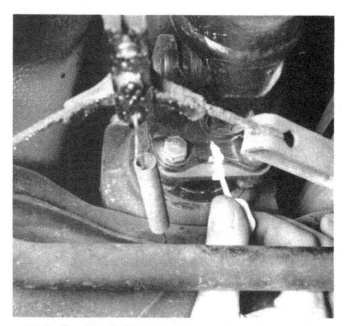

15.7 The driveshaft is balanced so mark its relation to the carrier flange so it can be reinstalled in its original location

suspension crossmember.

Because of the complexity and critical nature of the differential adjustments, as well as the special equipment needed to perform the operations, we recommend any disassembly of the differential be done by a Nissan dealer or other qualified mechanic.

The power is transferred from the differential to the rear wheels by the driveaxles, which rotate through two universal joints, one attached to the differential and the other to the rear hub. The type of driveaxle used on Turbo ZX models is of the constant velocity Double offset-Birfield universal joint type, while the axle used on non-Turbo models is of the Tripod-Tripod universal joint type.

The rear hub/axle stub assemblies incorporate two rear wheel bearings, and are mounted on the trailing ends of the suspension arms.

---

15 Driveshaft — removal, installation and balancing

Refer to illustrations 15.3, 15.7 and 15.8

1   Raise the vehicle and support it securely on jackstands.
2   Remove the catalytic converter (Chapter 4).
3   Remove the heat shield above the catalytic coverter (see illustration) and remove the rear cross bar.
4   Loosen the emergency brake cable adjuster and remove the adjuster from the cable (Chapter 9).
5   Disconnect the emergency brake cable spring from the rear crossmember.
6   Remove the rear sway bar-to-frame hold down nuts and let the sway bar hang down (Chapter 10).
7   Mark the driveshaft and pinion flanges with paint (see illustration) so the shaft will be installed in its original position. **Note:** *The driveshaft and companion flange are balanced at the factory and must be installed in their original alignment.*
8   Loosen the rear flange nuts. It is not necessary to hold the bolt heads since they are held by the flange casting (see illustration). Remove the bolts and lower the driveshaft down while sliding it rearward from the transmission extension housing.
9   Installation is the reverse of the removal procedures. Be sure to match up the aligning marks on the flanges.

Correcting unbalanced shaft

10 Remove the driveshaft from the vehicle following the procedure

15.8 The flange head bolts wedge against the flange, making removal a one wrench operation

above and remove any undercoating or foreign material which could affect driveshaft balance.
11 Reinstall the driveshaft and road test. If the vibration still occurs, disconnect the driveshaft at the differential carrier companion flange and rotate the flange 180° and reinstall the driveshaft.
12 Install the driveshaft and road test the vehicle. If the vibration still exists, replace the driveshaft assembly.

---

16 Differential carrier — general information

Refer to illustration 16.1

The main rear axle component is the hypoid final drive and differential unit (see illustration), which is fixed to the body at the rear using a

16.1 Exploded view of the differential carrier

cross-bracket located in rubber mountings. The front of the differential is mounted to the suspension crossmember.

Because of the complexity and critical nature of the differential adjustments, as well as the special equipment needed to perform the operations, we recommend any disassembly of the differential be done by a Nissan dealer or other qualified mechanic.

The power is transferred from the differential to the rear wheels by the driveaxles, which rotate through two universal joints, one attached to the differential and the other to the rear hub.

17 Carrier assembly — removal and installation

Refer to illustration 17.8

1   Raise the rear of the vehicle, support it securely on jackstands and drain the differential oil (Chapter 1).
3   Disconnect the driveshaft from the differential pinion flange (Section 15).
4   Disconnect the driveaxles and remove them from the vehicle (Section 19).
5   Remove the catalytic converter heat shield.
6   Position a jack under the differential carrier and raise it just enough to support it.
7   Remove the carrier mount.
8   Remove the bolts securing the front of the differential unit to the suspension crossmember (see illustration). The differential unit can now be carefully lowered and removed from the vehicle.

17.8   Remove the through bolts securing the carrier to the suspension crossmember (arrows)

9   Install the carrier using a floor jack to raise it into place.
10 Install the through bolts and attach the carrier to the suspension crossmember. Tighten all fasteners to specifications.
11 Install the carrier mount.
12 Install the catayltic converter heat shield.
13 Install the driveaxles.
14 Install the driveshaft.
15 Remove the jackstands and lower the vehicle.

18 Carrier front oil seal — replacement

Refer to illustrations 18.4, 18.6, 18.7 and 18.8

1   Raise the vehicle and support it securely on jackstands.
2   Place a container beneath the differential, then remove the drain plug and allow the oil to drain from the unit. Install the drain plug.
3   Disconnect the rear of the driveshaft and wire the shaft to the body out of the way (Section 15).
4   Mark the relation of the pinion flange retaining nut to the carrier with paint (see illustration).
5   Use a large adjustable wrench to hold the flange while you loosen the pinion nut. Count the number of turns it takes to remove the nut and record it for installation.
6   Using a puller, withdraw the pinion flange from the differential unit (see illustration).
7   Remove the oil seal by driving in one side of the seal and levering it out or use a seal puller tool ST33290001 (see illustration).

18.4   Mark the relation of the pinion nut to the carrier input shaft and count the number of turns it takes to remove the nut. Record the number of turns for installation

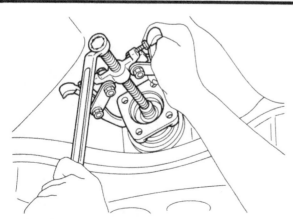

18.6   Use a puller to remove the carrier pinion flange

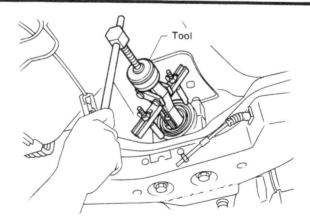

18.7   Using a seal removal tool to remove the differential carrier front seal

8   Install the new oil seal. The lips of the oil seal must face inwards. Using a suitable socket or Nissan Tool KV38100500 (see illustration), carefully drive the new oil seal into the axle housing recess until the face of the seal is flush wirh the housing.

9   Install the pinion flange and thrust washer onto the drive pinion and thread on the pinion nut. Count the turns of the nut while tightening it. Turn the nut the exact number of turns it took to remove it, then turn the nut one-quarter turn more to insure proper tightness.

10   Connect the driveshaft to the carrier companion flange.

11   Refill the differential with the proper grade and quantity of lubricant.

12   Remove the jackstands and lower the vehicle.

19   Carrier side oil seal — replacement

Refer to illustrations 19.4 and 19.5

1   Raise the rear of the vehicle and support it securely on jackstands. Place the jackstands under the suspension arms to compress the suspension system when the vehicle is lowered onto the jackstands.

2   Remove the driveaxle from the differential (Section 19).

3   Use a pry bar or large screwdriver to pry the side flange out of the differential. While prying, keep one hand on the flange to keep it from dropping out and possibly being damaged.

4   Pry the oil seal out with a screwdriver (see illustration), being careful not to scratch or damage the surface of the bore.

5   Drive in a new seal using Nissan Tool KV38100200 or a suitable socket (see illustration).

6   Apply grease to the cavity between the oil seal lips.

7   Install the side flange.

8   Reinstall the driveaxle.

9   Open the filler plug on the carrier housing and check the fluid level (Chapter 1).

10   Remove the jackstands and lower the car.

20   Driveaxle — general information

Refer to illustration 20.1

The differential transfers power to the rear wheels by driveaxles which rotate through two universal joints (see illustration).

Two types of constant velocity (CV) joints are employed, and the type used on your vehicle depends on whether it is a Turbo model or not. The Double offset-Birfield type is used on Turbo models and consists of caged ball bearings placed between the inner race and the cage. This assembly is then housed in a slide joint housing and capped off with a plug and covered with a rubber boot. The non-Turbo models are equipped with a "Tripod-Tripod" type of bearing. This type employs needle bearings which are part of a spider assembly, splined onto the shaft and housed inside a slide joint housing. The bearing is held in place by a spring and end cap. The shaft end is covered with a rubber boot filled with lubricant.

21   Driveaxle boot replacement

As of the writing of this manual no boot replacement parts could be obtained. It was necessary to buy the complete driveaxle from a Nissan dealer. Check with your local parts store for replacement parts availability.

22   Driveaxle — removal and installation

Refer to illustrations 22.2a, 22.2b, 22.3a, 22.3b, 22.4 and 22.5

1   Raise the vehicle and secure it on jackstands. Place the jackstands under the suspension arms so the suspension will be compressed.

2   Remove the bolts securing the driveaxle to the companion flange (see illustration). A back up wrench will be necessary to hold the bolts (see illustration).

3   Place a jack under the spring seat stay, remove the spring seat stay bolts (see illustration) and lower the jack (see illustration). **Caution:** *Use a screwdriver to pry the parking brake cable over the spring seat stay.*

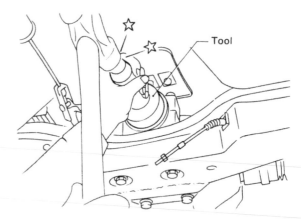

18.8   Special seal replacing tool being used to drive in the differential carrier front seal

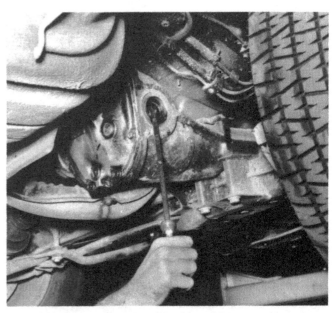

19.4   Insert a screwdriver into the seal and pry the seal from the housing

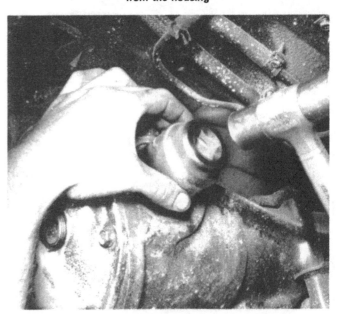

19.5   Use a socket to drive the seal evenly into the carrier recess

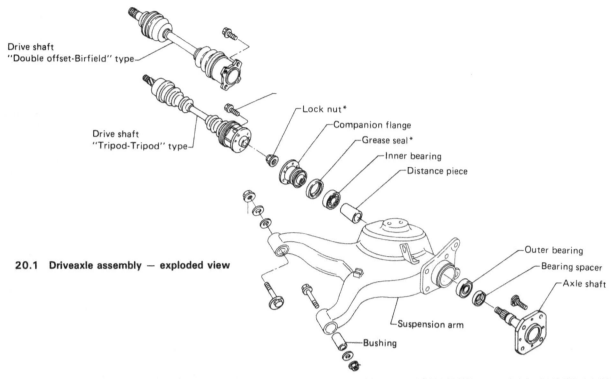

Drive shaft "Double offset-Birfield" type

Drive shaft "Tripod-Tripod" type

Lock nut*
Companion flange
Grease seal*
Inner bearing
Distance piece

Outer bearing
Bearing spacer
Axle shaft

20.1   Driveaxle assembly — exploded view

Suspension arm

Bushing

22.2a   Remove the bolts securing the driveaxle (Double offset-Birfield type shown) to the wheel companion flange

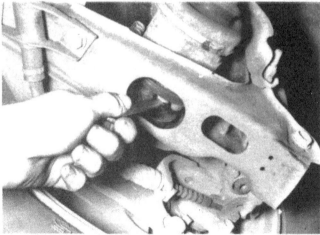

22.2b   Use a back up wrench to remove the driveaxle-to-wheel bolts

22.3a   With a jack under the spring seat stay, remove the spring seat stay attaching bolts

22.3b   Lower the spring seat stay to allow the removal of the rear axle

22.4   Push the driveaxle towards the differential carrier and pull down to remove it from the wheel flange

22.5   Give the pry bar a quick rap with a hammer to release the driveaxle from the carrier housing

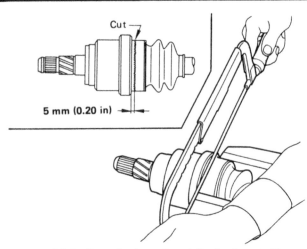

23.3   Use a hacksaw to cut the boot assembly as indicated

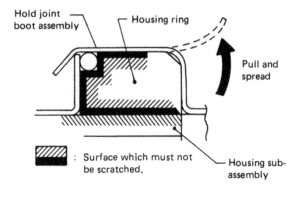

23.4   Carefully pry the boot assembly back from the axle. Be careful not to scratch the housing sub-assembly

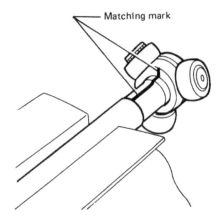

23.5   Make an alignment mark to note the relation of the spider gear assembly to the shaft

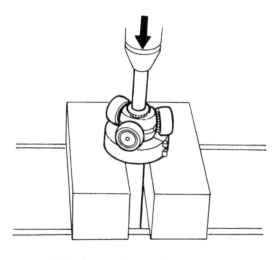

23.6   Press off the spider assembly

4   Carefully pry the driveaxle towards the differential carrier to get enough clearance to remove it from the wheel side flange. With the driveaxle pushed back gently force it down from the wheel flange (see illustration). **Caution:** *Keep the slide joint housing grease cap located next to the wheel companion flange in place to prevent grease spillage.*

5   Remove the driveaxle from the differential carrier by placing a small pry bar between the driveaxle and carrier and giving the bar a quick rap with a hammer (see illustration).

6   Install the driveaxle by inserting the splined shaft into the differential carrier.

7   Work the wheel side of the driveaxle into its companion flange.

8   Place a jack under the spring seat stay and raise it up to install the retaining bolts. **Caution:** *Carefully pry the brake cable over the end of the spring seat stay when jacking it up.*

9   Install the driveaxle to wheel companion flange bolts and tighten.

10   Remove the jack from under the spring seat stay.

11   Remove the safety jackstands and lower the vehicle.

---

23 Driveaxle (Tripod-Tripod type) — disassembly and assembly

Refer to illustrations 23.3, 23.4, 23.5, 23.6, 23.9, 23.10, 23.11, 23.13, 23.14, 23.15, 23.20, 23.21 and 23.22

Note: *A press is required to remove the spider gear assembly from the shaft. Refer to steps 6 and 17.*

1   Remove the driveaxle (Section 22).

2   Place the driveaxle assembly in a vise.

3   Use a hacksaw to cut the boot assembly (see illustration) near the joint housing. **Caution:** *Before cutting, ensure that the shaft is pushed into the housing sub-assembly to prevent the spider gear assembly from being scratched.*

4   Pry the boot assembly back from the axle (see illustration) and be careful not to scratch the housing sub-assembly. Cut the remaining part of the boot from the axle assembly.

5   Make an alignment mark across the shaft and spider gear assembly (see illustration).

6   Place the shaft in a suitable press (see illustration) and press off the spider assembly. **Note:** *Do not disassemble the spider gear assembly.*

7   Inspect the bearing assembly, joint housing and housing sub-assembly for cracks or other damage and replace either component if necessary.

8   Place the shaft in a vise between two wooden blocks.

9   Install the spider assembly and match up alignment marks made during removal. Drive the assembly onto the shaft using a suitable tool (see illustration).

10   Stake the serration portion evenly at three places around the assembly (see illustration) and avoid areas that have been previously staked. **Note:** *Always stake two or three teeth at each spot.*

11   Install the boot assembly, fold back the boot band and crimp down the retaining dogs (see illustration).

12   Pack the boot with the specified amount and type of grease.

13   Set the boot length by holding the shaft straight out in relation to the joint assembly and crimping down the boot band at the length specified (see illustration).

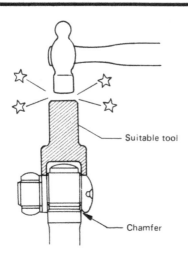

23.9   Drive the spider assembly onto the shaft after matching the alignment marks made during disassembly

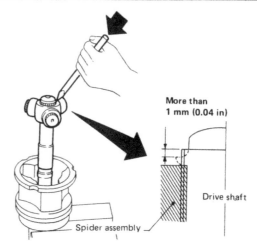

23.10   Stake the serration portion evenly at three places around the assembly. Do not stake areas previously staked

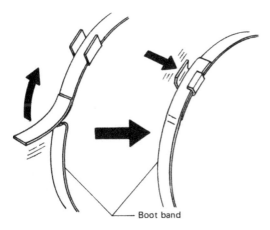

23.11   Install the hold joint assembly and fold back the boot band, then use a small hammer to bend back the dogs

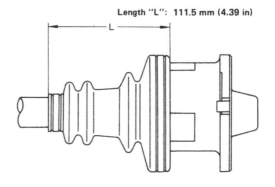

23.13   Hold the boot out straight and set the boot length as specified

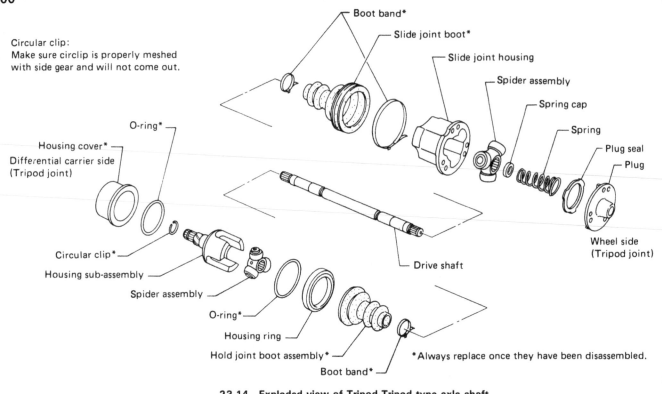

Circular clip:
Make sure circlip is properly meshed with side gear and will not come out.

Boot band*
Slide joint boot*
Slide joint housing
Spider assembly
Spring cap
Spring
Plug seal
Plug

Housing cover*
Differential carrier side (Tripod joint)

O-ring*

Circular clip*
Housing sub-assembly
Spider assembly
O-ring*
Housing ring
Hold joint boot assembly*
Boot band*

Drive shaft

Wheel side (Tripod joint)

*Always replace once they have been disassembled.

23.14   Exploded view of Tripod-Tripod type axle shaft

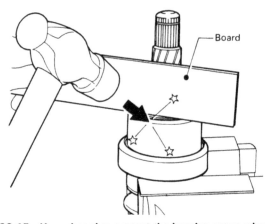

Board

23.15   Use a board to protect the housing cover when crimping the edge of the housing ring

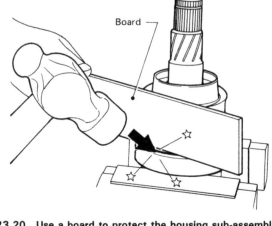

Board

23.20   Use a board to protect the housing sub-assembly when bending the edge of the hold joint assembly

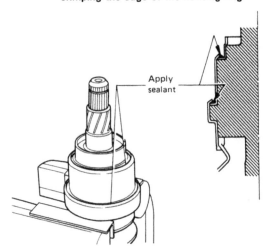

Apply sealant

23.21   Apply sealant at the locations noted

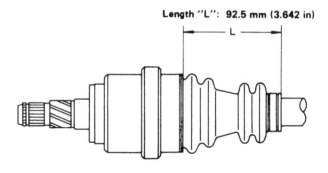

Length "L": 92.5 mm (3.642 in)

L

23.22   Hold the shaft out straight and set the boot length before crimping down the boot band

Circular clip:
Make sure circular clip "A" is properly meshed with side gear and that circular clip "B" is also meshed with joint assembly, and will not come out.

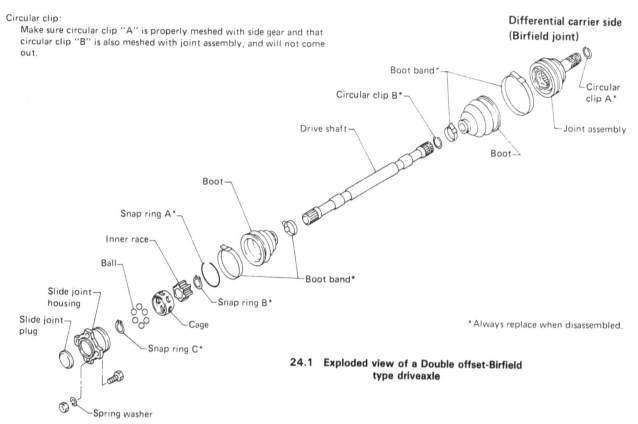

* Always replace when disassembled.

24.1 Exploded view of a Double offset-Birfield type driveaxle

Wheel side (Double offset joint)

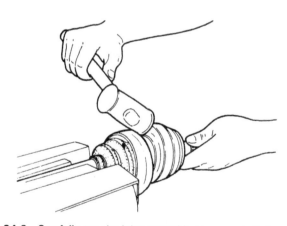

24.3 Carefully tap the joint assembly to separate it from the shaft

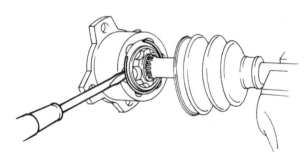

24.6 Remove the snap ring and pull the slide joint housing out

14 Install the housing ring and housing sub-assembly onto the carrier side of the shaft (see illustration). **Note:** *Always replace housing ring and sub-assembly as a set.*
15 Bend the edge of the housing ring over along the entire circumference. Start by bending the edge at two positions 180° apart. Place a board on the housing cover (see illustration) so you will not damage it.
16 Install a new boot band and boot assembly onto the driveaxle.
17 Install the spider assembly.
18 Pack the boot with grease.
19 Insert the boot assembly so that only the flange is retained in a vise.
20 Insert the housing sub-assembly into the boot assembly and bend the edge (see illustration) in the manner described in Step 15.
21 Apply sealant to the edges of the hold down joint assemblies (see illustration).
22 Set the boot length by holding the shaft straight out in relation to the joint assembly and crimping down the boot band at the length specified (see illustration).
23 After the shaft has been assembled, be sure that it moves smoothly over the entire range without binding.
24 Install the driveaxle (Section 22).

---

24 Driveaxle (Double offset-Birfield type) — dissassembly and assembly

---

Refer to illustrations 24.1, 24.3, 24.6, 24.8, 24.14, 24.16, 24.17 and 24.23

Carrier side disassembly

1 Remove the driveaxle (see illustration) from the vehicle (Section 22) and place it in a vise between two boards so not to damage the shaft.
2 Before disassembly, place alignment marks on the shaft and the joint assembly for alignment during installation.
3 Lightly tap on the joint assembly to separate it from the shaft (see illustration). **Note:** *The joint on this side can not be disassembled.*

Wheel side

4 Remove the boot bands.
5 Scribe matching marks on the slide joint housing and the inner race.
6 Pry off the snap ring (see illustration) with a screwdriver and pull the slide joint housing out.

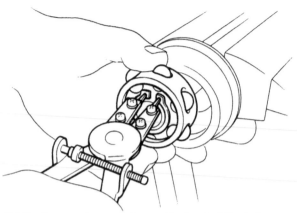

24.8   Remove the snap ring on the front of the ball cage
and remove the cage, balls and inner race assembly

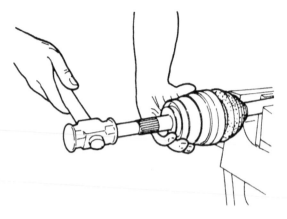

24.14   Gently tap the joint assembly onto the shaft

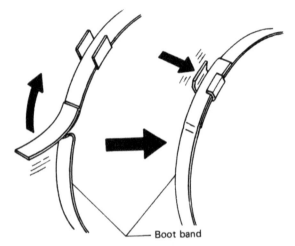

24.16   Install the large boot band and flatten the tabs
down against the band

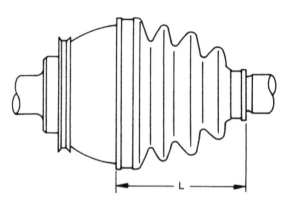

Length "L": 90.8 mm (3.575 in)

24.17   Set the carrier side boot length and crimp down
the boot band

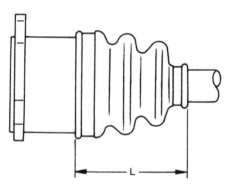

Length "L": 90.4 mm (3.559 in)

24.23   Set the wheel side boot length and crimp down the
boot band

7   Scribe matching marks on the shaft and the inner race.
8   Use snap ring pliers and remove the snap ring (see illustration) on the front of the ball cage. Remove the ball cage, balls and inner race as an assembly.
9   Pry off the snap ring and lay the cage assembly in a container.
10   Remove the boot from the shaft.

Inspection

11   Inspect the shaft for cracks or distortion. Check the assemblies for burns, rust, wear or excessive play. Check the boots for fatigue or worn spots. Replace any of the components if damaged.

Carrier side assembly

12   Install the boot from the wheel side. Be careful not to damage the boot on the edge of the driveshaft.
13   Install a new small boot band on the shaft.
14   Install shaft in a vise protected with wooden blocks and gently tap the joint assembly onto the shaft (see illustration). Make sure the alignment marks are properly aligned.
15   Pack the joint/boot assembly with grease.
16   Install the large boot band and secure the retaining tabs by flattening them against the band (see illustration).
17   Set the boot length by holding the shaft straight out in relation to the joint assembly and crimping down the boot band at the length specified (see illustration).

Wheel side assembly

18   Install the boot, being careful not to damage the boot on the edge of the shaft.
19   Install the large boot band and secure the band retainer by flat-

tening it against the band.
20   Install the ball cage assembly and secure with the snap rings.
21   Install the ball cage assembly into the slide joint housing and retain with the snap ring.
22   Pack the boot/joint assembly with grease.
23   Set the boot length by holding the shaft straight out in relation to the joint assembly and crimping down the boot band at the length specified (see illustration).

25.3   Wedge a pry bar between the carrier and its mount and pry it back to see if the rubber has separated from the mount

26.3   Remove the heat shield retaining bolts and note ground wire attached to one of them

24   After the driveaxle has been assembled, ensure that it moves smoothly over its entire range without binding.
25   Install the driveaxle (Section 22).

25   Carrier mount — inspection

Refer to illustration 25.3

1   Raise the vehicle and support it securely on jackstands.
2   Visually check the rubber part of the insulator to see if it has torn away from the metal backing.
3   Insert a pry bar into the mount (see illustration) and pry back to see if the rubber has separated from the metal backing. If the rubber has parted, replace the mount (Section 26).
4   Remove the jackstands and lower the vehicle.

26   Carrier mount — removal and installation

Refer to illustration 26.3

1   Raise the vehicle and support it securely on jackstands.
2   Place a jack under the carrier assembly.
3   Remove the heat shield and note ground wire attached to the body (see illustration).
4   Remove the carrier mount retaining nuts and bolts.
5   Lower the jack slightly until the mount clears the studs and pull the mount rearward to remove it from the studs on the carrier.
6   Insert the mount onto the carrier studs and raise the jack until the mount can be secured to the body.
7   Install and tighten the carrier mount nuts and bolts.
8   Replace the heat shield and be sure and install the ground wire.
9   Remove the jack from under tha carrier.
10 Remove the jackstands and lower the vehicle.

Chapter 9 Brakes

Contents

Specifications

Recommended brake fluid	Dot 3

Brake pedal

free height
  automatic transmission . . . . 184 to 194 mm (7.24 to 7.64 in)
  manual transmission . . . . 182 to 192 mm (7.17 to 7.56 in)
depressed height
  1984 thru 1986 . . . . 90 mm (3.54 in) or more
  1987 on . . . . 80 mm (3.15 in) or more
stop lamp switch clearance . . . . 0.3 to 0.0 mm (0.012 to 0.039 in)
ASCD switch clearance . . . . 0.3 to 1.0 mm (0.012 to 0.039 in)
free play . . . . .10 to 3.0 mm (0.039 to 0.118 in)

Torque specifications

	Kg-m	Ft-lbs
Brake hose connector to caliper	1.7 to 2.0	12 to 14
Brake pedal		
input rod lock nut	1.6 to 2.2	12 to 16
fulcrum bolt nut	3.1 to 4.1	22 to 30
stop lamp and ASCD switch	1.2 to 1.5	9 to 11
Master cylinder nut	0.8 to 1.1	5.8 to 8.0
Brake booster nut	0.8 to 1.1	5.8 to 8.0
Caliper		
torque member fixing bolt		
1984 thru 1986	7.3 to 9.9	53 to 72
1987 on	3.9 to 5.3	28 to 38
pin bolt		
1984 thru 1986	2.2 to 3.2	16 to 23
1987 on	3.2 to 4.2	23 to 30
air bleeder	0.7 to 0.9	5.1 to 6.5
Emergency brake		
lever retaining bolts	1.6 to 2.1	12 to 15
cable adjuster locking nut	0.32 to 0.44	2.3 to 3.2
front cable retaining bolts	0.8 to 1.1	5.8 to 8.0

1 General information

Refer to illustration 1.1

1   The braking system in the 300ZX is a split system design (see illustration). It incorporates two separate circuits; one for the front brakes and one for the rear brakes. With this system if one circuit fails, the other circuit will still function.
2   The master cylinder is designed for the split system and incorporates a primary piston for one circuit and a secondary piston for the other.

3   A vacuum servo unit is used which draws vacuum from the intake manifold to add power assistance to the normal brake pressure.
4   A proportioning valve inside the master cylinder regulates the fluid pressure in the brake lines so that all wheels receive equal pressure.
5   All four wheels are equipped with disc brakes. These consist of a rotor which is attached to the axle and wheel. Around the rotor is mounted a stationary caliper assembly which houses two hydraulically-operated disc brake pads. The inner pad is mounted to a piston facing the inner surface of the rotor, while the outer pad is mounted either to a moveable yoke or to the caliper body so that it faces the outer surface of the rotor. When the brake pedal is applied, brake fluid

pressure forces both pads against the rotor.

6   The rear brakes are also equipped with cable-operated parking brake mechanisms, which lock the rear pads against their rotors.

7   After completing any operation involving the dismantling of any part of the brake system always test drive the car to check for proper braking performance before resuming normal driving. When testing the brakes, perform the tests on a clean, dry, flat surface. Conditions other than these can lead to inaccurate test results. Test the brakes at various speeds with both light and heavy pedal pressure. The car should brake evenly without pulling to one side or the other.

2   Brake pads — removal, inspection and installation

Refer to illustrations 2.2a, 2.2b, 2.3, 2.4, 2.9, 2.10, 2.11, 2.12, 2.17, 2.18, 2.19, 2.20, 2.21, 2.24, 2.26, 2.27 and 2.28

Front pad replacement

1   Raise the vehicle and support it securely on jackstands. Remove the front wheels.

2   Remove the lower pin retaining bolt from the front brake caliper (see illustrations).

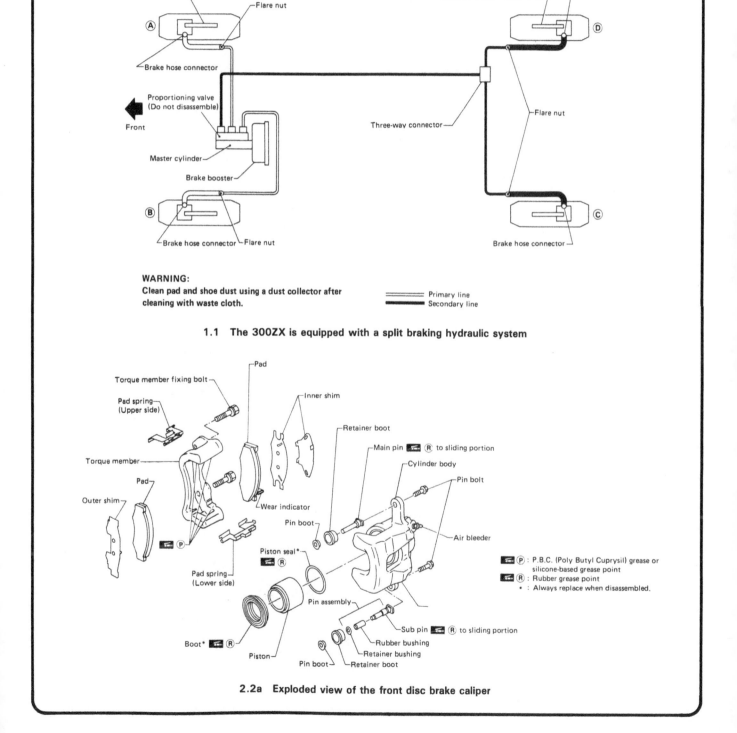

WARNING:
Clean pad and shoe dust using a dust collector after cleaning with waste cloth.

Primary line
Secondary line

1.1   The 300ZX is equipped with a split braking hydraulic system

:::: P.B.C. (Poly Butyl Cuprysil) grease or silicone-based grease point
:::: (R) : Rubber grease point
  * : Always replace when disassembled.

2.2a   Exploded view of the front disc brake caliper

2.2b   Remove the caliper lower pin retaining bolt (arrow)

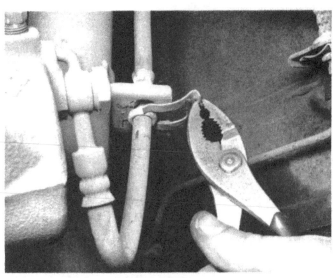

2.3   Disconnect the brake line hose retaining spring clip to allow enough hose slack for the caliper to swing up

2.4   Pivot the caliper body up to gain acess to the brake pads

2.9   Arrows indicate the contact surfaces requiring silicone-based grease

2.10   Install pad retainers (arrows) onto the torque member

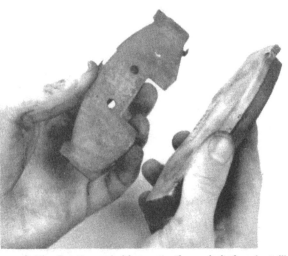

2.11   Put the pad shims onto the pads before installing the pads

3   Remove the brake line retaining clip at the hose bracket located on the strut (see illustration). Remove the hose from the bracket. This will allow enough hose slack for the caliper to swing up.
4   Lift the caliper body upward to allow access to the brake pads (see illustration).
5   Remove the inner and outer shims and note their installed positions.
6   Lift out the brake pads. **Note:** *After removing the pads, do not depress the brake pedal, as this will force the piston out of the cylinder.*
7   If the pads are glazed, damaged, fouled with oil, grease or worn beyond the limit (see Specifications), they should be replaced. **Note:** *Always replace all four pads on the axle (two in each brake assembly) at the same time, and do not mix different pad materials.*
8   Prior to installation, use brake fluid to clean the end of the piston and the pin bolts. Do not use mineral based solvents.
9   Before installation of the pad retainers raise the cylinder body and apply a silicone-based grease to the contact surfaces between the torque member and the pads (see illustration). During installation, be careful not to allow grease to get on the friction surfaces of the pads.
10 Reinstall the retainers before installing the pads (see illustration).
11 Install the the pad shims onto the pads before installing them in the caliper (see illustration). Install the pads in the caliper. **Caution:**

Always replace the shims when replacing the pads.
12 Use a small C-clamp to compress the piston (see illustration). **Caution:** *Make sure the reservoir fluid level will allow the piston to be compressed without overflowing the reservoir.*
13 Lower the caliper body and install the lower pin bolt, tightening it to specifications.
14 Depress the brake pedal several times to settle the pads into their proper positions.
15 Remove the jackstands and lower the vehicle after reinstalling the wheels.

Rear pad replacement

16 Raise the vehicle and support it securely on jackstands.
17 Loosen the emergency brake cable at the adjusting turnbuckle (see illustration) located in front of the differential carrier housing.
18 Remove the cable retaining spring clip located on the bottom of the caliper (see illustration).
19 Remove the emergency brake cable end from the retaining bracket by pushing the actuating arm back and sliding the cable end from the lever (see illustration).

2.12   Use a C-clamp to compress the piston

2.17   Loosen the emergency cable adjusting bolt to release tension from the cable

2.18   Use pliers to remove the cable retaining spring clip

2.19   Push the actuating arm back and lift the cable end from the lever

2.20   Remove the lower retaining pin bolt (arrow) to allow the caliper to be raised

20 Remove the lower caliper retaining pin (see illustration) and raise the caliper. **Note:** *Due to the short length of the hydraulic brake hose the caliper will not flip all the way up, but will go up enough to allow pad removal.*

21 Remove the pads with the shims attached (see illustration).

22 Lift out the pad retainers.

23 Clean the components with brake fluid and let them dry.

24 Before installation of the pad retainers raise the cylinder body and apply a silicone-based grease to the contact surfaces between the torque member and the pads (see illustration). During installation, be careful not to allow grease to get on the friction surfaces of the pads.

25 Examine the pads for wear and condition.

26 Install the pad retainers (see illustration).

27 Retract the piston into the cylinder body by using needle nose pliers to turn the piston clockwise until the pad has enough room to be installed (see illustration).

28 Install the shims onto the pads (see illustration). **Note:** *Always replace the shims when replacing the pads.*

29 Reinstall the pads and lower the caliper.

30 Install the lower caliper retaining pin.

31 Install the emergency cable into its retaining bracket and secure it with the spring clip.

32 Push the emergency brake acutating lever back and insert the emergency cable end into its recess.

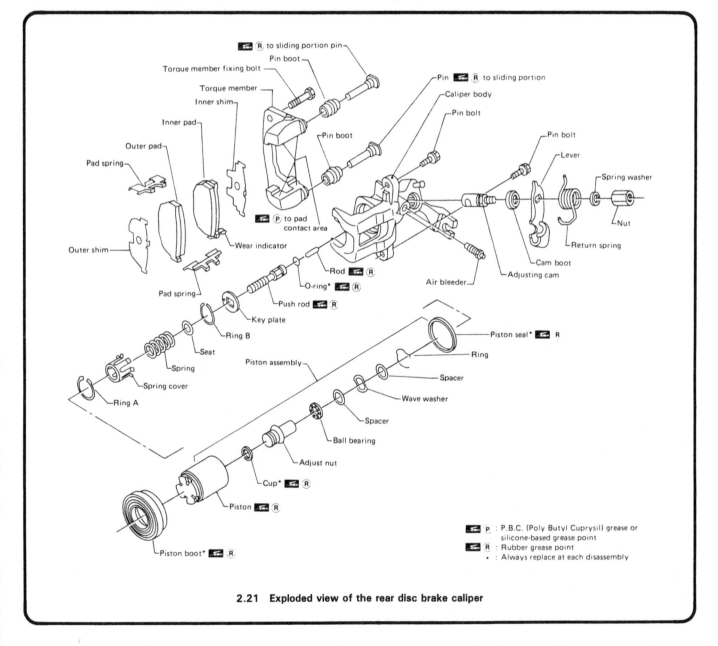

2.21   Exploded view of the rear disc brake caliper

33 Depress the brake pedal several times to settle the pads into their proper positions.
34 Adjust the emergency brake cable (Section 10).
35 Remove the jackstands and lower the vehicle.

---

3   Caliper — removal and installation.

Refer to illlustrations 3.3 and 3.13

Front caliper

1   Raise the vehicle and support it securely on jackstands.
2   Remove the wheel.
3   Disconnect the hydraulic hose by disconnecting the banjo bolt from the caliper (see illustration). Be prepared to catch a small amount of brake fluid.
4   Remove the two caliper retaining bolts and lift off the caliper and pads as an assembly.
5   Inspect the caliper for leaking seals. Inspect the caliper and torque member for cracks, wear or other damage and replace if necessary.
6   Install the caliper and secure with the top retaining pin bolt.
7   Install the pads (Section 2).
8   Reconnect the hydraulic hose.
9   Bleed the hydraulic system (Section 9).

2.24   Apply silicone based grease to the contact surfaces (arrows) between the torque member and the pads

2.26   Install pad retainers (arrows) into the torque member

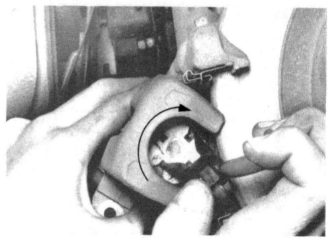

2.27   Arrow shows rotation direction of the piston to create enough room to install the new pads

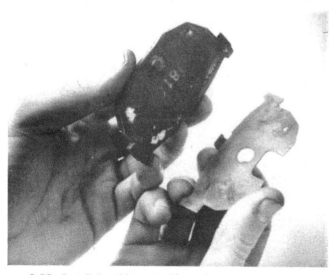

2.28   Install the shims onto the pads before installation

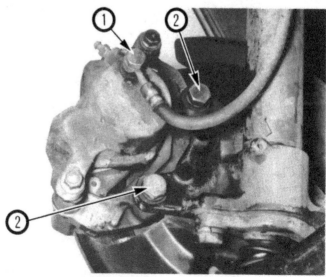

3.3   Disconnect the brake hose by removing the Banjo bolt (1). The caliper is held on by two bolts (2)

Rear caliper

10 Raise the vehicle and support it securely on jackstands.

11 Remove the wheel.

12 Disconnect the emergency brake cable by loosening the cable adjuster then removing the cable from the parking brake lever at the caliper.

13 Use a flare nut wrench to disconnect the hydraulic hose where the metal line connects to the rubber hose (see illustration).

14 Pull the hose retaining spring clip from the hose retaining bracket and remove the hose from the bracket.

15 Remove the caliper retaining bolts and remove the caliper.

16 Inspect the caliper for leaking seals and inspect the caliper and torque member for cracks, wear or other damage.

17 To make caliper installation easier remove the pads and place the caliper onto the rotor and secure it with the two retaining bolts.

18 Install the brake pads (Section 2).

19 Reconnect the hydraulic line.

20 Bleed the hydraulic system (Section 9).

21 Replace wheel, remove the jackstands and lower the vehicle.

4   Disc brake rotor – removal and installation

Refer to illustrations 4.6, 4.7, 4.9, 4.10, 4.11, 4.12 and 4.20.

Front rotor

Note: *The front brake rotors on 1988 and later turbo models can be pulled off the hub after the caliper is removed. There is no need to remove the hub from the spindle.*

1   Raise the vehicle and support it securely on jackstands.

2   Remove the wheel.

3   Remove the caliper. Leave the hydraulic brake hose connected during caliper removal and wire the caliper to the spring to get it out of the work area.

4   Remove the dust cap from the rotor (see illustration).

5   Remove the cotter pin and discard the pin (see illustration).

6   Pull the wheel bearing locking nut cap off the wheel bearing locking nut.

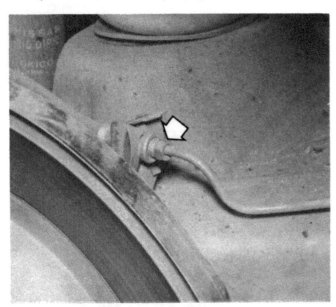

3.13   Use a flare nut wrench to disconnect the metal brake line where it joins the rubber hose (arrow)

4.4   Pry the dust cap off the rotor

4.5   Remove the cotter pin and replace it with a new one during installation

4.7   Remove the wheel bearing locking nut

7   Remove the wheel bearing nut and washer from the end of the spindle (see illustration).
8   Remove the hub/rotor components from the spindle by grasping it with your hands (see illustration) and wiggling it from the spindle.
9   To remove the rotor from the hub, lay the rotor face down and remove the bolts (see illustration).
10 To install, carefully place the hub assembly onto the spindle and push the outer bearing into position (see illustration).
11 Install the washer and adjuster/locking nut. Snug the adjusting nut down and spin the rotor in both directions to seat the bearings.
12 Adjust wheel bearing pre-load (Chapter 10).
13 Replace the brake caliper (Section 2).
14 Replace the wheel, remove the jackstands and lower the vehicle.

Rear rotor

15 Raise the vehicle and support it securely on jackstands.
16 Remove the wheel.
17 Remove the caliper (Section 3). Leave the hydraulic brake hose connected during caliper removal and wire the caliper to the spring

to get it out of the work area.
18 The rotor should pull off. If it is rusted on screw two bolts into the threaded holes in the rotor (see illustration) and turn the bolts to press the rotor from the flange.
19 Install by sliding the rotor onto the wheel lugs.
20 Install the caliper (Section 2).
21 Adjust parking brake cable (Section 10).
22 Replace the wheel, remove the jackstands and lower the vehicle.

---

5   **Master cylinder – removal and installation**

Refer to illustrations 5.3, 5.4 and 5.5

1   Disconnect the negative cable at the battery. Place the cable out of the way so it cannot accidentally come in contact with the negative terminal of the battery.
2   Place newspapers or rags under the master cylinder to catch any leaking brake fluid.

4.8   Place your thumbs on the spindle to keep from dropping the outer bearing

4.9   Separate the rotor from the hub by removing the four bolts attaching the two components

4.10   Carefully push the rotor onto the spindle and insert the outer bearing

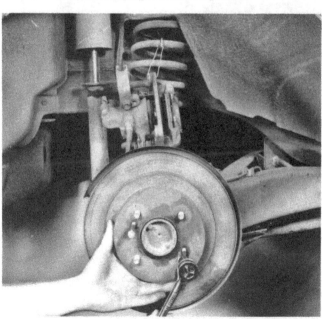

4.18   Install two bolts into the threaded holes provided and turn the bolts to force the rotor from the flange

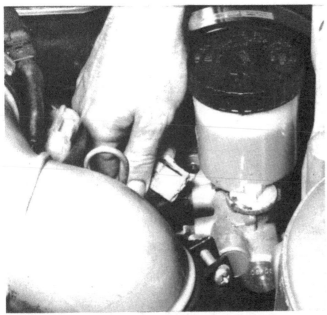

5.3   Disconnect the low fluid level electrical connector
from the master cylinder

5.4   Loosen the metal lines attached to the master cylinder
(arrows)

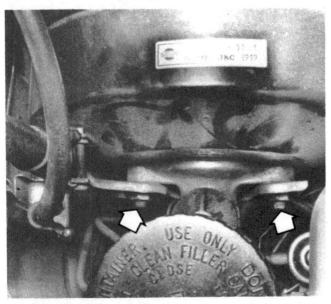

5.5   Remove the nuts attaching the master cylinder to the
servo unit (arrows)

3   Disconnect the low fluid level sensor electrical connector (see illustration).
4   Use a flare nut wrench to loosen the nuts securing the brake lines to the master cylinder (see illustration).
5   Remove the two nuts securing the master cylinder to the vacuum servo unit (see illustration).
6   Carefully lift the cylinder off the mounting studs, remove the brake lines from the cylinder, and immediately place your fingers over the holes to prevent leakage of fluid. Lift the cylinder out of the engine compartment.
7   Plug the fluid lines to prevent further leakage of fluid.

Bench bleeding procedure

8   Before installing the new master cylinder it should be bench bled. Because it will be necessary to apply pressure to the master cylinder piston and, at the same time, control flow from the brake line outlets, it is recommended that the master cylinder be mounted in a vice. Use caution not to clamp the vice too tightly, or the master cylinder body might be cracked.

9   Insert threaded plugs into the brake line outlet holes and snug them down so that there will be no air leakage past them, but not so tight that they cannot be easily loosened.
10   Fill the reservoir with brake fluid of the recommended type (see *Recommended fluids and lubricants*).
11   Remove one plug and push the piston assembly into the master cylinder bore to expel the air from the master cylinder. A large Phillips screwdriver can be used to push on the piston assembly.
12   To prevent air from being drawn back into the master cylinder the plug must be replaced and snugged down before releasing the pressure on the piston assembly.
13   Repeat the procedure until only brake fluid is expelled from the brake line outlet hole. When only brake fluid is expelled, repeat the procedure with the other outlet hole and plug. Be sure to keep the master cylinder reservoir filled with brake fluid to prevent the introduction of air into the system.
14   Since high pressure is not involved in the bench bleeding procedure, an alternative to the removal and replacement of the plugs with each stroke of the piston assembly is available. Before pushing in on the piston assembly, remove the plug as described in Step 12. Before releasing the piston, however, instead of replacing the plug, simply put your finger tightly over the hole to keep air from being drawn back into the master cylinder. Wait several seconds for brake fluid to be drawn from the reservoir into the piston bore, then depress the piston again, removing your finger as brake fluid is expelled. Be sure to put your finger back over the hole each time before releasing the piston, and when the bleeding procedure is complete for that outlet, replace the plug and snug it before going on to the other port.
15   Install the master cylinder onto the vacuum servo retaining studs and fasten it to the servo.
16   Attached the fluid lines to the master cylinder using a flare nut wrench.
17   Connect the low fluid level sensor electrical connector.
18   Top up the fluid level in the master cylinder reservoir until the level is between the Max and Min marks.
19   Bleed the entire brake system as described in Section 9.
20   Remove the newspapers from underneath the master cylinder. Be careful not to drip any fluid on painted surfaces.
21   Connect the negative ground cable at the battery.

---

6   Vacuum servo booster unit — general information and testing

Refer to illustration 6.1

1   A vacuum servo unit (see illustration) is part of the brake hydraulic circuit.
2   The unit operates by vacuum obtained from the intake manifold.

6.1   Vacuum servo unit and vacuum hose

The servo unit and hydraulic master cylinder are connected together so that the servo unit piston rod acts as the master cylinder pushrod.
3   The controls are designed so that assistance is given under all conditions and, when the brakes are not required, vacuum in the rear chamber is established when the brake pedal is released. All air entering the rear chamber is passed through a small air filter.
4   Under normal operating conditions the vacuum servo unit is very reliable and does not require replacement except at very high mileage.
5   To check for a satisfactory vacuum servo unit, depress the brake pedal several times. The distance which the pedal travels on each depression should not vary.
6   Hold the pedal fully depressed and start the engine. The pedal should be felt to move down slightly when the engine starts.
7   Depress the brake pedal and switch off the engine, holding the pedal down for about 30 seconds. The position of the pedal should not alter.

7   Vacuum servo booster unit — removal and installation

1   Remove the master cylinder (Section 5).
2   Disconnect the vacuum line from the servo unit.
3   Remove the dash panel lower trim piece (Chapter 11).
4   Working under the dash, disconnect the servo pushrod from the brake pedal by removing the clevis pin.
5   Remove the nuts from the servo unit mounting studs. Return to the engine compartment and withdraw the unit from the car.
6   Install the vacuum servo unit and working from inside the vehicle install the nuts securing the unit to the bulkhead.
7   Connect the brake pedal to the servo unit pushrod and install the clevis pin.
8   Connect the vacuum line to the servo unit.
9   Replace the master cylinder.
10   Check the brake pedal height and free play and adjust them if necessary as described in Section 12.
11   Replace the dash panel lower trim piece.

8   Hydraulic brake hose and lines — inspection and replacement

1   Every six months the flexible hoses which connect the steel brake lines to the calipers should be inspected for cracks, chafing of the outer cover, leaks, blisters, and other damage. These are vulnerable parts of the brake system and inspection should be complete. A light and mirror will prove helpful for a thorough check. If a hose exhibits any of the above conditions, replace it with a new one.
2   Prefabricated brake lines are available from dealer parts departments and auto parts stores. **Caution:** *Do not, under any circumstances, use lines that have been formed from copper or aluminum.*
3   When installing the brake line, leave at least 0.75 in (19 mm) clearance between the line and any moving or vibrating parts. Be sure to bleed the brakes (Section 9) after the line is installed.

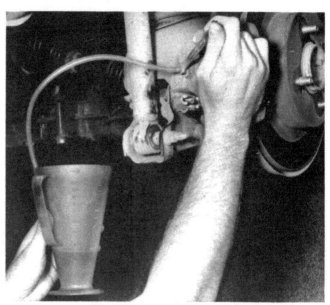

9.7   Place one end of the tubing over the bleeder valve and submerge the other end in a container partially filled with clean brake fluid

9   Hydraulic system — bleeding

Refer to illustration 9.7
Caution: *Spilled brake fluid can damage the car's paint.*
1   Any time any part of the brake system is disassembled or develops a leak, or when the fluid in the master cylinder reservoir runs low, air will enter the system. To eliminate this air the brakes must be bled.
2   If air has entered the system because the master cylinder has been disconnected, or the master cylinder reservoir has been low or empty of fluid, or if a complete flushing of the system is needed, all four brakes should be bled. If a brake line serving only one brake is disconnected, then only that brake need be bled.
3   Before beginning, have an assistant on hand, as well as a good supply of new brake fluid, an empty clear container such as a glass jar, a length of 3/16-inch plastic, rubber or vinyl tubing to fit over the bleeder valve and a brake line wrench to open and close the bleeder valve. The vehicle will have to be raised and placed on jackstands for clearance.
4   If the vehicle is equipped with power brakes, remove the vacuum reserve in the system by applying the brake several times with the engine off.
5   Check that the master cylinder reservoir is full of fluid and be sure to keep it at least half full during the entire operation. If the reservoir runs low of fluid the entire bleeding procedure must be repeated.
Caution: *Do not mix different types of brake fluid, and do not re-use any old fluid, as this could deteriorate brake system components.*
6   Begin with the left rear wheel. Loosen the bleeder valve slightly to break it loose then tighten it to a point where it is snug but can still be loosened quickly and easily.
7   Place one end of the tubing over the bleeder valve and submerge the other end in brake fluid in the container (see illustration).
8   Have your assistant pump the brakes a few times to build pressure in the system. On the last pump have him hold the pedal firmly depressed.
9   While the pedal is held depressed, open the bleeder valve just enough to allow a flow of fluid to leave the valve. Watch for air bubbles to exit the submerged end of the tube. When the fluid flow slows after a couple of seconds, close the valve and have your assistant release the pedal.
10   Repeat the procedure until no more air is seen in the fluid leaving the tube. Fully tighten the bleeder valve and proceed to the right rear wheel, the right front wheel, and the left front wheel, in that order, and peform the same operation. Be sure to check the fluid in the master cylinder reservoir frequently.
11   Refill the master cylinder with fluid at the end of the operation.

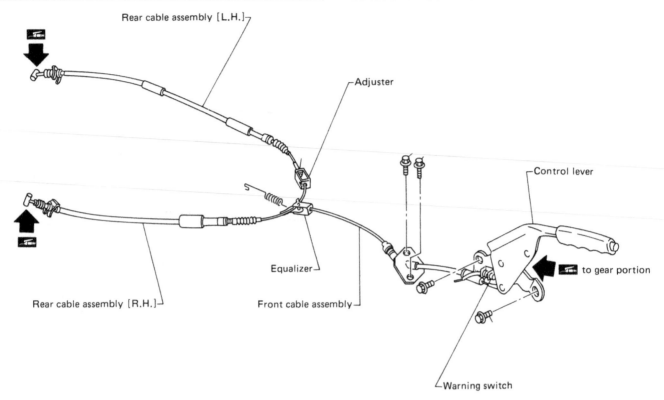

10.1   Parking brake system components

10.4   To adjust the emergency cable, back off on the lock
nut (1) and turn the adjuster (2) until the correct tension
is achieved

10   Parking brake — adjustment

Refer to illustrations 10.1, 10.4 and 10.9

1   The parking brake (see illustration) does not need routine maintenance, but the cable may stretch over a period of time necessitating
adjustment. Also, the parking brake should be checked for proper adjustment whenever the rear brake cables have been disconnected to
perform other procedures.

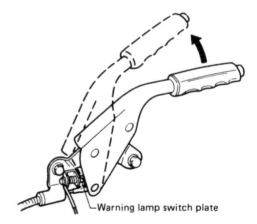

10.9   Bend the parking brake warning lamp switch plate
so the brake warning light comes on when the lever is
pulled up one notch and goes out when fully released

2   Pull up on the parking brake lever with heavy pulling force. When
correctly adjusted, the lever should move upward 8-10 notches.
3   If adjustment is necessary, raise the vehicle and support it securely
on jackstands.
4   Loosen the locknut (see illustration) at the parking brake cable
adjuster.
5   Turn the adjuster to tighten the cable so the parking brake lever,
when pulled, moves the proper distance.
6   Tighten the adjuster locknut.
7   Check that both rear wheels turn freely with the parking brake
released and that there is no brake drag in either direction.
8   Remove the jackstands and lower the vehicle.

Warning lamp adjustment

9   Bend the parking brake warning lamp switch plate, located at the
base of the brake handle (see illustration), so that the brake warning
light comes on when the lever is pulled one notch and goes out when
fully released.

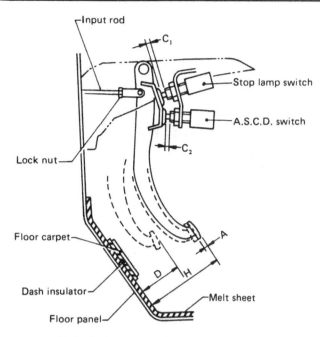

12.1 Brake pedal and switch adjustments

A 1.0 to 3.0 mm
    (0.039 to 0.018 in)

C1 0.3 to 1.0 mm
    (0.012 to 0.039 in)

C2 0.3 to 1.0 mm
    (0.012 to 0.039 in)

D 90 mm (3.54 in)

H Automatic transmission
    184 to 194 mm
    (7.24 to 7.64 in)
    Manual transmission
    182 to 192 mm
    (7.17 to 7.56 in)

---

11 Stop light switch — removal, installation and adjustment

1   The brake light switch is the top switch mounted to the top of the brake pedal.
2   Disconnect the wires leading to the switch.

3   Loosen the locknut located on the pedal side of the switch.
4   Unscrew the switch from the retaining bracket.
5   Installation is the reverse of the removal procedure.
6   Once installed, adjust the switch-to-pedal clearance.

Adjustment

7   Refer to Section 12 and confirm that brake pedal free play is within specifications.
8   Use a feeler gauge inserted between the pedal stopper and the threaded end of the stop lamp switch. Turn the stop lamp switch adjusting nut until the distance between the switch and stopper is as specified.
9   Tighten adjusting nut.
10   Adjust the Automatic Speed Control Device (ASCD) switch in the same manner as the stop lamp switch.
11   Press the brake pedal and check that the brake lights come on when depressed and go off when released.

---

12 Brake pedal — adjustment

Refer to illustration 12.1

1   You will find two switches attached to the brake pedal near the pivot point. The top switch is the stop lamp switch and the bottom is the Automatic Speed Control Device (ASCD) switch (see illustration).
2   Adjust the brake pedal free height by loosening the locknut on the booster input rod.
3   Turn the input rod until the free pedal height is as listed in the Specifications. **Caution:** *Be sure that the tip of the input rod stays inside the booster.*
4   Tighten the locking nut on the booster input rod.
5   Check the clearance between both switches where they contact the brake pedal. See the Specifications for the correct clearance. If the clearance does not meet the Specifications, adjustment is made by loosening the locking nut on the switch and turning the adjusting nut.
6   Check the pedal free play and compare it to the Specifications.
7   Check to see that the stop lamp is off when the pedal is released.
8   Check the brake pedal depressed height with the engine running. Depress the pedal and check the distance from the face of the pedal to the floor board and compare with the Specifications. If height is below the specified value, check the brake system for leaks, accumulation of air or damage to the hydraulic system components.

Chapter 10 Suspension and steering

Contents

Specifications

Front suspension	Strut with coil spring
Rear suspension.............................	Independent rear suspension, semi-trailing arm
Steering overall gear ratio.......................	17.1
Power steering fluid	
type	Dexron
capacity	0.9 liter (1-7/8 US pints)

Torque specifications	Kg-m	Ft-lbs
Front suspension		
wheel hub bearing lock nut	2.5 to 3.0	18 to 22
wheel hub to disc rotor	5.0 to 7.0	36 to 51
wheel nut................................	8.0 to 10.0	58 to 72
knuckle arm to side rod	5.5 to 10.0	40 to 72
knuckle arm to knuckle spindle	7.3 to 9.9	53 to 72
torque member fixing bolt	7.3 to 9.9	53 to 72
knuckle spindle to baffle plate	0.33 to 0.44	2.4 to 3.2
tie rod lock nut	8 to 10	58 to 72
lower balljoint to knuckle arm	9.8 to 12.2	71 to 88
strut assembly		
strut mounting insulator fixing bolt	3.2 to 4.3	23 to 31
piston rod lock nut......................	7 to 9	51 to 65
transverse link to suspension member	9.5 to 11.5	69 to 83
tension rod to tension rod bracket	4.5 to 5.5	33 to 40
tension bracket to body	3 to 4	22 to 29
tension rod to transverse link...................	4.3 to 6.0	31 to 43
stabilizer bar clamp to body....................	3 to 4	22 to 29
stabilizer bar to transverse link	1.6 to 2.2	12 to 16
suspension member to body	7 to 9	51 to 65
Rear suspension		
wheel nut................................	8.0 to 10.0	58 to 72
wheel bearing lock nut	30 to 40	217 to 289
three way connector mounting bolt	0.5 to 0.7	3.6 to 5.1
connector to brake tube	1.5 to 1.8	11 to 13
shock absorber		
lower end fixing bolt	6 to 8	43 to 58
upper end fixing bolt	3.2 to 4.3	23 to 31
locking nut (adjustable shock absorber)	5.0 to 6.5	34 to 46
suspension member to suspension member stay........	8 to 11	58 to 80
suspension member stay to body	2.0 to 2.6	14 to 19
suspension member to suspension arm	10 to 12	72 to 87
spring seat stay to suspension arm		
front	6 to 8	43 to 58
rear	7 to 9	51 to 65
stay to parking cable clamp	1.6 to 2.1	12 to 15
stabilizer bar to suspension arm	1.6 to 2.1	12 to 15
stabilizer bar clamp to suspension member............	3.2 to 4.3	23 to 31

Steering column		
steering wheel nut	5.0 to 6.0	36 to 43
column joint fixing bolt		
1984 thru 1986	3.3 to 3.9	24 to 28
1987 on	2.4 to 3.0	17 to 22
Steering gear and linkage		
tie rod lock nut	8.0 to 10.0	58 to 72
tie rod to knuckle arm	5.5 to 10.0	40 to 72
Oil pump		
mounting bracket to engine	1.4 to 1.8	10 to 13
oil pump casing to sub bracket	3.2 to 4.3	23 to 31
adjusting bar bracket to mounting bracket	1.6 to 2.1	12 to 15
sub bracket to adjusting bar	1.6 to 2.1	12 to 15
rear cover fixing bolt	3.9 to 5.3	28 to 38
suction pipe to casing		
1984 thru 1987	0.9 to 1.2	6.5 to 8.7
1988 on	1.4 to 1.8	10 to 13
low pressure pipe to steering gear	2.8 to 4.0	20 to 29
high pressure pipe to steering gear	1.5 to 2.5	11 to 18
high pressure pipe connector bolt on pump	5.0 to 7.0	36 to 51
oil pressure switch	1.6 to 2.4	12 to 17

1   Suspension system — general information

Refer to illustrations 1.1 and 1.4

This vehicle features an independent front suspension (see illustration) of MacPherson strut design. The strut uses a combination shock absorber and spring assembly which is mounted to the steering knuckle.

The front spindles, on which the front hubs are mounted, are also part of the strut assembly.

A lower control arm (transverse link) which pivots on the front crossmember is attached to the steering knuckle by way of a balljoint. The balljoint allows for free movement between the transverse link and the front wheel when the strut compresses during normal driving.

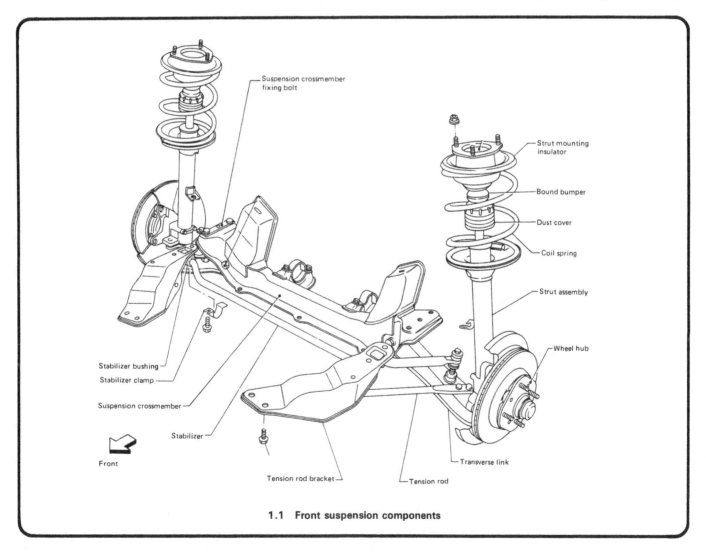

1.1   Front suspension components

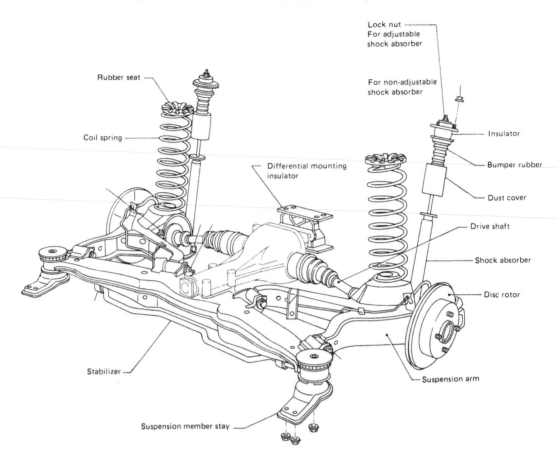

Rubber seat

Coil spring

Differential mounting insulator

Lock nut
For adjustable shock absorber

For non-adjustable shock absorber

Insulator

Bumper rubber

Dust cover

Drive shaft

Shock absorber

Disc rotor

Stabilizer

Suspension arm

Suspension member stay

1.4   Rear suspension components

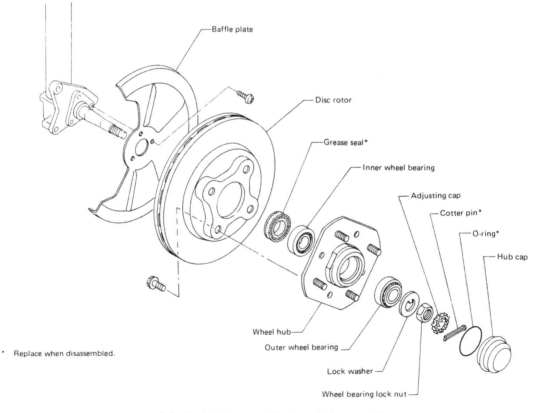

Baffle plate

Disc rotor

Grease seal*

Inner wheel bearing

Adjusting cap

Cotter pin*

O-ring*

Hub cap

Wheel hub

Outer wheel bearing

Lock washer

Wheel bearing lock nut

* Replace when disassembled.

2.4   Exploded view of the front hub assembly

2.9   Pry the inner bearing grease seal out of the hub

2.13   Use a brass drift to drive the race from the hub

2.14   Use a socket to drive the race into the hub

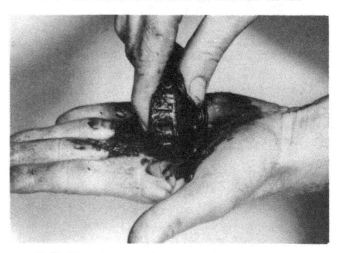

2.15   Work the grease into the bearings and be sure to pack enough between the rollers and cage

Front and rear forces are controlled by tension rods, and a front stabilizer bar is installed to control body roll in corners.

The rear suspension (see illustration) is also independent, made up of a crossmember assembly, suspension arm assemblies, and coil springs which are independent of the shock absorbers. A rear stabilizer bar is installed to help control body roll.

The suspension member is bolted directly to the body and supports the front of the differential. The wheels and lower mounts of the shock absorbers are mounted to the suspension arms, which pivot on the suspension member.

Standard non-adjustable shocks found on some models are sealed hydraulic units. They are non-adjustable, non-refillable and cannot be disassembled. They are mounted at the bottom to a bracket on the axle housing and at the top to the body. Some models have electronically actuated adjustable damping of the shocks. A switch located on the console allows for three settings and transfers the information to the actuator mounted on top of the shock assembly. The adjustable shock absorbers are rebuildable, but due to the complexity of the job it is advisable to have your Nissan dealer or other specialist service the unit.

Never attempt to heat or straighten any suspension part, as this can weaken the metal or in other ways damage the part.

---

2   Wheel bearings — removal, inspection and adjustment

Refer to illustrations 2.4, 2.9, 2.13, 2.14, 2.15, 2.18, 2.19, 2.20 and 2.21

1   Raise the vehicle and support it securely on jackstands.

2   Remove the wheel.
3   Remove the brake caliper (Chapter 9).
4   Pry off the dust cap (see illustration) and remove the O-ring.
5   Remove the cotter pin and discard it.
6   Pull the wheel bearing locking cap off of the wheel bearing nut.
7   Remove the wheel bearing nut and washer from the spindle.
8   Remove the hub and rotor components from the spindle. Be prepared to catch the outside bearing.
9   Lay the rotor face down on a work bench and use a screwdriver to pry out the inner bearing grease seal (see illustration).
10   Remove the inner bearing.
11   Use parts solvent to remove all traces of old grease from the bearings, hub and spindle.
12   Inspect the bearings for cracks, heat discoloration, bent rollers, etc. and replace if necessary.
13   If the bearing race shows cracks, scoring or uneven surface wear it will have to be replaced along with the bearing. Use a brass drift to carefully drive the race out of the hub (see illustration).
14   Install a new race using a socket to carefully drive the race into its seat (see illustration). **Caution:** *Wheel bearings and bearing races are replaced as a matched set. Never use an old bearing in a new race or a new bearing in an old race.*
15   Use high temperature front wheel bearing grease to pack the bearings. Work the grease fully into the bearings, forcing the grease between the rollers and cage (see illustration).

2.18   Apply a small amount of grease to the inside of the seal lip

2.19   Carefully drive the grease seal into position using a proper size socket so the seal is driven in evenly

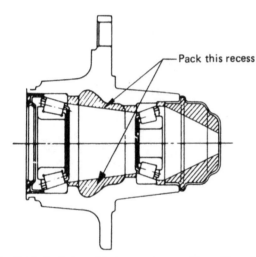

Pack this recess

2.20   Pack the hub and hub grease cap with multi-purpose grease up to the shaded areas shown

16 Apply a thin coat of grease to the spindle at the outer bearing seat, inner bearing seat, shoulder and seal seat.
17 Install the inner bearing into the back of the hub and put a little more grease outboard of the bearing.
18 Apply a small amount of grease to the inner part of the seal lip (see illustration).
19 Place the new seal over the inner bearing and tap the seal with a socket (see illustration) until it is flush with the hub.
20 Carefully place the hub assembly onto the spindle and push the grease packed outer bearing into position. Be sure to pack the hub with grease before installation (see illustration).

Bearing pre-load

21 Install the washer and adjuster nut. Snug the adjuster nut down and turn the rotor in both directions to seat the bearings. Tighten the adjuster nut to the specified torque and then back off the adjuster nut 60° (see illustration).
22 Install the adjusting cap and install a new cotter pin.
23 Install the dust cap with a new O-ring.
24 Remove the jackstands and lower the vehicle.

---

3   Front spring and strut components — removal and installation

Refer to illustrations 3.5, 3.6, and 3.8

1   Raise the vehicle and support it securely on jackstands.

2.21   Once the adjusting nut is tightened to the specified torque, back off the nut 60° (arrow) to set pre-load

2   Remove the wheel.
3   Remove the brake caliper (Chapter 9). It will not be necessary to remove the brake line from the caliper. Just disconnect the caliper and wire the caliper to the stabilizer bar out of the work area.
4   Remove the rotor (Chapter 9).
5   If replacing the strut assembly, remove the baffle plate by removing the three Phillips head screws (see illustration).
6   Remove the bolts (three bolts on some models) securing the knuckle arm to the transverse link (see illustration).
7   Remove the bottom stabilizer retaining nut to allow the transverse link to lower enough to allow the strut removal.
8   Remove the upper retaining nuts and remove the strut assembly from the vehicle (see illustration).
9   Install the upper end of the strut and secure it with the retaining nuts.
10 Place the bottom of the strut in place over the balljoint and secure the knuckle arm to the transverse link.

3.5   If replacing the strut or baffle plate, remove the screws and separate the two components

3.6   Remove the bolts (arrows) securing the knuckle arm to the transverse link

3.8   Remove the nuts (arrows) securing the top of the strut to the vehicle

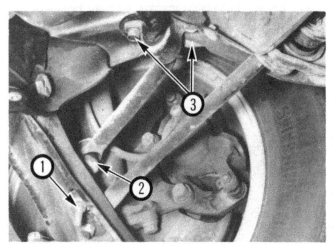

4.3   To disconnect the stabilizer bar first remove the nut (1) at the end of the bar attaching the bar to the transverse link. Remove the bar from the transverse link and note sequence of the bushings (2) removed. Disconnect the bolts (3) to remove the stabilizer support brackets from the body

11   Install and tighten the bottom front stabilizer retaining nut.
12   If removed, replace the baffle plate and secure with the Phillips head screws.
13   Install the rotor (Chapter 9).
14   Install the brake caliper (Chapter 9).
15   Install the wheel.
16   Remove the jackstands and lower the vehicle.

---

4   Front stabilizer bar — removal and installation

Refer to illustration 4.3

1   Raise the front of the vehicle and support it securely on jackstands.
2   Remove the splash shield (Section 20).
3   Remove the lower nut (see illustration) from the end of the stabilizer.
4   Remove the stabilizer links from the transverse links. Remove the bushings, washers and spacers, being sure to keep them in their installed order, and then remove the stabilizer links from the stabilizer bar.
5   Remove the bolts and nuts holding the stabilizer support brackets to the body and lift out the stabilizer bar.
6   If the stabilizer bar bushings need to be replaced spread them apart

and slide them off the bar.
7   Inspect the stabilizer link bolts and bushings for any cracks or other damage and replace as necessary.
8   Fit the stabilizer support brackets onto the stabilizer bar. If marked, the bracket marked with an L goes on the left side of the bar.
9   Raise the stabilizer bar into position and loosely install both nuts and mounting bolts.
10   Install the links through the transverse link and stabilizer bar.
11   Install the stabilizer link nuts and torque them to the specified torque.
12   Lower the car to the ground so that the full weight of the car is on the wheels, then tighten the stabilizer bar mounting bolts and nuts to the specified torque.

---

5   Tension rod — removal and installation

Refer to illustration 5.3

1   Raise the vehicle and support it securely on jackstands.
2   Remove the wheel.

5.3   Remove the bolts securing the rod to the transverse link (arrow) and remove the nut at the other end of the rod (arrow)

3   Remove the two nuts securing the tension rod to the transverse link (see illustration).
4   Remove the nut securing the tension rod to the tension rod bracket. Note the sequence of the bushings when removing the rod from the bracket.
5   Insert the rod into the bracket, assembling bushing in their original locations.
6   Install the other end of the rod on the transverse link.
7   Install the wheel.
8   Remove the jackstands and lower the vehicle.

---

6   Balljoint and transverse link — removal and installation

Refer to illustrations 6.4 and 6.6

1   Raise the front of the vehicle and support it securely on jackstands.
2   Remove the wheel.
3   Remove the splash shield (Section 20).
3   Remove the front stabilizer bar. Note the position of the washers and spacers for reinstallation.
4   Remove the nuts attaching the tension rod (see illustration) to the transverse link.
5   Remove the cotter pin and nut attaching the side rod to the steering knuckle and separate the two. Remove the bolts attaching the steering knuckle to the strut.
6   Remove the pivot bolt holding the inner end of the transverse link to the crossmember (see illustration).
7   Remove the transverse link.
8   Place the transverse link in a vise and remove the bolt attaching

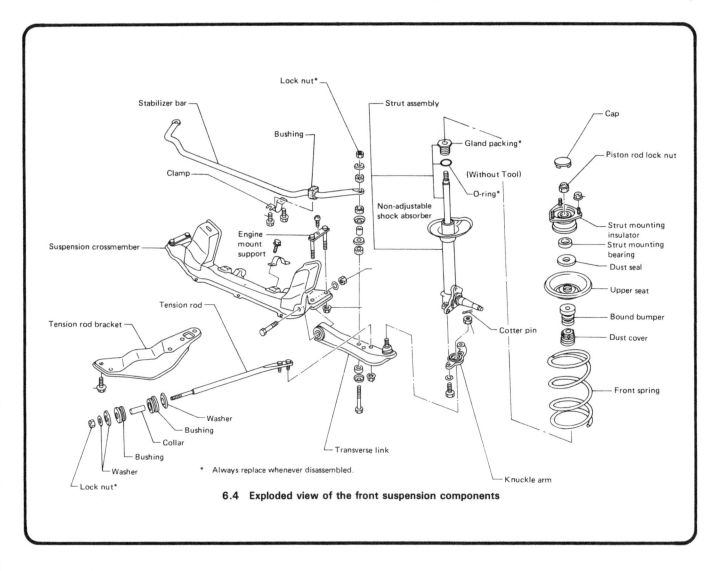

6.4   Exploded view of the front suspension components

Stabilizer bar
Bushing
Clamp
Suspension crossmember
Engine mount support
Tension rod
Tension rod bracket
Washer
Bushing
Collar
Bushing
Washer
Lock nut*
* Always replace whenever disassembled.
Transverse link
Knuckle arm
Cotter pin
Lock nut*
Strut assembly
Gland packing*
(Without Tool)
O-ring*
Non-adjustable shock absorber
Cap
Piston rod lock nut
Strut mounting insulator
Strut mounting bearing
Dust seal
Upper seat
Bound bumper
Dust cover
Front spring

the steering knuckle and balljoint to the transverse link. Replace the transverse link if the balljoint is damaged.

9   If the transverse link bushing needs to be replaced it will be necessary to replace the entire transverse link.

10 Install the transverse link to the crossmember but do not tighten the pivot bolt until the car has been lowered to the ground.

11 Install the nut and cotter pin attaching the side rod to the steering knuckle.

12 Attach the steering knuckle to the the strut.

13 Attach the tension rod to the transverse link.

14 Install the front stabilizer bar.

15 Install the splash shield.

16 Replace the wheel.

17 Remove the jackstands and lower the vehicle.

18 Tighten the transverse link pivot bolt.

7   Rear stabilizer bar — removal and installation

Refer to illustration 7.2 and 7.3

1   Raise rear of the vehicle and support it securely on jackstands.

6.6   Remove the transverse link pivot bolt (arrow) to remove the link

7.3   Remove the stabilizer bar mounting brackets

2   Remove the stabilizer bar link bolts that pass through the suspension arms (see illustration).

3   Remove the bolts attaching the stabilizer bar mounting brackets to the suspension crossmember (see illustration) and remove the bar.

4   If the bushings exhibit any hardening or cracking they should be replaced.

5   Install the stabilizer bar with the bushings and brackets in place and install the retaining bolts. **Note:** *Final tightening of these bolts should be carried out once the vehicle is lowered to the ground under curb weight.*

6   Install the stabilizer bar link bolts that pass through the suspension arms.

7   Remove the jackstands and lower the vehicle.

8   Tighten the retaining bolts to the specified torque.

8   Rear shock absorber — removal and installation

Refer to illustrations 8.1, 8.3, 8.4 and 8.5

1   Raise the vehicle and support it by placing jackstands under the spring seat stays (see illustration).

7.2   Remove the stabilizer bar link bolts (arrow) that pass through the suspension arm

8.1   Place the jackstand under the spring seat stay to support the weight of the suspension arm once the shock absorber is removed

2   Remove the wheel.
3   Remove the bolt (see illustration) securing the shock absorber to the spring seat stay.
4   From inside the rear compartment of the vehicle remove the speaker cover (Chapter 11) to gain access to the top of the shock absorber and remove the shock asorber mounting insulator nuts (see illustration). Push down on the studs to lower the shock mounting insulator which will gain access to the shock piston rod holding point inside the fender well. On adjustable shock models, disconnect the electrical connector.
5   Use a wrench to hold the shock absorber piston rod hold point (see illustration) and use a socket to remove the shock absorber top nut.
6   Work the bottom of the shock absorber out of the spring seat stay and lower it out the bottom of the vehicle. Note the sequence of the insulator dust cover and washers. If it is an adjustable shock absorber note the sub-harness location.
7   Install the shock absorber by making sure the insulator and cover

are in the proper sequence and insert the top of the shock absorber through the opening. If equipped, install the electrical connector. Let the mounting insulator hang down and while placing a wrench on the piston hold point. Install the rod top bolt.
8   Install the lower shock absorber to spring seat stay bolt and loosely tighten the nut.
9   Replace the wheel.
10   Remove the jackstands and lower the vehicle.
11   Tighten the bottom shock absorber retainer.
12   Replace the speaker cover.

***

9   Rear spring — removal and installation

Refer to illustrations 9.2, 9.5 and 9.8
1   Remove the driveaxle (Chapter 8).

8.3   Remove the lower shock absorber through bolt

8.4   Remove the shock absorber insulator nuts (arrows) and push down on the stud bolts to release the insulator

8.5   Use a wrench wedged against the fender well to hold the shock absorber piston while you remove the top nut

9.2   Remove the lower shock absorber-to-spring seat stay bolt

2   With the jackstands under the suspension arms so the springs are compressed, remove the lower shock absorber retaining bolt (see illustration).
3   Position a jack under the differential carrier and raise the vehicle which will relieve the tension from the springs. Place jackstands under the frame rails on both sides to support the vehicle.
4   Mark the spring to indicate the bottom position.
5   Press down on the suspension arm and lift the bottom of the spring off of the rubber spring seat insulator (see illustration). Note the installed location of the rubber insulator.
6   Inspect the spring for deformation or cracks and replace if necessary.
7   Check the upper and lower spring seat rubber insulators for damage and replace if necessary.
8   The lower spring seat insulator has to be set in its original position. Place the rubber seat onto the suspension arm with the notch facing the inside of the vehicle (see illustration).

9   Note the reference mark painted on the spring bottom and install the top spring seat insulator onto the top of the spring.
10 Install the top of the spring onto its retaining recess and press down on the suspension arm to install the bottom of the spring.
11 Place a jack under the spring seat stay and lift the suspension arm. Place a jackstand under the suspension arm and remove the jack.
12 Install the bottom of the shock absorber into the spring seat stay mounting bracket and secure the bolt.
13 Install the driveaxle.
14 Remove the jackstands and lower the vehicle.

10   Suspension arm — removal and installation

Refer to illustrations 10.4, 10.5a and 10.5b

1   Remove the rear spring (Section 9).
2   Remove the rear stabilizer bar (Section 7).
3   Remove the brake caliper and brake rotor (Chapter 9).
4   Use a flare nut wrench to disconnect the brake line hose where the metal hydraulic brake line-to-rubber hose connection bracket is attached to the body (see illustration). Use pliers to remove the spring clip to free the rubber hose.
5   Locate the two through bolts securing the suspension arm to the suspension crossmember (see illustration) and note the head of the innermost retaining bolt and the alignment marks notched in the special bolt head/washer component (see illustration). If the marks are not easily visible use paint and mark the bolt location in relation to the suspension arm. The special bolt/washer component is used to align the wheel assembly.
6   Remove the two retaining through bolts and note location and sequence of any bushings and washers before removing the suspension arm.
7   Inspect the suspension arm for deformation and structural cracks and replace if necessary. Check the bushing for wear or other damage and if replacement is necessary, take the suspension to your Nissan dealer or other qualified service center to have the bushing replaced with a press.
8   Install the suspension arm and match the alignment marks made during removal on the innermost through bolt. Final tightening of these bolts should be carried out once the wheels are mounted and the vehicle is on the ground under curb wieght.
9   Connect the rubber hydraulic hose to the bracket retaining it to the body and secure it with the spring clip.
10 Attach the metal hydraulic brake line to the rubber brake hose.

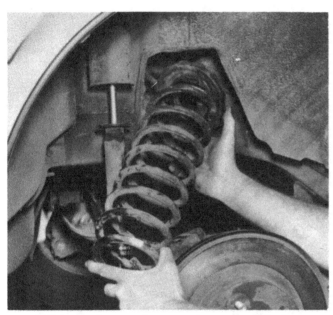

9.5   Press down on the suspension arm and lift the bottom of the spring off the rubber spring seat insulator

9.8   Install the bottom rubber spring seat insulator in its original location and with the notch facing inward

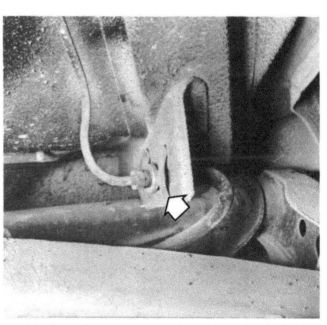

10.4   Use a flare nut wrench to disconnect the brake hose at the body mounting bracket and remove the spring clip (arrow) to free the hose

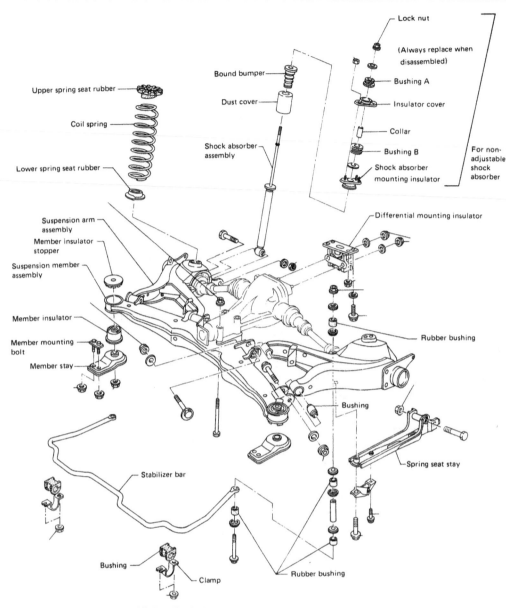

10.5a Exploded view of the rear suspension

11 Install the rear spring (Section 9).
12 Install the rear stabilizer bar (Section 7).
13 Install the rotor and brake caliper (Chapter 9).
14 Bleed the hydraulic system (Chapter 9).
15 Remove the jackstands and lower the vehicle.
16 Tighten suspension arm through bolts and the lower shock absorber and rear stabilizer bar mounting bolts to the specified torque.

11 Steering system — general information

Refer to illustrations 11.1a and 11.1b

   The components that make up the steering system are the steering wheel, steering column and the rack-and-pinion assembly (see illustrations).
   The steering system is a rack and pinion type which is power assisted. The steering action is transfered through either a tilt or standard steering column. Inside the rack and pinion housing the pinion gear works the rack, which operates the side rods that control the steering knuckles. If the power steering system loses its hydraulic pressure it will still function manually, though with increased effort.
   The steering column is of the collapsible, energy-absorbing type, designed to compress in the event of a front end collision to minimize

10.5b Note the alignment notch on the special bolt head/washer securing the suspension arm to the suspension crossmember

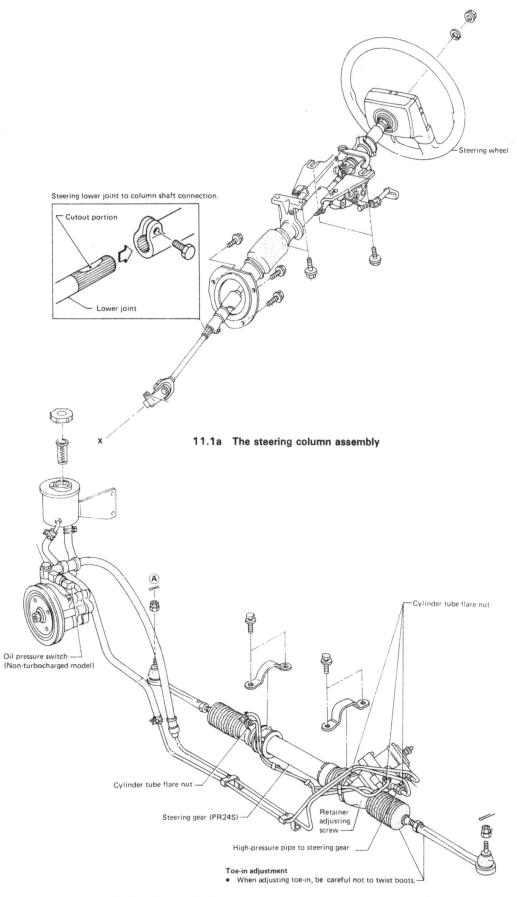

Steering lower joint to column shaft connection.

Cutout portion

Lower joint

Steering wheel

11.1a   The steering column assembly

X

Oil pressure switch
(Non-turbocharged model)

A

Cylinder tube flare nut

Cylinder tube flare nut

Steering gear (PR24S)

Retainer adjusting screw

High-pressure pipe to steering gear

Toe-in adjustment
• When adjusting toe-in, be careful not to twist boots.

11.1b   Rack and pinion steering gear with hydraulic power assistance

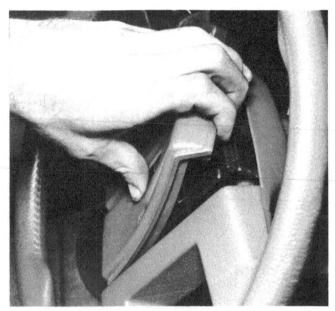

12.2 Carefully pry the steering wheel center trim piece from the steering wheel

12.3 Mark the relation of the steering wheel to the column shaft so the wheel can be installed in its original position

12.5a Install the steering wheel puller and turn the forcing bolt to pull the steering wheel from the shaft

12.5b If necessary, remove the horn assembly and disconnect the electrical connector

injury to the driver. The column also houses the ignition switch, steering column lock, headlight switch, turn signal control, headlight dimmer switch, windshield wiper and washer control and, on GL models, the cruise control switch.

Due to the column's collapsible design, it is important that only the specified screws, bolts and nuts be used as designated and that they be tightened to the specified torque. Other precautions particular to this design are noted in appropriate Sections.

12 Steering wheel — removal and installation

Refer to illustrations 12.2, 12.3, 12.5a and 12.5b

1 Disconnect the negative cable at the battery. Place the cable out of the way so it cannot accidentally come into contact with the negative terminal of the battery.

2 Carefully pry the center finish piece from the steering wheel (see illustration). The trim piece is fastened at the top (center) and on each side by pop-out type fasteners.

3 Use a dab of paint or suitable marker to mark the relationship of the steering wheel to the column shaft (see illustration).

4 Remove the steering wheel nut and washer.

5 Install a steering wheel puller (see illustration) and remove the steering wheel from the shaft. **Note:** *If necessary, remove the horn switch assembly by removing the screws and disconnecting the electrical connector (see illustration).* **Note:** *Springs are located under the assembly on each of the screws.*

6 Align the marks and slide the steering wheel onto the shaft.

7 Install the steering wheel retaining washer and nut and tighten the nut to the specified torque.

8 Align the horn pad and press it onto the steering wheel.

9 Connect the battery negative cable.

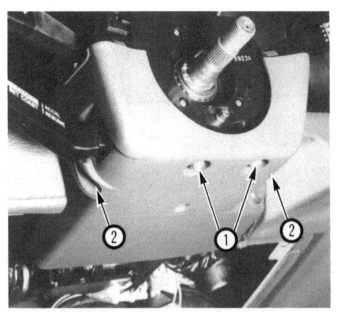

13.3   Remove the steering wheel trim piece cover screws and lift the pieces off. (1) removes the bottom piece and (2) removes the top piece

13.6   The shear-head screws (arrows) will have to be drilled out

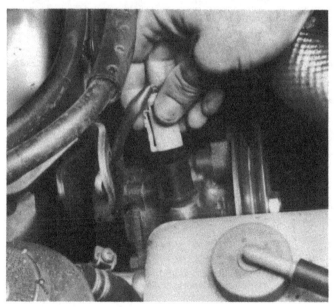

15.3   Disconnect the electrical connector from the power steering pump by pressing in on the connector tab and pulling the connector up

13   Steering lock — removal and installation

Refer to illustrations 13.3 and 13.6

1   Obtain two self-shear screws from your Nissan dealer, needed to reinstall the steering lock.
2   Remove the steering wheel as described in Section 12.
3   Remove the steering column trim covers (see illustration).
4   Remove the two cluster panel switches (Chapter 12).
5   Remove the dash panel trim piece by removing the screws holding it to the dash.
6   The steering lock is secured to the steering shaft with two self-shear screws (see illustration). Removing the shear screws will involve very careful drilling of the screws. Start by center punching the center of the screw. Select a drill that is about the diameter of the root of

the thread. Very carefully drill out the center of the screw until a screw extractor can be used to remove the screw.
7   Disconnect the ignition switch wires.
8   To install, align the mating surface of the steering lock with the hole in the steering column tube. Check the operation of the lock using the key.
9   Install the two self-shearing screws in their appropriate holes, making sure the tops snap off.
10   Install the dash panel trim piece.
11   Install the two cluster panel switches.
12   Install the steering column trim covers.
13   Install the steering wheel.
14   Connect the negative battery cable.

14   Power steering system — maintenance and adjustment

Refer to illustrations 14.1a and 14.1b

1   The hydraulic components used in the power steering system include a belt driven pump with a remote reservoir tank, attached to the rack-and-pinion assembly by hoses and metal lines.
2   Normal maintenance of the power steering system consists of periodically checking the fluid level in the reservoir, keeping the pump drivebelt tension correct and visually checking the hoses for any evidence of fluid leakage. It will also be necessary, after a system component has been removed, to bleed the system as described in Section 16.

15   Power steering pump — removal and installation

Refer to illustrations 15.3, 15.7, 15.8, 15.9, 15.18a and 15.18b

Pump removal

1   Remove the power steering pump drivebelt (Chapter 1).
2   Disconnect the negative cable at the battery. Place the cable out of the way so it cannot accidentally come in contact with the battery negative terminal.
3   Disconnect the power steering pump electrical connector (see illustration) by pressing in on the tab and pulling up on the connector.
4   Disconnect the cooling system reservoir by removing the bolts securing it to the fender well and the low fluid level electrical connector, using a small screwdriver to release the tang on the connector. It is

15.7   Remove the pump bracket to engine bolts (arrows)

15.8   The banjo bolt (1) secures the high pressure line to the pump and the low pressure line (2) is held on by a hose clamp

15.9   Use a bolt to plug the power steering hose

15.18a   Remove the cap and lift the screen from the power steering fluid reservoir

not necessary to disconnect the coolant hose. Simply set the reservoir on the fender well out of your way.

5   Place a catch pan under the vehicle to catch any leaking hydraulic fluid.

6   Remove the bolts retaining the engine lifting bracket and remove the bracket.

7   Remove the bolts securing the pump to the engine block (see illustration).

8   Remove the banjo bolt (see illustration) connecting the high pressure line to the pump and raise the hose up and secure it in the up position.

9   Pull the bottom hose off the pump and cover the end of the hose with your thumb. Lift the pump out of the engine compartment. Use a bolt (see illustration) to plug the end of the hose to prevent further leakage.

10   Install the return hose to the pump and secure with a hose clamp.

11   Install the pump on the engine block and tighten the two bolts.

12   Install the banjo bolt fitting to the pump and tighten the bolt.

13   Install the engine lifting bracket.

14   Connect the power steering pump electrical connector.

15.18b   Use a small hand pump to siphon the fluid from the reservoir

17.4   A separating tool is required to separate the tie rod from the steering knuckle

17.6   Mark the location of the tie rod locking nut so toe-in will be close during installation

22 Attach the hoses to the bottom of the reservoir and secure with hose clamps.
23 Replace the screen and fill the reservoir with power steering fluid.

---

16   Power steering system — bleeding

Note: *Whenever a hose in the power steering system has been disconnected, it is quite probable, no matter how much care was taken to prevent air entering the system, that the system will need bleeding.*

1   Raise the front of the vehicle until the front wheels are just clear of the ground and secure on jackstands.
2   While adding fluid to the reservoir, quickly turn the steering wheel all the way to the right lock and then the left lock. Do not allow the lock stoppers to be struck hard. This operation should be repeated about ten times or until correct fluid level is achieved in the reservoir.
3   Start the engine and allow it to idle. Repeat Step 2.
4   If the air bleeding is insufficient, the oil reservoir will be extremely foamy and the pump will be noisy. In this case, allow the foam in the reservoir to disperse, recheck the level and repeat the entire bleeding process.
5   If it becomes obvious, after several attempts, that the system cannot be satisfactorily bled, there is probably a leak in the system. Visually check the hoses and their connections for leaks. If no leaks are evident, the problem could be in the steering system itself and the only solution is to have the entire system checked by a Nissan dealer or other specialist.

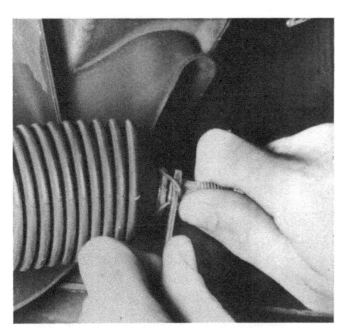

17.7   Use a small screwdriver to remove the O-ring on the small end of the boot

15 Connect the coolant reservoir electrical connector and secure the reservoir to the fender well.
16 Check the power steering fluid reservoir and add power steering fluid if necessary.
17 Connect the negative cable to the battery.

Power steering fluid reservoir

18 Remove the cap and lift out the fluid screen (see illustration). Use a small hand pump (see illustration) to siphon the fluid out of the reservoir.
19 Disconnect the hoses attached to the bottom of the reservoir, held on by hose clamps.
20 Remove the bolts securing the the reservoir to the fender well and lift the reservoir from the vehicle.
21 Install the reservoir onto the fender well.

---

17   Tie rod and boot — removal and installation

Refer to illustrations 17.4, 17.6, 17.7 and 17.8

1   Raise the front of the vehicle and support it securely on jackstands.
2   Remove the splash shield retaining bolts and remove the shield.
3   Remove the cotter pin and nut securing the tie rod to the steering knuckle.
4   Use an appropriate separating tool (see illustration) to separate the tie rod from the steering knuckle.
5   Remove the clamp securing the rubber boot to the rack and pinion.
6   Use paint to mark the location of the tie rod locknut (see illustration). Loosen the tie rod locknut one-quarter turn and remove the tie rod by turning it counterclockwise to remove it.
7   Remove the rubber O-ring from the small end of the boot (see illustration).

17.8 Disconnect the breather hose (arrow) from the boot

18.3 Disconnect the hydraulic lines attached to the
front crossmember

18.4 Use a flare nut wrench to loosen the flare nuts
where the hydraulic lines enter the pinion assembly
(arrows)

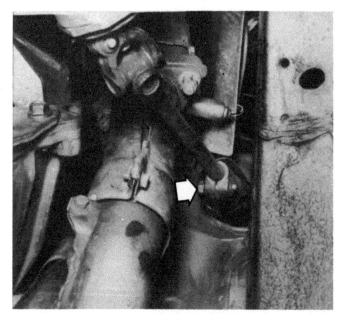

18.5 Remove the bolt securing the upper end of the lower
joint shaft to the steering column (arrow)

8   Remove the clamp securing the breather hose attached to the boot (see illustration).

9   Slide the boot off the rack and pinion unit.

10 When installing the boot be sure it is installed with the breather nipple facing upward.

11 Use a new O-ring on the small end of the boot.

12 Install the tie rod end. If the locknut was loosened only one-quarter turn the toe-in should be close when the tie rod is brought up to the locknut.

13 Install the nut and cotter pin securing the tie rod to the steering knuckle.

14 Install the splash shield.

15 Remove the jackstands and lower the vehicle.

16 Have the toe-in set by your Nissan dealer or other front end specialist.

18   Rack and pinion unit — removal and installation

Refer to illustrations 18.3, 18.4, 18.5, 18.6 and 18.8

1   Raise the front of the vehicle and support it securely on jackstands.

2   Remove the splash shield.

3   Remove the bolts attaching the hydraulic lines to the crossmember (see illustration).

4   Place a drain pan under the rack and pinion and disconnect the flare nuts where the hydraulic lines enter the rack and pinion (see illustration). Allow the fluid to drain into the pan. Plug all the openings to prevent the entry of foreign matter into the system.

5   Remove the bolt securing the upper end of the lower joint shaft to the steering column (see illustration).

6   Disconnect the bolt securing the lower joint to the rack and pinion

18.6 Remove the through bolt and separate the lower steering column joint shaft from the pinion gear. Arrow indicates the alignment mark to the match groove on the lower joint coupling

18.8 Loosen but do not remove the bolts securing the rack and pinion to the crossmember

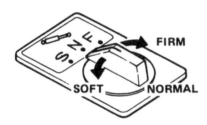

20.1 The adjustable shock asborber control switch is located on the console and offers three selections of damping force

gear and separate the two pieces (see illustration). Note that the alignment spacer is aligned with the lower joint coupling retraction groove.

7   Remove the cotter pins and nuts securing the tie rods to the steering knuckles. Using a suitable separating tool, separate the tie rods from the knuckles.

8   Loosen, but do not remove, the bolts securing the rack-and-pinion to the crossmember (see illustration).

9   Remove the nuts attaching the front engine mounting insulators to the crossmember (Chapter 2A).

10 Securely connect an engine hoist to the lifting eyes on the engine and lift the engine slightly until the weight is taken off the engine mounts.

11 Lift the engine slowly and keep a careful check on clearances around the engine while it is being lifted. Check especially the fan shroud, vacuum lines leading to the engine and the rubber ducts connected to the air flow meter.

12 Remove the rack and pinion mounting bolts and lift the unit from the crossmember.

13 If the rack and pinion unit needs to be overhauled, this should be done by a Nissan dealer or other qualified mechanic.

14 Install the rack and pinion onto the crossmember and install the mounting bolts.

15 Carefully lower the engine, making sure the engine mounts align with their insulators and install the insulator nuts.

16 Install the tie rods to the steering knuckles and secure with the nuts and cotter pins.

17 Install the top of the lower steering column joint shaft but do not

install the through bolt until the lower end is installed.

18 Install the lower end of the steering column lower joint shaft, making sure the groove in the coupling is aligned with the notch on the pinion housing. Install the upper end through bolt.

19 Connect the metal hydraulic fluid lines to the pinion housing.

20 Connect the fluid lines to the crossmember.

21 Replace the splash shield.

22 Remove the jackstands and lower the vehicle.

19 Steering wheel play — adjustment

Due to the special tools and knowldege required, it is advised that you have your Nissan dealer or other qualified specialist make adjustments to the steering rack unit.

20 Adjustable suspension (turbo models) — general description

Refer to illustration 20.1

Some models come equipped with a three-way adjustable suspension system. The system is electronically controlled by a switch located on the console (see illustration). The dampening force of the shock absorbers is set by turning the switch to the S, N, or F settings. The letters denote S for soft, N, for normal and F, gives a firm ride.

The switch can only be activated when the ignition switch is on. An indicator light will come on below the letter selected. If there is an abnormality in the system all the indicator lights will flicker. If this should occur the system should be checked by your Nissan dealer or other service facility.

21 Tire changing procedure

Refer to illustration 21.3, 21.4 and 21.12

1   The car should be on level ground with the transmission in Park (P), if equipped with an automatic transmission, or in Reverse (R) on manual equipped transmissions.

2   Set the parking brake and block the front and rear of the wheel of the wheel diagonally opposite of the wheel to be changed to prevent the car from rolling.

21.3   Remove the spare tire and jack from its storage area by removing the plastic trim cover and freeing the wheel and jack retaining wing nut

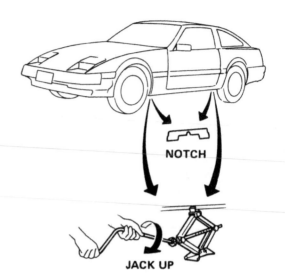

21.4   Position the jack near the wheel to be changed and be sure the jack head notch engages in the body rail

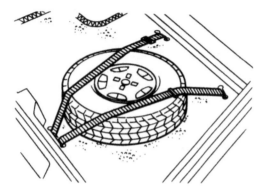

21.12   Secure the flat tire before traveling to prevent the tire from becoming a dangerous projectile in the event of an accident

3   Remove the spare tire and jack from the storage area (see illustration) by removing the two retainers holding the plastic cover to the side of the luggage compartment. Turn the center wing nut counterclockwise to release the spare tire and lift out the jack.
4   Position the jack at the appropriate jacking point (see illustration) and insert the jack head notch into the body rail.
5   Use the wheel nut wrench to loosen the the wheel nuts one or two turns. **Caution:** *Do not remove wheel nut until the tire is off the ground.*

6   Slowly turn the jack handle in a clockwise motion and be attentive to any shifting in the weight of the vehicle.
7   Once the wheel has cleared the ground, remove the wheel nuts and carefully remove the the wheel by pulling it towards you. **Warning:** *Never get under a vehicle supported only by the jack. Do not start or run the engine while the vehicle is on the jack.*
8   Lay the flat tire on the ground near you and mount the spare tire by aligning the holes in the wheel with the wheel lugs and sliding it onto the wheel hub.
9   Replace all the wheel nuts by hand, making sure to replace the washers between the nuts and the wheel. **Caution:** *If mounting the spare wheel onto the hub, use the special steel wheel lug nuts provided in the tool kit located under the carpet in the luggage compartment.*
10 Use the wheel nut wrench and slightly tighten each wheel nut alternately and evenly until there is no wheel play.
11 Lower the vehicle slowly until the wheel touches the ground, then tighten the wheel nuts. **Caution:** *As soon as possible after changing the wheel it is advisable to have a technician tighten the wheel nuts to the specified torque.*
12 Lower the vehicle the rest of the way to the ground. **Warning:** *Always secure the flat tire (see illustration) and jacking equipment after use to prevent them from becoming dangerous projectiles in the event of an accident.*

Chapter 11 Body

Contents

1 General information

During the years 1984 through 1986 the 300ZX was available in two basic body styles: the 2-seater coupe and the 2 + 2 style.

The 300ZX is constructed with a unitized body, in which the body is designed to provide vehicle rigidity so that a separate frame is not necessary.

Certain body panels which are particularly vulnerable to accident damage can be replaced by unbolting them and installing replacement parts. These panels include the fenders, inner fender skirts, grille front apron, headlamp case, hood and hatch.

2 Body — general maintenance

1   The condition of your vehicle's bodywork is of considerable importance, as it is on this that the resale value will mainly depend. It is much more difficult to repair neglected bodywork than mechanical components. The hidden portions of the body, such as the wheel arches, the underframe and the engine compartment, are equally important, although obviously not requiring such frequent attention as the immediately visible paint.

2   Once a year or every 12,000 miles it is a good idea to have the underside of the body steam cleaned. All traces of dirt and oil will be removed and the underside can then be inspected carefully for rust, damaged hydraulic lines, frayed electrical wiring and similar trouble areas. The front suspension should be greased on completion of this job.

3   At the same time, clean the engine and the engine compartment either using a steam cleaner or a water-soluble cleaner.

4   The wheel arches and fender skirts should be given particular attention, as undercoating can easily come away here and stones and dirt thrown up from the wheels can soon cause the paint to chip and flake, and so allow rust to set in. If rust is found, clean down to the bare metal and apply an anti-rust paint.

5   Use a mild detergent and soft sponge to wash the exterior of the car and rinse immediately with clear water. Owners who live in coastal regions and where salt or chemicals are used on the roads should wash the finish often to prevent damage to the finish. Do not wash the car in direct sunlight or when the metal is warm. To remove road tar, insects or tree sap use a tar remover rather than a knife or sharp object, which could scratch the surface.

6   A coat of wax or polish may be your best protection against the elements. Use a good grade of polish or wax suitable for a high-quality synthetic finish. Do not use a wax or polish which contains large amounts of abrasives, as these will scratch the finish.

7   Bright metal parts can be protected with wax or a chrome preservative. During winter months or in coastal regions apply a heavier coating or, if necessary, use a non-corrosive compound such as petroleum jelly for protection. Do not use abrasive cleaners, strong detergents or materials such as steel wool on chrome or anodized aluminium parts, as these may damage the protective coating and cause discoloration or deterioration.

8   Interior surfaces can be wiped clean with a damp cloth or with cleaners specifically designed for car interior fabrics. Carefully read the manufacturer's instructions and test any commercial cleaners on an inconspicuous area first. The carpet should be vacuumed regularly and can be covered with mats.

9   Cleaning the mechanical parts of the car serves two functions. First, it focuses your attention on parts which may be starting to fail, allowing you to fix or replace them before they cause problems. Second, it is much more pleasant to work on parts which are clean. You will still get dirty on major repair jobs, but it will be less extreme. Large areas, such as the firewall and inner fender panels, should be brushed with detergent, allowed to soak for about 15 minutes and then carefully rinsed clean. Cover ignition and carburetor parts with plastic to prevent moisture from penetrating these critical components.

3 Vinyl trim — maintenance

Under no circumstances try to clean any external vinyl trim or roof covering with detergents, caustic soap or petroleum based cleaners. Plain soap and water is all that is required, with a soft brush to clean dirt that may be ingrained. Wash the covering as frequently as the rest of the vehicle.

After cleaning, application of a high quality rubber and vinyl protectant will help prevent oxidation and cracking. This protectant can also be applied to all interior and exterior vinyl components, as well as to vacuum lines and rubber hoses, which often fail as a result of chemical degradation.

4   Upholstery — maintenance

1   Every three months remove the carpets or mats and clean the interior of the vehicle (more frequently if necessary). Vacuum the upholstery and carpets to remove loose dirt and dust.
2   If the upholstery is soiled, apply upholstery cleaner with a damp sponge and wipe it off with a clean, dry cloth.

5   Body damage — minor repair

Refer to the color photo sequence on pages 242 and 243

Repair of minor scratches

If the scratch is superficial, and does not penetrate to the metal of the body, repair is simple. Lightly rub the area of the scratch with a fine rubbing compound to remove loose paint from the scratch and to clear the surrounding paint of wax buildup. Rinse the area with clean water.

Apply touch-up paint to the scratch, using a small brush. Continue to apply thin layers of paint until the surface of the paint in the scratch is level with the surrounding paint. Allow the new paint at least two weeks to harden, then blend it into the surrounding paint by rubbing with a very fine rubbing compound. Finally, apply a coat of wax to the scratch area.

Where the scratch has penetrated the paint and exposed the metal of the body, causing the metal to rust, a different repair technique is required. Remove any loose rust from the bottom of the scratch with a pocket knife, then apply rust inhibiting paint to prevent the formation of rust in the future. Using a rubber or nylon applicator, coat the scratched area with glaze type filler. If required, this filler can be mixed with thinner to provide a very thin paste, which is ideal for filling narrow scratches. Before the glaze filler in the scratch hardens, wrap a piece of smooth cotton cloth around the tip of a finger. Dip the cloth in thinner and then quickly wipe it along the surface of the scratch. This will ensure that the surface of the filler is slightly hollowed. The scratch can now be painted over as described earlier in this section.

Repair of dents

When denting of the vehicle's body has taken place, the first task is to pull the dent out until the affected area nearly attains its original shape. There is little point in trying to restore the original shape completely, as the metal in the damaged area will have stretched on impact and cannot be reshaped fully to its original contours. It is better to bring the level of the dent up to a point which is about 1/8-inch below the level of the surrounding metal. In cases where the dent is very shallow, it is not worth trying to pull it out at all.

If the underside of the dent is accessible, it can be hammered out gently from behind using a mallet with a wooden or plastic head. While doing this, hold a block of wood firmly against the metal to absorb the hammer blows and prevent a large area of the metal from being stretched.

If the dent is in a section of the body which has double layers, or some other factor making it inaccessible from behind, a different technique is in order. Drill several small holes through the metal inside the damaged area, particularly in the deeper sections. Screw long self-tapping screws into the holes just enough for them to get a good grip in the metal. Now the dent can be pulled out by pulling on the heads of the screws with locking pliers.

The next stage of repair is the removal of paint from the damaged area and from an inch or so of the surrounding metal. This is accomplished most easily by using a wire brush or sanding disk in a drill motor, although it can be done just as effectively by hand with sandpaper. To complete the preparation for filling, score the surface of the bare metal with a screwdriver or the tang of a file (or drill small holes in the affected area). This will provide a very good grip for the filler material. To complete the repair, see the Section on filling and painting.

Repair of rust holes or gashes

Remove all paint from the affected area and from an inch or so of the surrounding metal using a sanding disk or wire brush mounted in a drill motor. If these are not available, a few sheets of sandpaper will do the job as effectively. With the paint removed, you will be able to determine the severity of the corrosion and therefore decide whether to replace the whole panel, if possible, or repair the affected area. New body panels are not as expensive as most people think and it is often quicker to install a new panel than to attempt to repair large damaged areas.

Remove all trim pieces from the affected area (except those which will act as a guide to the original shape of the damaged body such as headlamp, shells etc.). Using metal snips or a hacksaw blade, remove all loose metal and any other metal that is badly affected by rust. Hammer the edges of the hole inwards to create a slight depression for the filler material.

Wire brush the affected area to remove the powdery rust from the surface of the metal. If the back of the rusted area is accessible, treat it with rust-inhibiting paint.

Before filling can be done it will be necessary to block the hole in some way. This can be accomplished with sheet metal riveted or screwed into place, or by stuffing the hole with wire mesh. Once the hole is blocked off, the affected area can be filled and painted.

Filling and painting

Many types of body fillers are available, but generally speaking, body repair kits which contain filler paste and a tube of resin hardener are best suited for this type of repair work. A wide, flexible plastic or nylon applicator will be necessary for imparting a smooth and contoured finish to the surface of the filler material.

Mix up a small amount of filler on a clean piece of wood or cardboard (use the hardener sparingly). Follow the instructions on the package, otherwise the filler will set incorrectly.

Using the applicator, apply the filler paste to the prepared area. Draw the applicator across the surface of the filler to achieve the desired contour and to level the filler surface. As soon as a contour that approximates the original one is achieved, stop working the paste. If you continue, the paste will begin to stick to the applicator. Continue to add thin layers of filler paste at 20 minute intervals until the level of the filler is just above the surrounding metal.

Once the filler has hardened the excess can be removed using a body file. From then on, progressively finer grades of sandpaper should be used, starting with a 180-grit paper and finishing with 600-grit wet-or-dry paper. Always wrap the sandpaper around a flat rubber or wooden block, otherwise the surface of the filler will not be completely flat. During the sanding of the filler surface, the wet-or-dry paper should be periodically rinsed in water. This will ensure that a very smooth finish is produced in the final stage.

At this point, the repair area should be surrounded by a ring of base metal, which in turn should be encircled by the finely feathered edge of the good paint. Rinse the repair area with clean water until all of the dust produced by the sand operation is gone.

Spray the entire area with a light coat of primer. This will reveal any imperfections in the surface of the filler. Repair these imperfections with fresh filler paste or glaze filler and once more smooth the surface with sandpaper. Repeat this spray-and-repair procedure until you are satisfied that the surface of the filler and the feathered edge of the paint are perfect. Rinse the areas with clean water and allow it to dry completely.

The repair area is now ready for painting. Paint spraying must be carried out in a warm, dry, windless and dustfree atmosphere. These conditions can be created if you have access to a large indoor working area, but if you are forced to work in the open, you will have to pick the day very carefully. If you are working indoors, dousing the floor in the work area with water will help settle dust which would otherwise be in the air. If the repair area is confined to one body panel mask off the surrounding panels. This will help to minimize the effects of a slight mismatch in paint color. Trim pieces such as chrome strips, door handles, etc., will also need to be masked off or removed. Use masking tape and several thicknesses of newspaper for the masking operations.

Before spraying, shake the paint can thoroughly, then spray a test area until the spray painting technique is mastered. Cover the repair area with a thick coat of primer. The thickness should be built up using several thin layers of primer rather than one thick one. After the primer has dried, use 600-grit wet-or-dry sandpaper, rub down the surface of the priomer until it is very smooth. While doing this, the work area should be thoroughly rinsed with water and the wet-or-dry sandpaper periodically rinsed as well. Allow the primer to dry before spraying additional coats.

Spray on the top coat, again building up the thickness by using several thin layers of paint. Begin spraying in the center of the repair area and then, using a circular motion, work out until the whole repair area and about two inches of the surrounding original paint is covered. Remove all masking material 10 to 15 minutes after spraying on the final coat of paint. Allow the new paint at least two weeks to harden, then, using a very fine rubbing compound, blend the edges of the new paint into the existing paint. Finally, apply a coat of wax.

6   Body repair — major damage

1   Major damage must be repaired by an auto body/frame repair shop with the necessary welding and hydraulic straightening equipment.
2   If the damage has been serious, it is vital that the unibody structure be checked for correct alignment, as the handling of the vehicle will be affected. Other problems, such as excessive tire wear and wear in the transmission and steering, may also occur.

7   Hinge and locks — maintenance

Every 3000 miles or three months, the door, hood and rear hatch hinges should be lubricated with a few drops of oil and the locks treated with dry graphite lubricant. The door and rear hatch striker plates should be given a thin coat of grease to reduce wear and ensure free movement.

8   Hood — removal and installation

Refer to illustration 8.2
1   Use blankets or fender covers to protect the fenders from scratches during removal.
2   Use paint or scratch an alignment mark (see illustration) to outline the hinges in relation to the hood. This will greatly aid alignment when reinstalling.
3   Have an assistant support the hood while you remove the bolts securing the hydraulic stays to the hood.
4   Remove the hinge-to-hood bolts on both sides of the hood.
5   Carefully lift the hood from the vehicle.
6   Installation is the reverse of the removal procedure, noting the alignment marks made during removal.

9   Hood — raising while servicing the engine

Refer to illustration 9.2
1   The hood may not have to be removed when servicing the vehicle and desiring extra head room. The hood is designed so that it may be opened further than normal during servicing procedures. Have an assistant hold the hood up while you remove the two bolts securing the hood hydraulic stays.
2   Raise the hood up farther and replace the bolts in the holes located below the two regular attachment holes (see illustration). **Caution:** *Do not attempt to shut the hood with the stays in the raised position. The hood could be damaged.*
3   After servicing of the vehicle is complete, have an assistant support the hood while you remove the hydraulic stay bolts. Have the assistant lower the hood until the bolts again fit into their original locations.

10   Hood latch release cable — removal and installation

Refer to illustrations 10.2, 10.3a, 10.3b, 10.3c, 10.5, 10.7 and 10.8
1   The hood latch is an assembly that includes the pull handle, control cable and housing.
2   To remove the cable, raise the hood and mark the location of the hood latch in relation to the front crossmember (see illustration).

8.2   Mark the hood to hinge relationship before removing the hood hydraulic stay and hinge-to-hood bolts

9.2   Remove the hydraulic stay bolts, then have an assistant raise the hood higher and install the stay bolts in the holes (arrow) provided

10.2   Mark the relationship of the latch to the crossmember

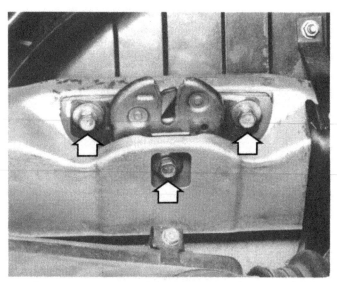

10.3a   Remove the retaining bolts (arrows) securing the
latch to the crossmember

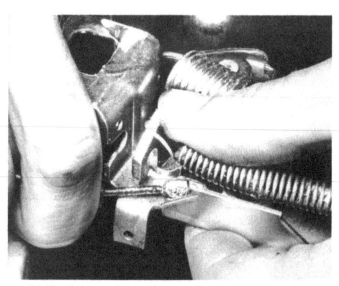

10.3b   Press the lever with your thumb and remove the
cable end from the lever

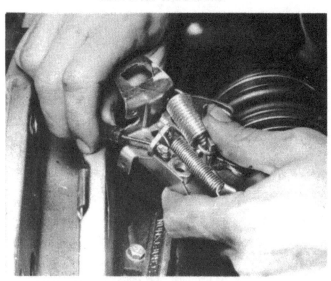

10.3c   Hold the latch in one hand and pull the cable
housing from the latch

10.5   Pry the plastic cable cover from the fender panel

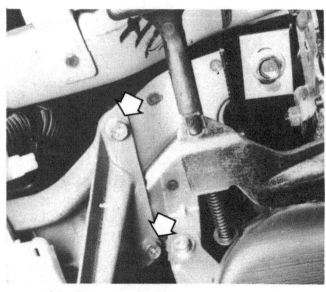

10.7   Remove the hood latch release handle bolts

10.8 Attach a strong piece of line to the cable end and
tape the connection to keep it from snagging when being
fed through the fender

3   Remove the bolts securing the latch to the crossmember (see illustration). Lift the latch out of the recess in the crossmember, and while holding the release lever back with your thumb work the cable end from the latch (see illustration). Pull the cable housing from the latch (see illustration).
4   Working towards the fender, remove the cable from the plastic retaining clips.
5   Pry the plastic cable cover from the fender panel (see illustration).
6   Remove the lower dash trim panel to gain access to the hood latch release handle and securing bolts.
7   Working under the dash, remove the cable assembly by removing the two retaining bolts (see illustration).
8   Before pulling the cable through the fender and dash, attach a strong piece of line to the end of the cable and tape the connection to prevent snags while pulling the cable out (see illustration).
9   From the driver's compartment pull the cable through the dash.
10   Disconnect the cable from the guide line.
11   Connect the guide line to the new cable and tape the connection.
12   Gently pull the guide line, feeding the new cable through the dash and fender.
13   Once the cable is pulled completely through the fender, install the cable into its retaining clips and install the plastic cover in the fender panel.
14   Insert the cable end into the hood latch by pushing back on the actuating lever and inserting the cable end into its retaining recess on the latch.
15   Install the latch and align it with the mark made during removal.

---

11   Door panel trim — removal and installation

Refer to illustrations 11.1, 11.2, 11.3, 11.4, 11.5, 11.6 and 11.7

1   Use a small screwdriver to pry out the arm rest retaining screw covers (see illustration).
2   Use a screwdriver to remove the arm rest retaining screws (see illustration).
3   Disconnect power window electrical connector by pressing down on the connector retaining tab and pulling the connector apart (see illustration).
4   Use a screwdriver to remove the screw attaching the door handle plastic cover (see illustration).

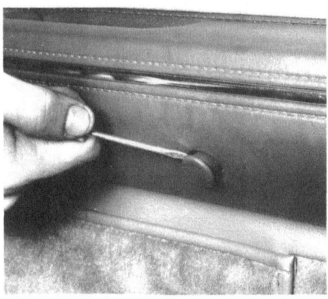

11.1   Pry out the arm rest retaining screw covers to gain acess to the screws

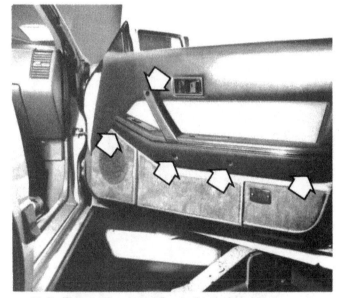

11.2   Remove the screws (arrows) attaching the arm rest to the door

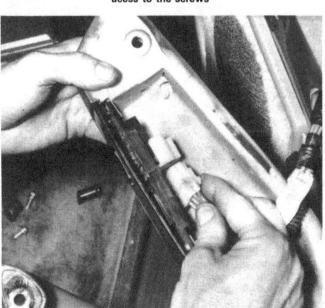

11.3   Disconnect the power window electrical connector

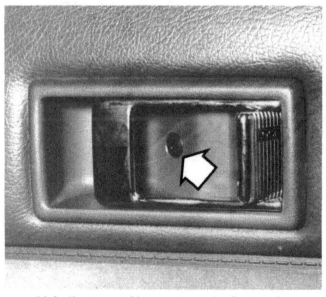

11.4   Use a screwdriver to remove the door handle retainer plastic cover

11.5   Remove the screws (arrows) holding the door safety light lens to the door

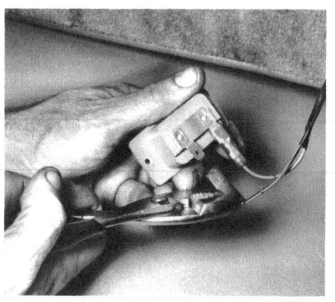

11.6   Unplug the electrical connectors from the the safety light switch

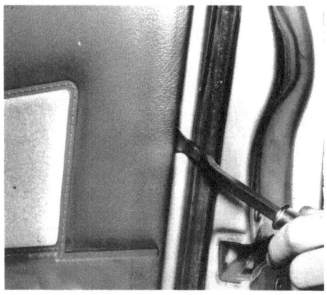

11.7   Pry the panel away from the door, being careful not to pull the plastic panel fasteners through the panel

12.1   Mark the alignment of the striker to the door before removing it

5   Use a screwdriver to remove the two screws attaching the door safety light lens to the door (see illustration).
6   Disconnect the electrical connectors from the safety light (see illustration).
7   Using a flat screwdriver or a panel removal tool, pry the panel away from the door, being careful not to pull the plastic panel fasteners through the panel (see illustration).
8   Remove the panel by sliding the panel away from the vehicle while lifting it up from the door.
9   Install the panel by pulling the electrical harnesses through their proper openings in the panel and sliding the inner most part of the panel into the rubber recess and aligning the panel fasteners. Press the fasteners into the door.
10   Connect the safety light electrical connectors.
11   Install the safety light lens.
12   Install the door handle release plastic cover.
13   Hold the arm rest next to the panel and connect the power window electrical connector.
14   Place the arm rest onto the panel and secure with the five attaching screws.
15   Insert the arm rest retaining screw covers and press them in firmly.

12   Door latch striker — removal and installation

Refer to illustration 12.1
1   Make an alignment mark to show the relation of the striker to the door jamb to facilitate proper installation (see illustration).
2   Remove the two retaining screws.
3   Installation is the reverse of the removal procedure. Align the striker with the mark made during removal.

13   Door latch — removal and installation

Refer to illustrations 13.2 and 13.3
1   Remove the door trim panel (Section 11).
2   Disconnect the door latch control rods by pressing down on the yellow plastic fasteners, releasing the fastener from the rod (see illustration). Remove the rod by pulling it out of the fastener.
3   Remove the latch-to-door screws (see illustration) and work the latch from the door.

13.2 Press down on the yellow plastic fasteners (arrows) to release the rods

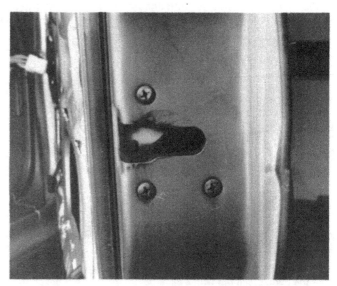

13.3 Remove the screws attaching the door latch to the door

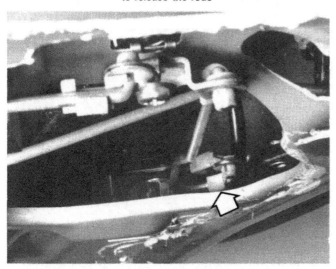

14.5 Use a long screwdriver to press down on the yellow plastic fastener (arrow) to release the rod from the latch actuator

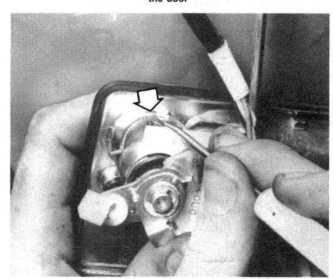

14.6 Pry the key cylinder retaining clip (arrow) up to release the cylinder lock

4   Install the latch by working it up through the inside of the door and placing it in the recess on the door end. Secure it with the three screws.

5   Connect the door latch control rods by pressing the ends of the rods into the yellow fasteners and securing it by turning the fastener up and snapping it to the rod.

6   Install the door trim.

---

14 Door lock cylinder — removal and installation

Refer to illustrations 14.5, 14.6 and 14.7

1   Remove the door trim panel (Section 11).

2   If equipped, disconnect the cylinder lock electrical connector.

3   Remove the two nuts securing the door handle assembly to the door.

4   Pull the anti-theft shield away from the door handle.

5   Use a screwdriver to disconnect the cylinder lock-to-latch actuating lever operating rod by pushing down on the yellow plastic rod fastener (see illustration) and pull the rod end from the latch actuator.

6   Lift the handle out of the door and pry the key cylinder retaining clip (see illustration) up to release the cylinder lock.

7   Pull the cylinder lock from the cylinder lock housing (see illustration).

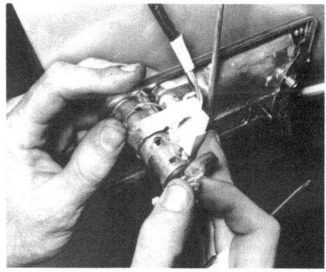

14.7 The cylinder lock can now be pulled from the housing

These photos illustrate a method of repairing simple dents. They are intended to supplement *Body repair - minor damage* in this Chapter and should not be used as the sole instructions for body repair on these vehicles.

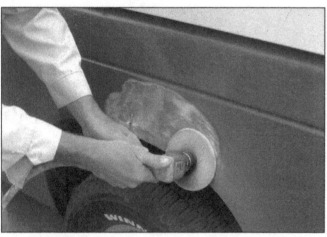

1   If you can't access the backside of the body panel to hammer out the dent, pull it out with a slide-hammer-type dent puller. In the deepest portion of the dent or along the crease line, drill or punch hole(s) at least one inch apart . . .

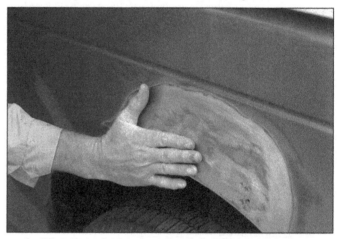

2   . . . then screw the slide-hammer into the hole and operate it. Tap with a hammer near the edge of the dent to help 'pop' the metal back to its original shape. When you're finished, the dent area should be close to its original contour and about 1/8-inch below the surface of the surrounding metal

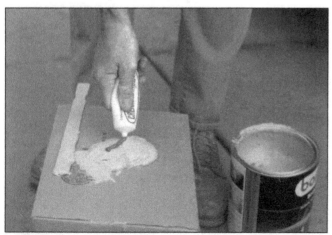

3   Using coarse-grit sandpaper, remove the paint down to the bare metal. Hand sanding works fine, but the disc sander shown here makes the job faster. Use finer (about 320-grit) sandpaper to feather-edge the paint at least one inch around the dent area

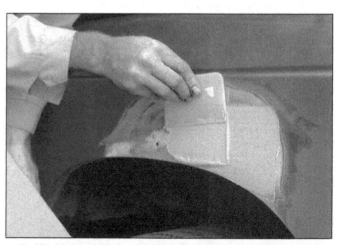

4   When the paint is removed, touch will probably be more helpful than sight for telling if the metal is straight. Hammer down the high spots or raise the low spots as necessary. Clean the repair area with wax/silicone remover

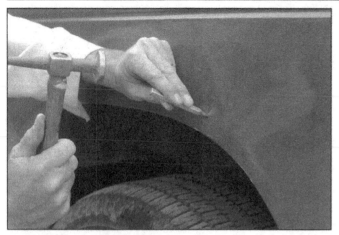

5   Following label instructions, mix up a batch of plastic filler and hardener. The ratio of filler to hardener is critical, and, if you mix it incorrectly, it will either not cure properly or cure too quickly (you won't have time to file and sand it into shape)

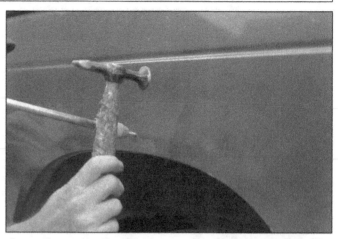

6   Working quickly so the filler doesn't harden, use a plastic applicator to press the body filler firmly into the metal, assuring it bonds completely. Work the filler until it matches the original contour and is slightly above the surrounding metal

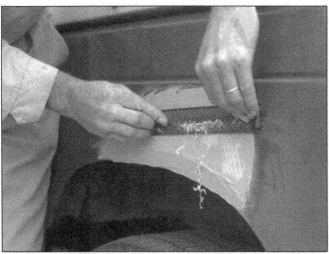

7   Let the filler harden until you can just dent it with your fingernail. Use a body file or Surform tool (shown here) to rough-shape the filler

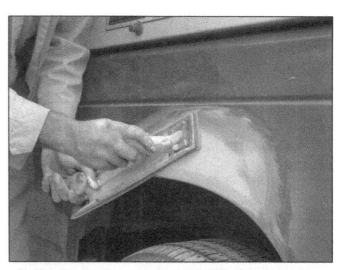

8   Use coarse-grit sandpaper and a sanding board or block to work the filler down until it's smooth and even. Work down to finer grits of sandpaper - always using a board or block - ending up with 360 or 400 grit

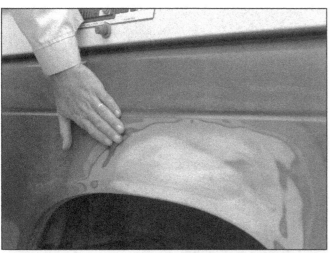

9   You shouldn't be able to feel any ridge at the transition from the filler to the bare metal or from the bare metal to the old paint. As soon as the repair is flat and uniform, remove the dust and mask off the adjacent panels or trim pieces

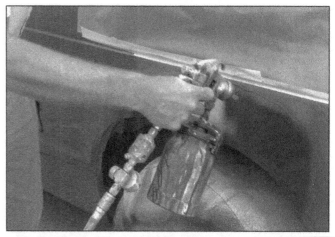

10   Apply several layers of primer to the area. Don't spray the primer on too heavy, so it sags or runs, and make sure each coat is dry before you spray on the next one. A professional-type spray gun is being used here, but aerosol spray primer is available inexpensively from auto parts stores

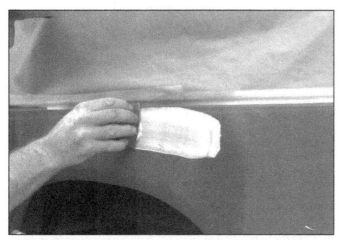

11   The primer will help reveal imperfections or scratches. Fill these with glazing compound. Follow the label instructions and sand it with 360 or 400-grit sandpaper until it's smooth. Repeat the glazing, sanding and respraying until the primer reveals a perfectly smooth surface

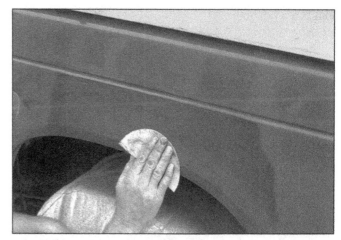

12   Finish sand the primer with very fine sandpaper (400 or 600-grit) to remove the primer overspray. Clean the area with water and allow it to dry. Use a tack rag to remove any dust, then apply the finish coat. Don't attempt to rub out or wax the repair area until the paint has dried completely (at least two weeks)

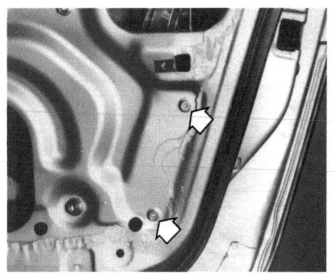

15.3   Remove the electric door lock bolts (arrows)

16.2   Remove the top apron bolts (left three bolts shown, three more on the right side)

16.3   From underneath the apron remove the screws attaching the bottom of the apron to the bumper (arrows)

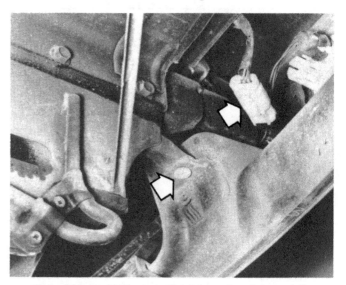

16.4   Disconnect the turn signal (arrow) and remove the screw located close to the bracket (arrow) on both sides of the vehicle

8   Install the cylinder lock into the housing and secure by pressing down on the key cylinder retaining clip.
9   Insert the handle into the door, making sure the door latch control rod slides into its actuating arm.
10   Connect the cylinder lock-to-latch operating lever and secure it by turning the yellow plastic fastener up and snapping it to the rod.
11   Install the anti-theft shield on the door handle.
12   Install and tighten the two handle retaining nuts.
13   Check the lock and door handle for proper operation.
14   Install the door trim panel.

15   Electric door lock actuating solenoid — removal and installation

Refer to illustration 15.3

1   Remove the door trim panel (Section 11).
2   Disconnect the door lock actuating rod by disconnecting the yellow plastic fastener. Press down on the fastener to release it from the rod then pull the rod end out of the fastener.
3   Remove the bolts securing the electric door lock to the door (see illustration).

4   Remove the solenoid and follow the electrical wire harness to the connector and disconnect the connector.
5   Install the solenoid by connecting the electrical connector and installing the solenoid inside the door.
6   Install and tighten the bolts securing the electrical door lock to the door.
7   Connect the door lock actuating rod by inserting the end of the rod into the yellow fastener attached to the actuating lever. Turn the end of the fastener up and snap the fastener onto the rod.
8   Check the electric door lock solenoid for proper operation.
9   Install the door trim panel.

16   Front apron — removal and installation

Refer to illustrations 16.2, 16.3, 16.4, 16.5 and 16.6

1   Raise the hood.
2   Remove the bolts (see illustration) securing the top of the apron to the body.
3   From underneath the apron, remove the screws attaching the bottom of the apron to the bumper (see illustration).
4   Disconnect the turn signal electrical connector and remove the

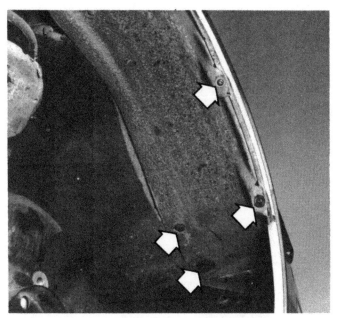

16.5   Remove the inner splash shield lower four screws

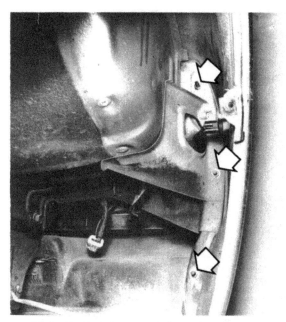

16.6   Remove the nuts retaining the apron to the fender (arrows)

17.2   Remove the headlight case bracket bolts (arrows)

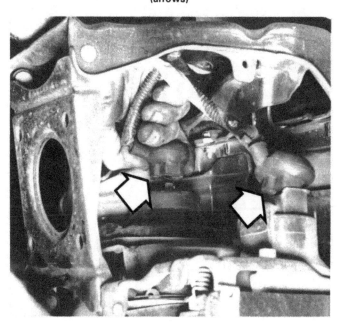

17.4   Reach inside the headlight case and disconnect the headlight electrical connectors (arrows)

screw close to it (see illustration) attaching the bottom of the apron to the frame bracket. Do this on both sides of the vehicle.

5   Remove the lowest four inner splash shield screws (see illustration). **Note:** *It will make this job easier if you raise the front of the vehicle, support it on jackstands and remove the front wheels.*

6   Remove the apron to fender mounting nuts on each side of the apron (see illustration). Lift the apron from the vehicle.

7   Install the apron by inserting the apron mounting studs through their respective mounting holes in the fender and install the nuts.

8   Install the inner splash shield and secure it with the four screws.

9   Connect the turn signal electrical connectors and install the apron lower mounting screws.

10   Install the five screws attaching the bottom of the apron to the bumper.

11   If removed, replace the wheels and remove the jackstands before lowering the vehicle.

12   Install the six bolts securing the top of the apron to the body.

17   Headlight case — removal and installation

Refer to illustrations 17.2 and 17.4

1   Remove the headlight motor (Chapter 12).

2   Remove the headlight case bracket bolts (see illustration).

3   Lift the headlight case up and disconnect the wire harness retaining clips.

4   Disconnect the two electrical connectors by pulling them from the lights (see illustration). Lift the case out.

5   Install the headlight box by connecting the two light electrical connectors.

6   Place the headlight case into the apron cavity, making sure the motor adjusting rod goes through the crossmember opening. Secure the box with the four bracket bolts.

7   Install and adjust the headlight motor.

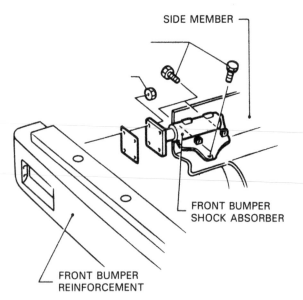

18.1a The front bumper energy absorber will compress
upon a five mph impact and return to its original position
after impact

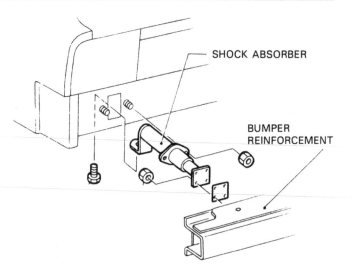

18.1b The rear bumper absorber is capable of handling
5 mph impacts

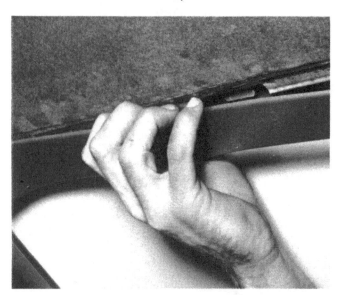

19.3a Pull the headliner trim piece rearward to release the
spring type clip retainers

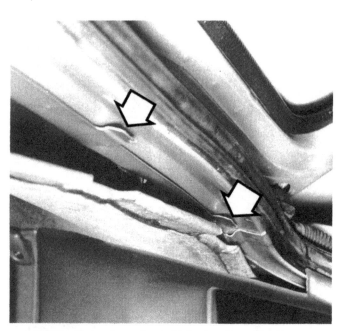

19.3b Spring type clips hold the headliner trim piece
in place

18 Bumpers and energy absorbers — general information

Refer to illusatrations 18.1a and 18.1b

1   Both the front and rear bumpers are equipped with gas/hydraulic
energy absorbers (see illustrations) designed to withstand a collision
into a fixed barrier at up to five mph without damage.
2   After any collision, the energy absorbers should be inspected. If
obvious physical damage or leaking oil is evident, the unit should be
replaced. These absorbers are not designed to be disassembled and,
if defective, should be replaced as a unit.
3   Due to the gas contained in the units, never apply heat to an energy
absorber or weld in its vicinity.

19 Rear hatch — removal, installation and adjustment

Refer to illustrations 19.3a, 19.3b, 19.7, 19.9, 19.10 and 19.17

1   Before opening the hatch, note alignment of the hatch in relation
to the body on all four sides.

2   Open the hatch.
3   Sitting inside the luggage compartment facing the front of the vehicle, grasp the rear headliner trim piece (see illustration) and pull backwards to release the five spring type retaining clips (see illustration)
attaching the trim piece to the body.
4   Remove the lower hatch trim panel.
5   Disconnect the three electrical connectors inside the bottom of
the hatch and release the wiring harness from the plastic retaining clips.
6   Attach a strong piece of line to the wiring harness and tape the
connection. Pull the harness through the hatch from the body side (top),
leave a few feet of line exposed at each end. As a precaution, tape
the ends of the guide line to the hatch to keep the line from falling out.
7   Use a small screwdriver to pry up on the washer nozzle to release
it from the exterior trim piece (see illustration).
8   Disconnect the nozzle from the rubber hose and attach a strong
piece of line to the rubber feed hose and tape the connection. Pull the
hose through the hatch from the body side. Leave a few feet of line
exposed at each end. As a precaution, tape the ends of the guide line
to the body to keep the line from falling out.
9   Use a piece of wood to support the hatch and remove the lower

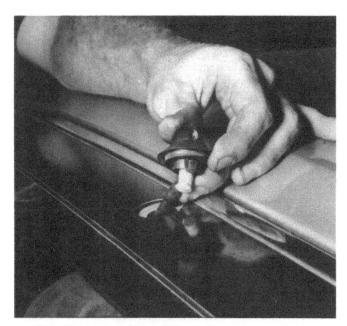

19.7   The rear window washer nozzle can be pryed out of
its hatch recess to disconnect the fluid feed hose

19.9   The bottom of the hydraulic stay can be
screwed out

19.10   Use a screwdriver to pop the stay off of the ball

19.17   Use a soft face hammer to knock the hydraulic
stay back onto the ball

attaching bolt securing the hydraulic stay (see illustration).

10   If removing only the stay, place the free end of the stay up near the hatch and use a screwdriver (see illustration) to pry the socket from the ball.

11   To remove the hatch, hold down on the headliner to expose the hatch hinge retaining nuts and use a ratchet and socket to remove the nuts.

12   Have an assistant help remove the hatch by lifting straight up to release the hinge bolts from the body mountings. Disconnect the guide line from the hatch.

13   Install the hatch with the help from an assistant. Insert the hinge bolts into the body mountings and loosely install the nuts.

14   Attach the guide line to the wiring harness and tape the connection. **Note:** *Spray the harness with silcone spray to reduce drag on the harness.*

15   Repeat Step 14 to install the rear washer hose and connect the hose to the spray nozzle.

16   Install the spray nozzle by pressing it into the recess on the outside

of the hatch.

17   Install and tighten the bottom hydraulic stay securing bolt. If the top was removed, use a soft face hammer to carefully force it onto the ball (see illustration).

18   Carefully lower the hatch until it latches. From the rear of the vehicle note the hatch alignment at all four corners and adjust by moving the hatch until it is squarely placed on the body.

19   From inside the vehicle, without disturbing the hatch alignment, tighten the hinge nuts.

20   Install the trim pieces.

20   Rear hatch latch — removal and installation

Refer to illustrations 20.3, 20.4 and 20.6

1   Open the hatch.

2   Remove the rear interior trim panel to expose the rear hatch latch.

20.3   Before removing the latch, mark its position in
relation of the body

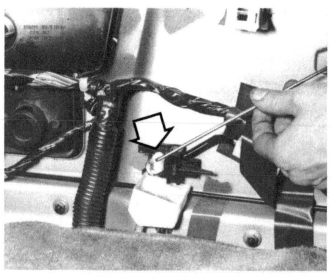

20.4   Press down on the white fastener (arrow) to
release the rod, then pull the end of the rod from
the fastener/actuating lever

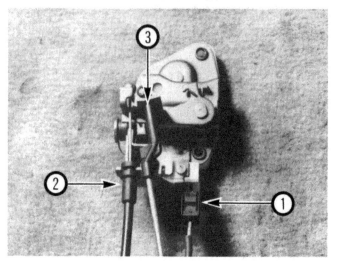

20.6   Disconnect the latch by unpluging the electrical
connector (1), removing the cable housing from the latch
recess (2) and disconnecting the cable end. Turn the
actuating rod sideways until the end (3) can be removed
from the latch

21.1   Pry the emblem on the front of the console up to
remove it

3   Using a marker, mark the position of the latch by outlining the relationship of the latch bolts to the body (see illustration).
4   Disconnect the cylinder lock actuating rod at the cylinder lock by
pressing down on the white plastic fastener to release the rod, then
pull the end of the rod from the fastener/actuating lever (see illustration).
5   Remove the three latch mounting bolts and work the latch out of
the body.
6   Disconnect the electrical connector (see illustration).
7   Lift the cable end from its retaining recess and pry the cable housing
from the latch.
8   Lift the cylinder lock actuating rod and turn the rod until the rod
can be removed from the latch.
9   Install the cable end into the retaining recess on the latch and press
the cable housing onto the latch. Place the end of the cylinder lock
actuating rod in the latch actuator and turn it to secure it to the latch.
Connect the electrical connector.
10 Place the latch into position and loosely install the latch bolts. Connect the actuating rod to the cylinder lock and fasten it with the plastic
fastener.
11 Align the latch retaining bolts with the marks made during removal
and tighten the bolts.
12 Check the operation of the latch and adjust if necessary by loosening the retaining bolts and moving the latch into correct alignment.
13 Install the rear trim piece.

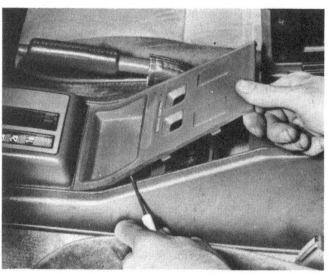

21.2   Pry up the switch cover plate

21 Console — removal and installation

Refer to illustration 21.1 and 21.2

1   Use a small screwdriver to pry up the 300ZX emblem at the front of the console (see illustration).

2   Open the console compartment lid and use a small screwdriver to pry up on the switch cover plate, which is held in place in eight places (see illustration).

3   Use a screwdriver to remove the console retainers. Two of the screws are located in the front, one at the back of the shifter and three inside the compartment.

4   Raise the console. Note and disconnect any electrical connectors.

5   Install by placing the console in position and connecting electrical connectors.

6   Install the console retainers.

7   Install the switch cover and snap it in place by pressing down.

8   Install the 300ZX emblem by snapping it in place.

22 T-bar roof — general description

Refer to illustration 22.1

1   The T-bar roof (see illustration) is made of glass and should be handled with care.

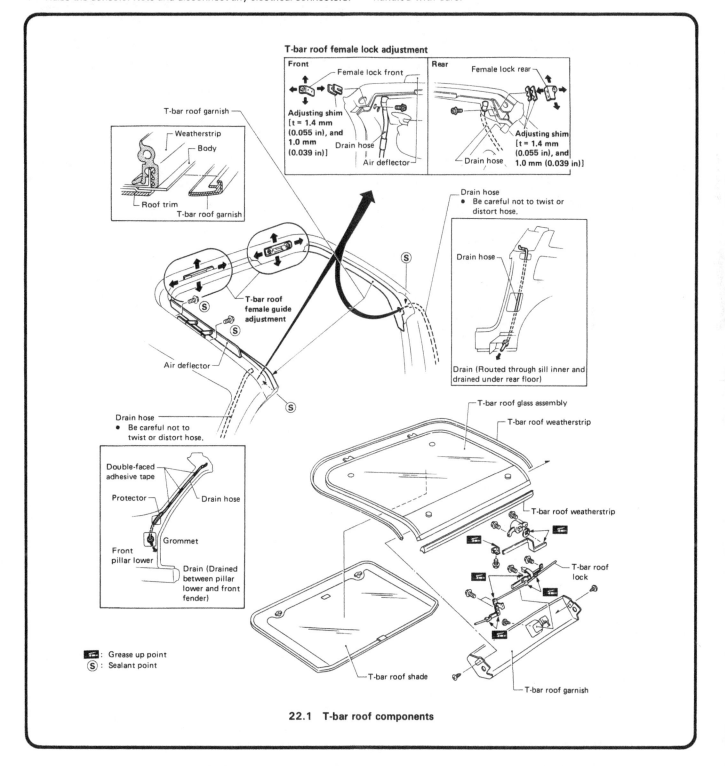

22.1   T-bar roof components

23.1   Press in the sides of the glove box to release the stops

23.2   To remove the glove box, remove the Phillips head screws (arrows)

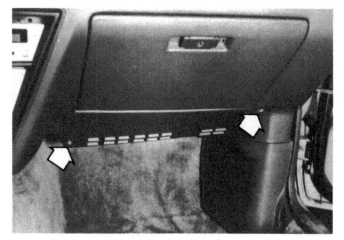

23.5   Remove the Phillips head screws to remove the passenger side lower trim panel

23.13   Disconnect the vacuum lines to free the panel

2   If necessary apply sealant to portions susceptible to water leakage.
3   The side molding, sash, lock basement and the glass of the T-bar roof constitute one unit which cannot be disassembled.
4   Latch adjustments can be made by loosening the fasteners and moving the latch into position, then securing the fasteners.
5   Apply a thin coat of grease to the areas indicated in illustration 22.1.

---

23   Interior trim — removal and installation

Refer to illustrations 23.1, 23.2, 23.5, 23.13, 23.19, 23.22, 23.25, 23.30 and 23.34

Glove box

1   Open the glove box and press the sides inward to release the stops on the outside of the box (see illustration).
2   Remove the four Phillips head screws (see illustration) and remove the glove box from the dash.
3   Install the box by placing it into the dash cavity and securing it with the four retaining screws.
4   Squeeze the sides of the glove box and push up to press it into place.

Passenger side lower trim panel

5   Remove the two Phillips head screws and lift the panel off (see illustration).
6   Disconnect the convience light by twisting and pulling the connector from the back of the light.
7   Reverse the removal procedure for installation.

Passenger side lower dash panel

8   Remove the Phillips head screw retaining the kick panel to the body.
9   Turn the plastic fastener located on the lower right corner a quarter turn to release it and lift the kick panel off.
10   Reverse the removal procedure for installation.

Driver side lower dash panel

11   Remove the three Phillips head screws, one on top near steering column and the other two on each corner of the panel, then lower the panel.
12   Disconnect the convenience light by twisting the light connector and pulling it from the light.
13   Remove the two vacuum hoses by pulling them from the nipples (see illustration).
14   Reverse the removal procedure for installation.

Rear luggage compartment trim panel

15   Turn the four plastic fasteners a quarter turn and pull the fasteners out. Unscrew the one knob head fastener, located on the top passenger side of the panel, by turning it counterclockwise.
16   Lean the trim piece forward and remove it from the vehicle.
17   Reverse the removal procedure for installation. Make sure all fasteners are locked in place.

Spare tire cover

18   Remove the rear luggage compartment trim panel.
19   Unscrew the knob fastener located on the top left corner of the panel by turning it counterclockwise. Lift out the panel (see illustration).
20   Reverse the removal procedure for installation.

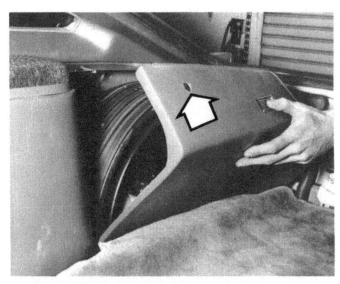

23.19 With the knob head fastener removed (arrow) tilt the panel forward and lift it out

23.22 Remove the tool kit to get at the panel fasteners

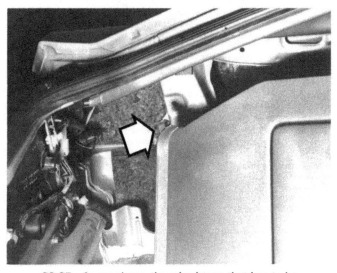

23.25 Arrow shows the raised tang that has to be inserted first during installation

23.30 Remove the hatch trim piece by prying down on it to release the pop-out type fasteners

Rear left trim piece

21 Remove the rear luggage compartment trim panel.
22 Fold back the carpet and lift out the tool kit (see illustration).
23 Remove the two trim piece retaining screws.
24 Lift the trim piece out from the bottom to release it from the three top plastic fasteners.
25 Install the trim piece by inserting the end with the raised tang (see illustration) first.
26 Install the two retaining screws.
27 Set the tool kit into position and lay the carpet over it.
28 Install the rear luggage compartment trim panel.

Rear hatch trim piece

29 Open the rear hatch.
30 Grasp the lower hatch trim panel and pull it down to release the nine plastic pop-out fasteners (see illustration).
31 Install the trim piece by matching the plastic fasteners to their holes, then press the trim piece into position.

Control panel trim piece

32 Remove the ash tray.
33 Pull the cigarette lighter from its retaining socket.
34 Remove the trim piece Phillips head screws (see illustration).
35 Pull the trim piece forward.

23.34 Remove the trim piece by removing the Phillips head screws (arrows)

Chapter 12 Chassis electrical system

Contents

Specifications

Bulbs

Headlight	H6054
Front combination light	1157
Front side marker light	158
Rear side marker light	158
Rear combination lights	
turn signal	1157
stop/tail	1157
back-up	1073

Fuses

10 Amp
  Headlight RH, headlight LH, room light, hazard light, horn, auto door lock, fuel pump, meter, turn signal, A/T control, theft warning, audio, RR wiper and air conditioning control

15 Amp
  Headlight motor RH, headlight motor LH, clearance, illumination, stop light, air conditioner blower, engine control, auxiliary driving light, FR wiper, fan motor (turbo models) and cigarette lighter

20 Amp
  RR defogger

Fusible links

Color	
black	Power supply to battery, accessory, EFI circuit and ignition switch
brown	Ignition switch
brown	Headlight circuit
green	Power window
brown	EFI control unit
brown	EFI injector

1 General information

This Chapter covers the repair and service procedures for the various lighting and electrical components not associated with the engine, as well as general information on various electrical circuits, including the theft warning system. Information on the battery, alternator, distributor and starter motor can be found in Chapter 5.

Electrical components located in the dashboard do not use ground wires or straps, but rather use grounding provisions which are integrated in the printed circuit mounted behind the instrument cluster.

It should be noted that whenever portions of the electrical system are worked on, the negative ground cable should be disconnected at the battery to prevent electrical shorts and/or fires.

2 Electrical troubleshooting — general information

A typical electrical circuit consists of an electrical component, any switches, relays, motors, etc. relevant to that component and the wiring and connectors that connect the components to both the battery and the chassis. To aid in locating a problem in any electrical circuit, complete wiring diagams of each model are included at the end of this Chapter.

Before tackling any troublesome electrical circuit, first thoroughly study the appropriate diagrams to get a complete understanding of what makes up that individual circuit. Trouble spots, for instance, can often be narrowed down by noting if other components related to that circuit are operating properly or not. If several components or circuits fail at one time, chances are the fault lies in the fuse or ground connection, as several circuits often are routed through the same fuse and ground connections. This can be confirmed by referring to the fuse box and ground distribution diagrams in this Chapter.

Often, electrical problems stem from simple causes, such as loose or corroded connections, a blown fuse or melted fusible link. Prior to any electrical troubleshooting, always visually check the condition of the fuses, wires and connections of the problem circuit.

If testing instruments are going to be utilized, use the diagrams to plan ahead of time where you will make the necessary connections in order to accurately pinpoint the trouble spot.

The basic tools needed for electrical troubleshooting include a circuit tester or voltmeter (a 12 volt bulb with a set of test leads can also be used), a continuity tester (which includes a bulb, battery and set of test leads) and a jumper wire, preferably with a circuit breaker incorporated, which can be used to bypass electrical components.

Voltage checks should be performed if a circuit is not functioning properly. Connect one lead of a circuit tester to either the negative battery terminal or a known good ground. Connect the other lead to a connector in the circuit being tested, preferably nearest to the battery or fuse. If the bulb of the tester goes on, voltage is reaching that point, which means the part of the circuit between that connector and the battery is problem-free. Continue checking along the circuit in the same fashion. When you reach a point where no voltage is present, the problem lies between there and the last good test point. Most of the time the problem is due to a loose connecton. Keep in mind that some circuits only receive voltage when the ignition key is in the Accessory or Run position.

A method of finding shorts in a circuit is to remove the fuse and connect a test light or voltmeter in its place to the fuse terminals. There should be no load in the circuit. Move the wiring harness from side to side while watching the test light. If the bulb goes on, there is a short to ground somewhere in that area, probably where insulation has rubbed off a wire. The same test can be performed on other components of the circuit, including the switch.

A ground check should be done to see if a component is grounded properly. Disconnect the battery and connect one lead of a self-powered testlight such as a continuity tester to a known good ground. Connect the other lead to the wire or ground connection being tested. If the bulb goes on, the ground is good. If the bulb does not go on, the ground is not good.

A continuity check is performed to see if a circuit, section of circuit or individual component is passing electricity properly. Disconnect the battery and connect one lead of a self-powered test light such as a continuity tester to one end of the circuit being tested and the other lead to the other end of the circuit. If the bulb goes on, there is con-

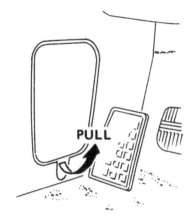

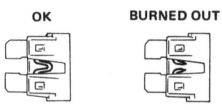

3.1 The fuse box is located behind the drivers kick panel. Pull the suspected fuse from the panel and check to see if it has blown

tinuity, which means the circuit is passing electricity properly. Switches can be checked in the same way.

Remember that all electrical circuits are composed basically of electricity running from the battery, through the wires, switches, relays, etc. to the electrical component (light bulb, motor, etc). From there it is run to the car body (ground) where it is passed back to the battery. Any electrical problem is basically an interruption in the flow of electricity from the battery to the component and back to the battery.

3 Fuses — general information

Refer to illustration 3.1

The electrical circuits of the car are protected by a combinaton of fuses and fusible links.

The fuse box is located in the side panel underneath the dash on the drivers' side of the car (see illustration). Each of the fuses is designed to protect a specified circuit, as identified on the use cover.

If an electrical component has failed, your first check should be the fuse. A fuse which has blown can be readily identified by inspecting the curved metal element inside the plastic housing. If this element is broken the fuse is inoperable and should be replaced with a new one. Fuses are replaced by simply pulling out the old one and pushing in the new one.

It is important that the correct fuse be installed. The different electrical circuits need varying amounts of protection, indicated by the amperage rating on the fuse. A fuse with too low rating will blow prematurely, while a fuse with too high a rating may not blow soon enough to prevent serious damage.

At no time should the fuse be bypassed by using metal or foil. Serious damage to the electrical system could result.

If the replacement fuse immediately fails, do not replace it with another until the cause of the problem is isolated and corrected. In most cases this will be a short circuit in the wiring system caused by a broken or deteriorated wire.

4 Fusible links — general information

Refer to illustration 4.1

In addition to fuses, the wiring system incorporates fusible links for

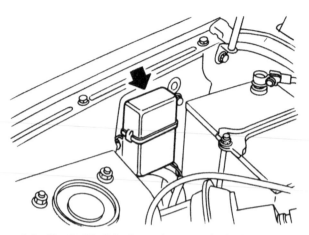

4.1   The fusible links located next to the battery are incorporated for overload protection

5.1   The electrical relays are grouped together (arrow) under the plastic cover next to the shock tower

overload protection (see illustration). These links are used in circuits which are not ordinarily fused, such as the ignition circuit and the Electronic Fuel Injection circuit.

The fusible links are located near the positive battery terminal, and are easily removed by unpluging the connectors at either end.

If an electrical failure occurs in one of the circuits covered by a fusible link, these should be the first check. If thé link is melted, the entire fusible link harness should be replaced, but only after checking and correcting the electrical fault that caused it.

5   Relays — general information

Refer to illustration 5.1

The various electrical relays are grouped together in the engine compartment (see illustration) for convenience in the event of needed replacement. If a faulty relay is suspected it can be removed and tested by a Nissan dealer or other qualified shop. Defective relays must be replaced as a unit.

6   Battery charging

1   If the battery is to remain in the vehicle during charging, disconnect the cables from the battery to prevent damage to the electrical system.
2   When the battery is being charged, hydrogen gas (which is very explosive) is produced. Do not smoke or allow an open flame near a charging or recently charged battery. Also, do not plug in the battery charger until the connections have been made at the battery posts.
3   The average time necessary to charge a battery at the normal rate is from 12 to 16 hours. Check the battery label and the charger instructions. Always charge the battery slowly. A quick charge or boost charge is hard on a battery and will shorten its life. Use a battery charger that is rated at no more than 1/10 the amp/hour rating of the battery.
4   Remove all of the vent caps (if equipped) and cover the holes with a clean cloth to prevent the spattering of electrolyte. Hook the battery charger leads to the battery posts (positive to positive, negative to negative), then plug in the charger. Make sure it is set at 12 volts if it has a selector switch.
5   Watch the battery closely during charging to make sure that it does not overheat. If the battery temperature rises above 60 °C (140 °F), stop charging.
6   The battery can be considered fully charged when it is gassing freely and there is no increase in specific gravity during three successive readings taken at hourly intervals. Overheating of the battery during charging at normal charging rates, excessive gassing and continual low specific gravity readings are indications that the battery should be replaced with a new one.

7   Turn signals and hazard flasher — general information

Small canister-shaped flasher units are incorporated into the elec-

trical circuits for the directional signals and hazard warning lights. These are located under the dash and just above the steering column.

When the units are functioning properly an audible click can be heard with the circuit in operation. If the turn signals fail on one side only and the flasher unit cannot be heard, a faulty bulb is indicated. If the flasher unit can be heard, a short in the wiring is indicated.

If the turn signal fails on both sides, the fault may be due to a blown fuse, faulty flasher unit or switch, or a broken or loose connection. If the fuse has blown, check the wiring for a short before installing a new fuse.

The hazard warning flasher is checked in the same manner as the turn signal flasher.

When replacing either of these flasher units it is important to buy a replacement of the same capacity. Check the new flasher against the old one.

8   Headlight – replacement

Sealed beam type
Refer to illustrations 8.2a, 8.2b, 8.3 and 8.4
1   Raise the headlights.
2   Remove the two screws attaching the plastic finisher plate to the headlight assembly (see illustration). Pull the top of the finisher plate forward and carefully pull the tight fitting finisher from the headlight assembly (see illustration).
3   Remove the screws attaching the lens retainer to the headlight assembly (see illustration).
4   Pull the headlight forward and disconnect the electrical connector from the back of the sealed beam unit (see illustration).
5   Install the the sealed beam unit by pressing the electrical connector onto the back of the unit and place it into the headlight box.
6   Install the lens retainer and secure with the four attaching screws.
7   Install the finisher plate and secure it with the two attaching screws.
8   Adjust the headlight (Sec 9).

Auxiliary driving light
Refer to illustration 8.10 and 8.11
9   If only replacing the auxiliary driving light, complete Steps 1 and 2.
10   Remove the top adjusting/retaining screw located on top of the light (see illustration).
11   Tilt the light forward and disconnect the spring attaching the bottom of the light to the holder (see illustration).
12   Unplug the connector and remove the light from the vehicle.
13   Plug the electrical connector into the back of the light.
14   Install the spring retainer onto the light and the headlight assembly.
15   Install the top adjusting/retaining screw.
16   Adjust the auxiliary headlight.

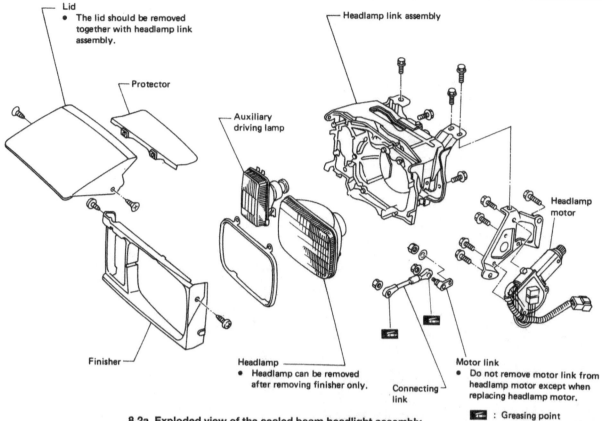

Lid
- The lid should be removed together with headlamp link assembly.

Protector

Auxiliary driving lamp

Headlamp link assembly

Headlamp motor

Finisher

Headlamp
- Headlamp can be removed after removing finisher only.

Connecting link

Motor link
- Do not remove motor link from headlamp motor except when replacing headlamp motor.

: Greasing point

8.2a Exploded view of the sealed beam headlight assembly

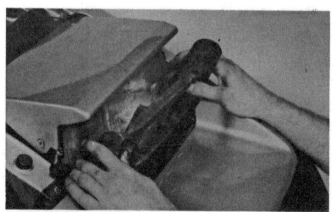

8.2b   The headlight finisher plate can be pulled from the headlight

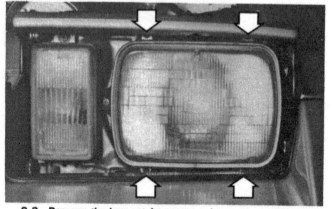

8.3   Remove the lens retainer screws (arrows) and lift the retainer out

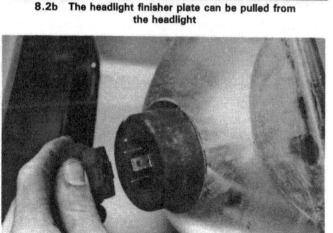

8.4   Pull the headlight forward and disconnect the electrical connector

8.10   Remove the auxiliary driving light by removing the screw located on top of the light (arrow)

8.11   Tilt the auxiliary driving light forward and disconnect the spring attaching the bottom of the light to the holder

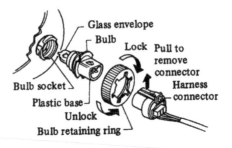

8.17   Halogen headlight bulb details

10.2   Remove the connecting link nut and washer (1) and remove the bolts (2) attaching the motor bracket to the body

Halogen type

Refer to illustration 8.17

17   Open the hood and locate the bulb retaining ring on the back of the headlight housing (see illustration).
18   Rotate the retaining ring counterclockwise (viewed from the rear).
19   Withdraw the bulb assembly from the headlight housing.
20   Unplug the holder from the electrical connector.
21   Without touching the glass with your bare fingers, insert the new bulb assembly into the headlight housing, then install and tighten the retaining ring. **Caution:** *If you do touch the glass with your fingers, clean it with rubbing alcohol and wipe it dry. The oils from your skin can cause the bulb to overheat and fail prematurely.*
22   Plug in the electrical connector.

---

9   Headlight — adjustment

Note: *It is important that the headlights be aimed correctly. If not ad-justed correctly they could blind an oncoming car and cause a serious accident or reduce your ability to see the road.*

Headlights have two spring loaded adjusting screws, one on the top controlling up and down movement and one on the side controlling left and right movement. The auxiliary lights have one adjusting screw located on the top that adjusts up and down movement.

10.6   Install the motor with the control arm in the down position

There are several ways to adjust the headlights. The simplest method is to set the vehicle on a level surface 25 feet from an empty wall. Adjustment should be made with the vehicle sitting level. Tire pressure should be correct, the gas tank should be full and the equivalent weight of the driver should be placed in the seat.

Preparation

1   Park the vehicle 25 feet in front of a light colored wall and turn on the lights.
2   Position masking tape vertically on the wall in reference to the center of the vehicle. Measure out from this line and place two vertical lines (one on each side) to represent the vertical center of each of the headlights. **Note:** *If the vehicle has a four headlight system, four vertical lines plus the vehicle centerline will be used.*
3   Position a horizontal tape line to reference the centerline of all the headlights. **Note:** *It may be easier to position the tape on the wall with the vehicle parked only a few inches away.*
4   Start adjustments with the low beam adjustment. Position the high-intensity zone so it is just below the horizontal line and four inches to the right of the headlight vertical line. Adjustment is made by turning the top adjusting screw clockwise to raise the beam and counterclockwise to lower the beam. The adjusting screw on the side should be used in the same manner to move the beam left or right.
5   With the high beams on, the high-intensity zone should be vertically centered four inches below the horizontal line. **Note:** *It may not be possible to position the headlight aim exactly for both high and low beams. If a comprise must be used, keep in mind that the low beams are the most used and have the greatest effect on driver safety.*
6   The auxiliary driving lights are preset horizontally but can be raised or lowered by turning the top adjusting screw.

---

10   Headlight motor — removal, installation and adjustment

Refer to illustrations 10.2, 10.6, 10.8 and 10.10

1   Disconnect the negative cable at the battery. Place the cable out of the way so it cannot accidentally come in contact with the negative terminal of the battery.
2   Remove the nut and washer holding the connecting link to the motor (see illustration).

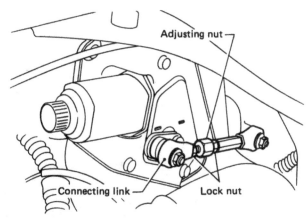

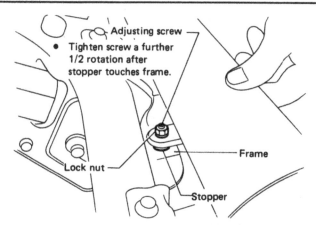

10.8 Adjust the headlight case connecting link by loosening the locking nut and turning the adjusting nut until it is properly adjusted

10.10 Adjust the stopper height by turning the adjusting screw until the stopper hits the frame, then tighten it 1/2-turn more

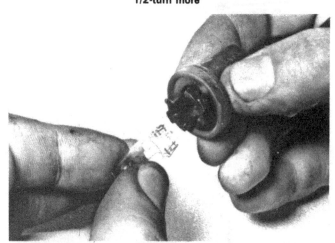

11.5 Grasp the side marker bulb connector assembly and twist the assembly clockwise to release it

11.6 Replace the side marker bulb by pulling it straight out of the connector

3   Remove the bolts attaching the motor bracket to the body.
4   Disconnect the electrical connector by pressing down on the tab and pulling the connector apart.
5   To install, connect the electrical connector.
6   Install the motor with the control arm in the down position (see illustration) and secure the motor to the body.
7   Install the connecting link to the motor and adjust if necessary.
8   With the headlight in the down position, adjust the connecting link so the lid is properly aligned with the hood and fender by loosening the locking nut and turning the adjusting nut until properly aligned (see illustration).
9   Tighten the locking nut.
10   Raise the headlight to the up position and adjust the stopper screw, if necessary.

---

11   Lenses and bulbs — replacement

Refer to illustrations 11.5, 11.6, 11.9, 11.13, 11.14, 11.15, 11.17, 11.21, 11.23 and 11.24

Exterior

Front turn signal
1   Remove the lens retainers and pull the lens from the apron.
2   Pull the assembly from the apron.
3   Replace the bulb by pushing in on the bulb and turning the bulb a quarter turn to release the bulb retaining tangs.
4   Install the bulb and assembly by reversing the removal procedure.

Side marker
5   Reach inside the front of the fender and grasp the bulb connector assembly (see illustration). Twist the assembly clockwise to release it from the lens assembly.

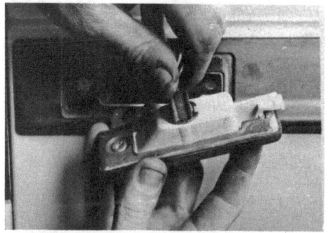

11.9 Twist the electrical connector a quarter turn to release it from the assembly

6   Replace the bulb by simply pulling it straight out of the connector (see illustration).
7   Install the bulb and assembly by reversing the removal procedure.

Rear side marker
8   Remove the screw securing the lens assembly to the body and pull the assembly from the body.
9   Grasp the connector and turn it a quarter turn to release it from the assembly (see illustration).
10   Replace the bulb by pulling it straight out from the connector.
11   Installation is the reverse of the removal procedure.

11.13　The cluster assembly has several light bulbs, so
determine which bulbs need to be replaced

11.14　The lens assembly has an arrow indicating the
connector/bulb locked position

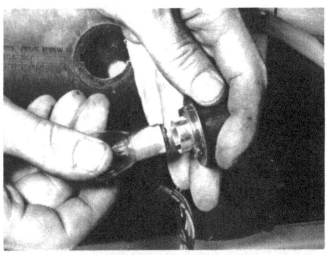

11.15　Remove the bulb by pressing in on the bulb while
turning it counterclockwise until the retaining tangs
are released

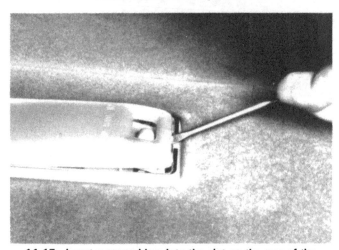

11.17　Insert a screwdriver into the slot on the rear of the
lens and pry the lens off

Tail lights
12　Remove the luggage compartment rear trim piece (Chapter 11).
13　Note the bulb assembly cluster and determine the location of the bulb to be replaced (see illustration).
14　Each connector is marked with an arrow to show the locked position (see illustration). Reverse the locking arrow position (counterclockwise) and release the electrical connector/bulb assembly.
15　Grasp the electrical connector/bulb assembly and push in on the bulb and turn the bulb counterclockwise to release the retaining tangs (see illustration).
16　Installation is the reverse of the removal procedure.

Interior

Dome light
17　Pry the lens off by inserting a screwdriver into the slot on the rear of the lens (see illustration).
18　Remove the bulb by pulling it from the spring contact/retainer.
19　Install the bulb by placing the bulb between spring contact/retainer and pressing it into place.
20　Install the lens by pressing it into place, making sure it the slot is facing rearward.

Convenience/map light
21　Grasp the convenience/map light fixture and pull down to release it from the overhead (see illustration).

11.21　Grasp the convenience/map light fixture and pull it
down from the overhead

11.23 Remove the screws (arrows) to separate the front face from the back of the fixture

11.24 The bulbs (arrows) can be replaced by pulling them from the springs/contacts

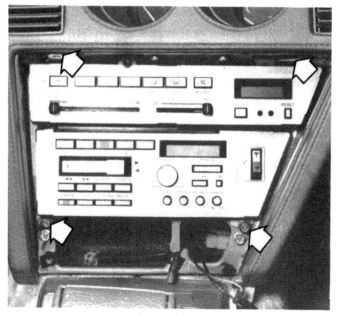

12.3 Remove the screws securing the radio and air control assembly to the dash (arrows)

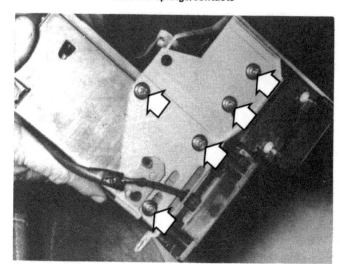

12.5 Remove the screws holding the radio to its bracket on the left side (arrows)

22 Disconnect the electrical connector by pressing down on the electrical connector retaining tab and pulling the connector apart.
23 Lay the convenience/map light face down and remove the screws to separate the face from the back of the fixture (see illustration).
24 Grasp the bulb and pull it from the spring retainer/contact (see illustration).
25 Reverse the removal procedure for installation.

Door safety light
26 Remove the screws and lift the lens off the door.
27 Grasp the bulb and pull it out of the spring contact/retainers.
28 Installation is the reverse of the removal procedure.

Floor convenience light — right and left side
29 Remove the appropriate side lower side dash panel trim piece (Chapter 11).
30 Grasp the electrical connector and turn counterclockwise.
31 Pull the bulb straight out of the connector.
32 Installation is the reverse of the removal procedure.

Control panel trim illumination
33 Remove the control panel trim piece (Chapter 11).
34 Pull the bulb from the electrical connector.
35 Installation is the reverse of the removal procedure.

12 Radio — removal and installation

Refer to illustration 12.3 and 12.5

1 Disconnect the negative cable at the battery. Place the cable out of the way so it cannot accidentally come in contact with the negative terminal of the battery, as this would once again allow power into the electrical system of the vehicle.
2 Remove the control panel trim piece (Chapter 11).
3 Remove the retaining screws (see illustration), two on top and two on bottom, securing the radio and air control assembly to the dash.
4 Slide the radio and air control assembly forward and disconnect the electrical connectors and the antenna wire.
5 Remove the screws retaining left side of the radio to the radio and air control assembly bracket (see illustration).
6 Disconnect the right side of the radio by removing the four screws located on the side and one screw located on the bottom of the radio bracket.
7 Lift the radio from the radio and air control bracket.
8 Install the radio by placing it into the radio and air control bracket and securing it to the bracket with the screws.
9 Connect all electrical connectors and the antenna cable, then place the radio and air control assemby in place and secure it with the four retaining screws.
10 Install the control panel trim piece.
11 Install the battery cable.

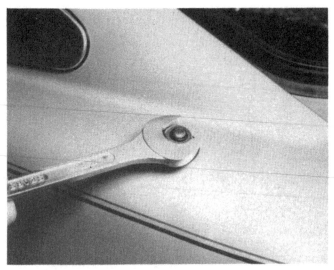

13.7   Remove the antenna mast retaining nut from the fender

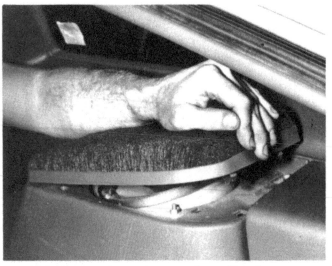

14.1   Remove the speaker cover by pushing the cover forward and lifting up on the back of the speaker cover

14.2   Remove the speaker screws (arrows) and lift the speaker up to remove the electrical leads attached to the back

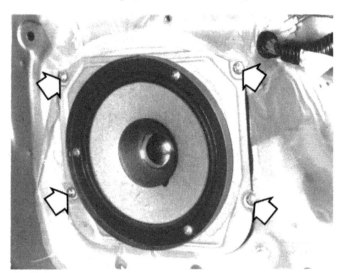

14.6   Remove the door speaker screws (arrows) to remove the speaker

13   Radio power antenna — removal and installation

Refer to illustration 13.7

1   Disconnect the negative cable at the battery. Place the cable out of the way so it cannot accidentally come in contact with the negative terminal of the battery, as this would once again allow power into the electrical system of the vehicle.
2   Remove the rear luggage compartment trim piece (Chapter 11).
3   Remove the drivers' side rear trim piece (Chapter 11).
4   Disconnect the antenna electrical connector by pressing down on the retaining tab and pulling the connector apart.
5   Disconnect the antenna cable by pulling it apart at the connection.
6   Remove the two screws holding the power antenna to the body.
7   Remove the antenna mast retaining nut (see illustration) from the fender panel.
8   Lower the power antenna from the fender and remove it from the vehicle.
9   Install the power antenna by placing the mast into the fender hole and loosely securing it with the mast retaining nut.
10 Loosely install the two lower retaining nuts securing the antenna to the body.
11 Tighten the mast retaining nut.
12 Tighten the two lower antenna retaining bolts.
13 Connect the antenna cable.

14 Connect the antenna electrical connector.
15 Install the side panel trim piece.
16 Install the rear luggage compartment trim piece.
17 Install the negative battery cable.

14   Speakers — removal and installation

Refer to illustrations 14.1, 14.2 and 14.6

Rear speakers

1   Push forward and pull up on the back of the speaker cover (see illustration).
2   Remove the three speaker screws (see illustration).
3   Raise the speaker up and disconnect the electrical connector.
4   Installation is the reverse of the removal procedure.

Door speakers

5   Remove the door panel (Chapter 11).
6   Remove the four screws (see illustration) holding the speaker to the door.
7   Lift the speaker out and disconnect the electrical connector.
8   Installation is the reverse of the removal procedure.

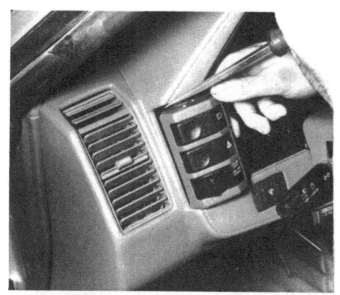

15.4 Carefully pry down on the top of the combination switch to release the tang on top of the switch

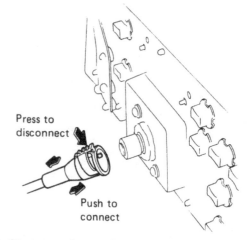

16.9 Disconnect the speedometer cable by pressing down on the tang and pulling the cable from the speedometer

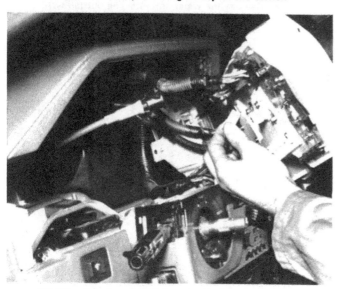

16.10a Pull the panel forward and disconnect all the electrical connectors

16.10b The back of the panel has each of the bulb/connectors labeled for indentification

15 Instrument panel combination switches — removal and installation

Refer to illustration 15.4

1   Disconnect the negative cable at the battery. Place the cable out of the way so it cannot accidentally come in contact with the negative terminal of the battery, as this would once again allow power into the electrical system of the vehicle.
2   Remove the lower trim panel on the driver's side (Chapter 11).
3   Look under the dash and locate the left instrument combination switch retaining nut. Remove the nut.
4   Carefully pry out the instrument panel combination switch by prying down on the top of the switch (see illustration) to release the tang on the switch. Pull the switch from the dash.
5   Push the tab on the side of the electrical connector and disconnect the connector.
6   Repeat Steps 3 and 4 for the right side combination switch.
7   Install the switches by placing the switch into its recess (bottom first) and start the retaining stud into its retaining bracket. Press the top of the switch into the panel recess.
8   Install the combination switch retaining nut.
9   Install the lower trim panel piece.
10 Connect the battery negative cable.

16 Instrument cluster panel — removal and installation

Refer to illustrations 16.9, 16.10a and 16.10b

1   Disconnect the negative cable at the battery. Place the cable out of the way so it cannot accidentally come in contact with the negative terminal of the battery, as this would once again allow power into the electrical system of the vehicle.
2   Remove the steering wheel (Chapter 10).
3   Remove the two instrument panel combination switches (Section 15).
4   Remove the screws retaining the upper steering column trim piece and remove it from the column.
5   Remove the four screws retaining the cluster panel trim piece to the dash. One is located on each of the lower corners and two are located on the top of the panel. Remove the trim piece.
6   Raise the vehicle and support it securely on jackstands.
7   Disconnect the speedometer cable from the transmission.
8   To gain enough room to release the speedometer cable, push the speedometer cable from the engine compartment towards the cluster.
9   Pull the cluster panel forward and tilt down the top of the panel. Reach behind the cluster panel and disconnect the speedometer by pressing the cable retaining tab and pulling the cable from the back of the speedometer (see illustration).
10 Pull the cluster panel from the dash and disconnect all the electrical connectors (see illustration) **Note:** *The back of the cluster panel has all of the bulbs labeled (see illustration). The large bulb connec-*

18.3 Grasp the cable housing and firmly pull it from the retaining bracket

19.2 Remove the wiper arm retaining nut (arrow)

tors are for the dash illumination and the small bulbs are for each of the sensors. To change the bulbs simply turn the connector and remove them from the panel.

11 Install the cluster panel by connecting all the electrical connectors disconnected during removal.

12 Reinstall the speedometer cable by pressing it into the back of the speedometer.

13 Install the cluster panel into the dash recess by first placing the bottom of the panel in and pressing the top of the panel into position. **Warning:** *Note the location of the trip meter reset knobs and do not break them off. Have an assistant pull the speedometer cable into the engine compartment so the cluster panel will settle into its dash recess.*

14 Install the cluster panel trim piece and secure it to the dash by installing the two lower corner screws, then press the top of the panel into place and install the two top screws.

15 Install the instrument panel combination switches.

16 Install the speedometer cable to the transmission and tighten the retaining ring.

17 Remove the jackstands and lower the vehicle.

18 Install the steering wheel.

19 Connect the negative battery cable to the battery.

20 Carefully check the operation of all instruments and warning lights before resuming normal vehicle operation

---

17 Speedometer — removal and installation

The speedometer is part of the instrument cluster panel and cannot be replaced as a separate unit. If failure occurs remove the instrument cluster panel and replace the entire unit.

Consult your local Nissan dealer regarding the mileage setting on the replacement speedometer, which must match the reading on the speedometer being replaced.

---

18 Speedometer cable — removal and installation

Refer to illustration 18.3

1 Remove the instrument cluster panel.

2 Working from the transmission end of the speedometer cable towards the speedometer head, open the hood and disconnect the cable plastic retainers.

3 From inside the vehicle, grasp the cable housing firmly (see illustration) and pull the cable and housing through the firewall.

4 Installation is the reverse of the removal procedure.

19.6 On the rear wiper, pry up the wiper arm retaining nut cover and remove the nut

---

19 Windshield wiper arm — removal and installation

Refer to illustrations 19.2 and 19.6

Front wiper arm

1 Open the hood.

2 Remove the wiper arm retaining nut (see illustration).

3 Note the location of the wiper arm and lift the arm off the wiper arm driveshaft.

4 Installation is the reverse of removal.

Rear wiper arm

5 Use a small screwdriver to pry up on the wiper arm retaining nut cover.

6 Remove the wiper arm retaining nut (see illustration).

7 Note the position of the arm for installation and lift up to remove it from the vehicle.

8 Installation is the reverse of the removal.

20.2 Remove the plastic drain screen by removing the
Phillips head screw (arrow)

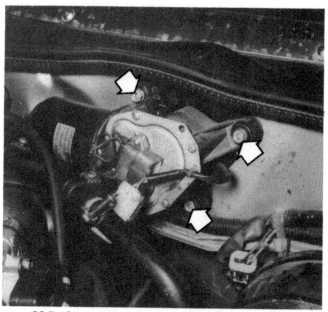

20.5 Remove the bolts (arrows) attaching the wiper
motor to the body

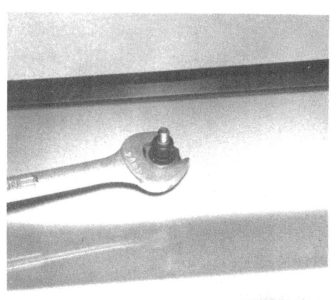

21.2 Remove the wiper motor large nut from the outside
of the hatch

21.5 Remove the wiper motor bolts (arrows)

20 Windshield wiper motor — removal and installation

Refer to illustrations 20.2 and 20.5

1 Raise the hood.
2 Remove the screw retaining the plastic drain screen located on the drain pan below the wiper arm (see illustration) and lift the screen out.
3 Look in the opening where the screen was removed and locate the linkage attached to the wiper motor. Use a screwriver to pry the linkage off.
4 Disconnect the two electrical connectors from the wiper motor.
5 Remove the three bolts attaching the wiper motor to the body (see illustration). Lift off the motor.
6 Install the wiper motor and secure the three retaining bolts.
7 Place the linkage onto the wiper motor driveshaft and press the linkage onto the shaft.
8 Connect the two electrical connectors.
9 Replace the plastic drain screen.

21 Rear wiper motor — removal and installation

Refer to illustrations 21.2 and 21.5

1 Remove the wiper arm.
2 Remove the large retaining nut on the outside of the hatch (see illustration).
3 Remove the rear hatch trim piece (Chapter 11).
4 Disconnect the electrical connector.
5 Remove the wiper motor retaining bolts (see illustration) and lift the motor from the hatch.
6 Install the motor and loosely install the large outside retaining nut.
7 Install the three wiper motor retaining bolts.
8 Connect the electrical connector.
9 Tighten the outside large retaining nut.
10 Install the wiper arm and secure the retaining nut.
11 Replace hatch trim piece.

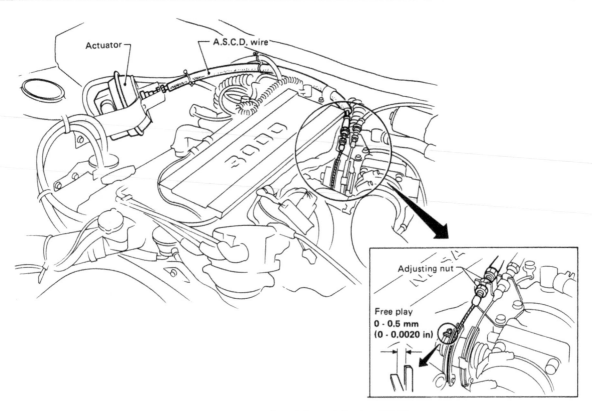

22.1 Automatic Speed Control Device (ASCD)

22 Cruise control — general information

Refer to illustration 22.1

1   The automactic speed control device (ASCD) is a cruise control system (see illustration) that allows the driver to maintain a constant highway speed without the necessity of continually applying foot pressure to the accelerator pedal.

2   The system utilizes a combination of electronic and vacuum controls. The ASCD controller senses the difference between the preset speed and the actual speed picked up by the speed sensor in the speedometer, and transmits an electrical signal to the servo valve. The servo valve then adjusts the vacuum which operates the actuator, which, in turn, adjusts the throttle valve opening to maintain the speed at the desired setting.

3   Due to the complexity of the ASCD system, if it begins to operate improperly, take the vehicle to your Nissan dealer or other qualified repair shop to have it diagnosed and corrected.

23 Steering column switches — removal and installation

Refer to illustrations 23.4, 23.8a and 23.8b

1   Remove the steering wheel (Chapter 10).

2   Remove the top steering column trim piece by removing the two retaining screws.

3   Remove the bottom steering column trim piece by removing the two retaining screws. Pull down on the trim piece to release the plastic retaining clip from the column. Disconnect the electrical connector from the trim piece.

4   Remove the screw retaining the switch assembly clamp to the column (see illustration).

5   Spread the switch assembly retaining clamp and push it as far to the left as possible to get the clamp off the retaining tab. Lift up on the switch assembly to release the retaining tab from the hole in the column and pull the assembly forward.

6   Unplug the electrical connectors from the back of the switch assembly.

7   Remove the lower trim panel (Chapter 11) and trace the small wires

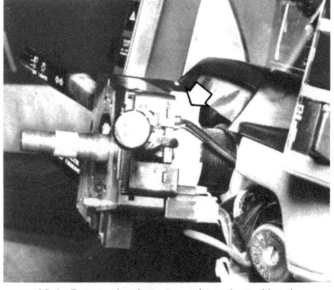

23.4   Remove the clamp screw (arrow) attaching the switch base to the column

attached to the switch assembly to their connectors under the dash and disconnect them.

8   To replace the individual switches remove the retaining screws located on the front of the switch assembly (see illustration). Slide the component from the assembly base (see illustration). Replacement is the reverse of the removal procedure.

9   Installation begins by connecting the large electrical connectors into the back of the switch assembly.

10 Place the switch assembly on the steering column shaft and connect the small electrical connectors.

11 Install the assembly clamp retaining screw. Make sure the assembly retaining tab is in the notch in the column and tighten the retaining screw.

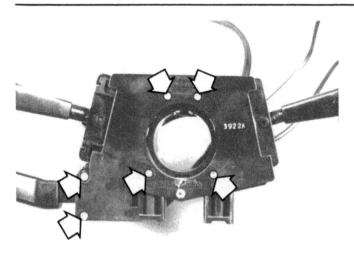

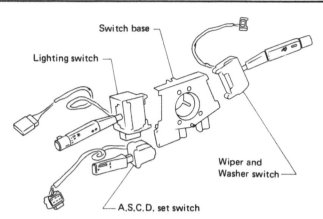

23.8a Remove the individual switches from the base by removing the component retaining screws (arrows) located on the front of the switch

23.8b The switch assembly is made up of four individual components

24.6 The steering lock/ignition switch self-shear screws (arrows) will have to be carefully drilled out

12 Replace the bottom column trim piece and secure with the two retaining screws.
13 Replace the top column trim piece and secure with the two retaining screws.
14 Replace the steering wheel.

24 Ignition switch — removal and installation

Refer to illustration 24.6

1 Obtain two self-shear screws from your Nissan dealer, needed to reinstall the steering lock.
2 Remove the steering wheel (Chapter 10).
3 Remove the top and bottom steering column covers attached to the column by two retaining screws each.
4 Remove the instrument panel combination switches.
5 Remove the trim panel from the cluster panel by removing the four retaining screws.
6 The steering lock/ignition switch assembly is secured to the steering shaft with two self-shear screws (see illustration). Removing the shear screws will involve very careful drilling of the screws.
7 Disconnect the ignition switch electrical connector.
8 To install, align the mating surface of the steering lock with the

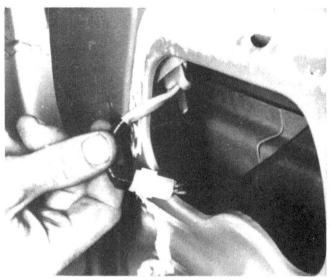

26.5 Disconnect the electrical connector and remove the wire body strap

hole in the steering column tube. Then loosely install the two self-shear screws. Once the operation of the lock is checked using the key, these screws can be tightened making sure the tops snap off.
9 Install the instrument cluster panel trim piece.
10 Install the instrument panel combination switches.
11 Install the top and bottom steering column trim pieces.
12 Install the steering wheel.

25 Remote control console switches — removal and installation

1 Remove the console (Chapter 11).
2 Turn the console over and press in on the two tangs securing the remote switch to the console and push the switch through the console.
3 Disconnect the switch electrical connector.
4 Install by pushing the switch into the console until the tangs catch.
5 Install the console.

26 Remote control door mirror — removal and installation

Refer to illustration 26.5

1 Lower the window.
2 Remove the door trim panel (Chapter 11).
3 Pry off the plastic cover that is held in place by three pop out type fasteners.
4 Remove the three screws and lift the mirror off.
5 Disconnect the electrical connector and remove the wire body strap (see illustration).

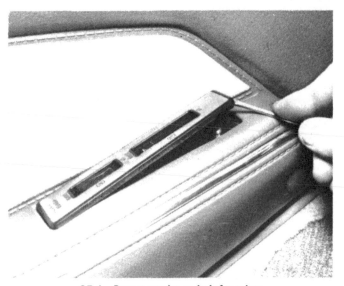

27.1   Pry up on the switch faceplate

6   Install the electrical connector and attach the wire body strap.
7   Secure the mirror to the door with the three screws.
8   Press the plastic face cover into place.
9   Install the door trim panel.

27   Power window switch — replacement

Refer to illustration 27.1

1   Open the door and use a small screwdriver to pry up the switch face plate (see illustration).
2   Remove the two switch screws.
3   Lift the switch assembly from the arm rest and disconnect the electrical connector.
4   Installation is the reverse of the removal procedure.

28   Electric door lock actuating solenoid — removal and installation

1   Raise the window.
2   Remove the door panel (Chapter 11).
3   Disconnect the door lock actuating rod by disconnecting the yellow plastic fastener.

4   Remove the two bolts securing the solenoid to the door.
5   Remove the solenoid and disconnect the electrical harness.
6   Installation is the reverse of the removal procedure. Make sure the yellow plastic fastener is secure on the actuating rod.

29   Windshield intermittent amplifier — general information

The windshield intermittent amplifier switch controls the speed of the windshield wiper while also controlling the windshield washer. Pull the lever toward you and hold it until there is sufficient water to wash the windshield. To operate the intermittent mode of the windshield wiping system, pull the lever down one notch. To adjust the intermittent time control knob, turn the end knob fully counterclockwise to activate the slowest intermittent operation of the wiper blade. This setting wipes the windshield every 12 seconds. For faster wiping action turn the knob clockwise. When the knob is turned fully clockwise the windshield will be wiped every four seconds.

30   Theft warning system — general information

The theft warning system has been designed to protect your vehicle and personal belongings when all the doors, hood and rear hatch are locked. If an attempt is made to open the doors, hood or rear hatch without using the key, the system will activate the headlights and horn while also making it impossible for the starter to operate. The headlights will flicker and the horn will sound intermittently. The alarm automatically stops in two to four minutes, but will activate again if the system is tampered with. If tampered with again the horn will sound and the starter will not work until the system is canceled by unlocking the door or hatch with a key. The retractable headlights will remain up after the alarm has stopped, making it necessary to retract the lights using the light retracting switch on the dash panel.
To arm the system, remove the key from the ignition switch, lock the doors and hatch and the warning light will flash for 30 seconds on the dash. If, during this time, the door is unlocked or ignition switch is turned to "On" or "Acc" the system will not arm. After the 30 second interval the system will be armed. The system will arm even if the window is left down. If this is the case it will be necessary to unlock the door using the key. If the door is unlocked by pulling up on the inside lock knob the alarm will go off.
Once the alarm is sounding the system can be stopped by unlocking the door or rear hatch with the key. The alarm will not go off by turning the ignition switch to "On" or "Acc".
If the system will not function properly, such as the warning light will not go off or the system will not arm itself, have the system checked by your Nissan dealer or other competent service facility.

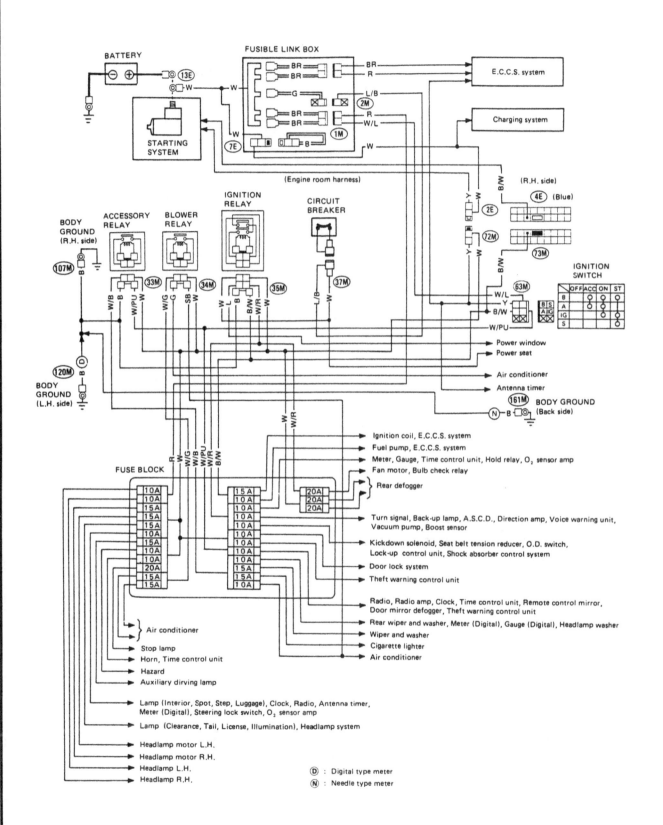

Power supply routing for Turbo models

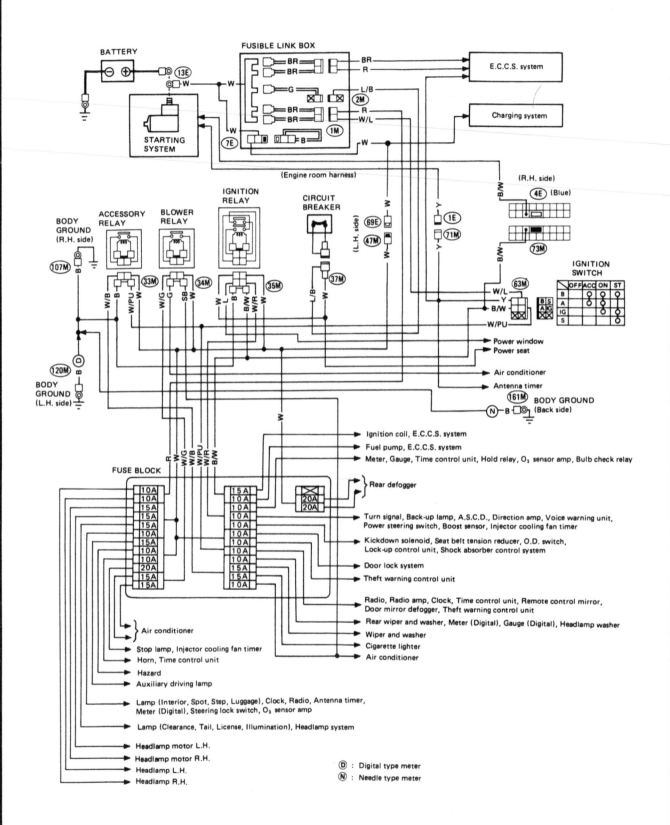

Power supply routing for Non-Turbo models

M/T MODELS

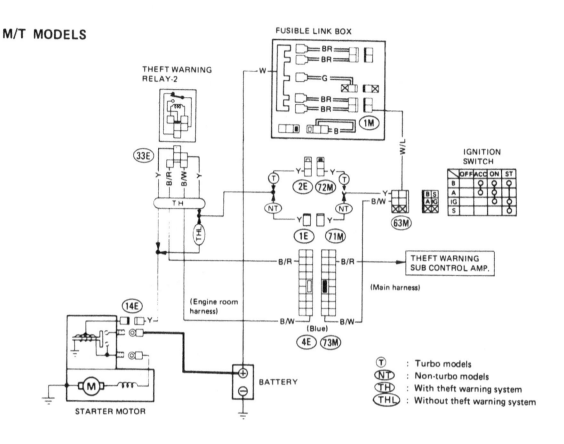

A/T MODELS

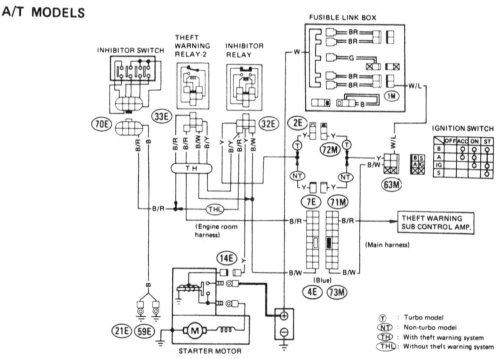

Starting system diagram with automatic and manual transmission models shown

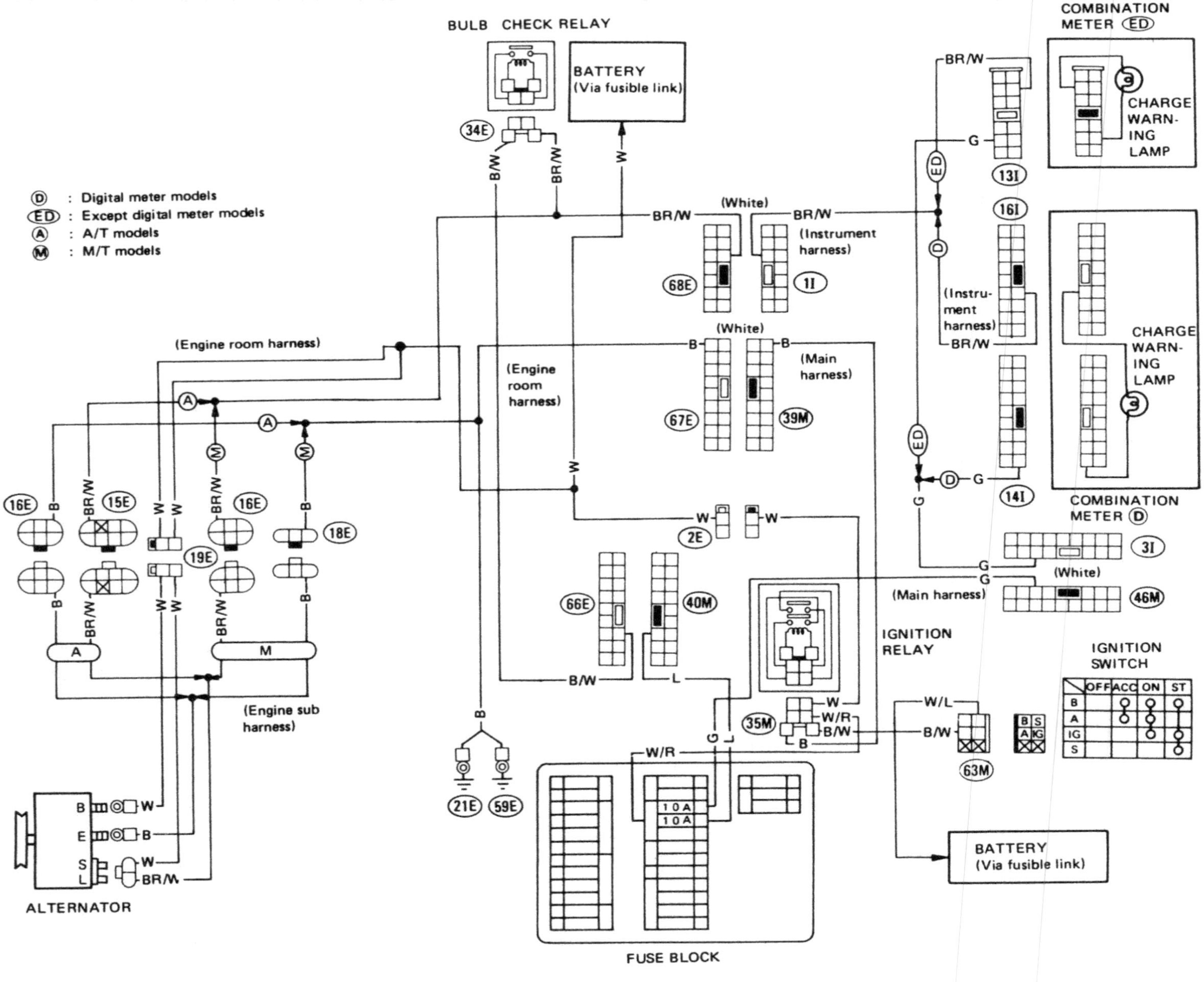

COMBINATION METER (ED)

CHARGE WARN-ING LAMP

CHARGE WARN-ING LAMP

COMBINATION METER (D)

31

46M

IGNITION SWITCH

	OFF	ACC	ON	ST
B				
A				
IG				
S				

B|S
A|IG
A|S

63M

BATTERY (Via fusible link)

131
161
141

BR/W
G

(Instrument harness)
BR/W

G

D
ED
G

(ED)
D

W/L
B/W

(Main harness)

IGNITION RELAY

W
W/R
B/W
B

35M

BR/W
(Instrument harness)

11

(White)

68E

BR/W

(Main harness)

B
39M

(White)

B
67E

W
W
2E

40M

66E

FUSE BLOCK

10A
10A

W/R

L

BATTERY (Via fusible link)

M

BULB CHECK RELAY

BR/W

B/W

34E

(Engine room harness)

M

B/W

B

59E

21E

(Engine room harness)

18E

M

B

B

M

(Engine sub harness)

16E

M
BR/W
BR/W

19E

W
W

M
M

15E

B
BR/W

B
BR/W

W
B
W
BR/W

A

16E

B

B

ALTERNATOR

B
E
S
L

Turbo model charging system

D : Digital meter models
ED : Except digital meter models
A : A/T models
M : M/T models

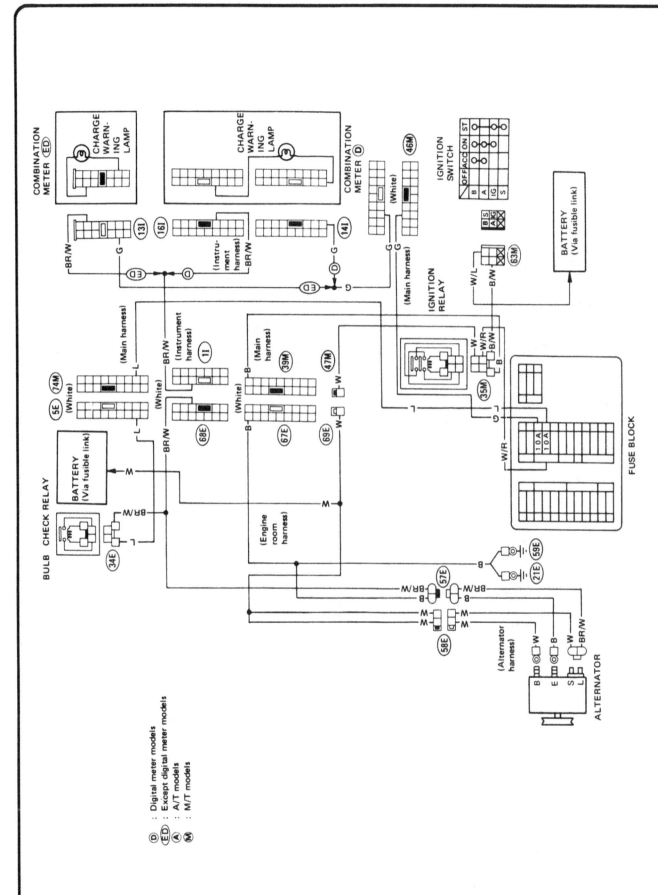

Non-turbo model charging system

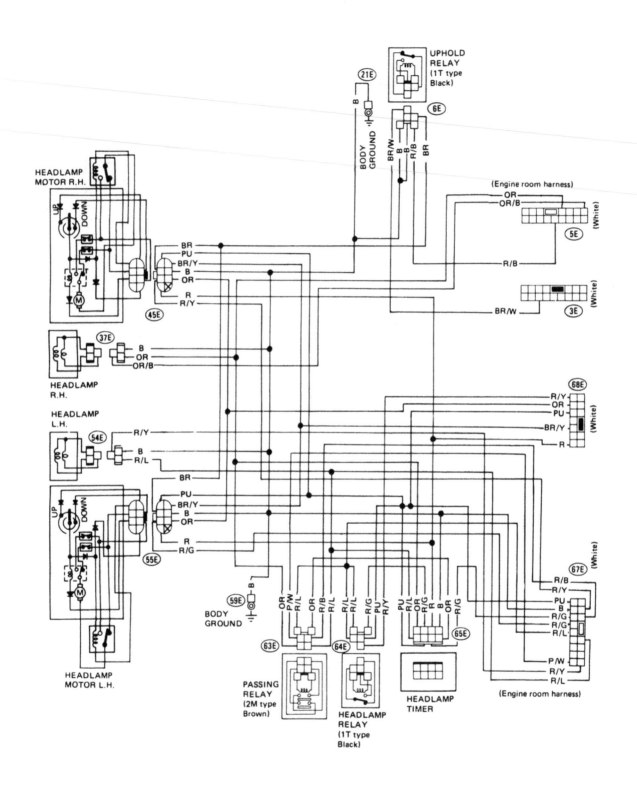

1985-86 Headlight diagram without headlamp sensor — SF model

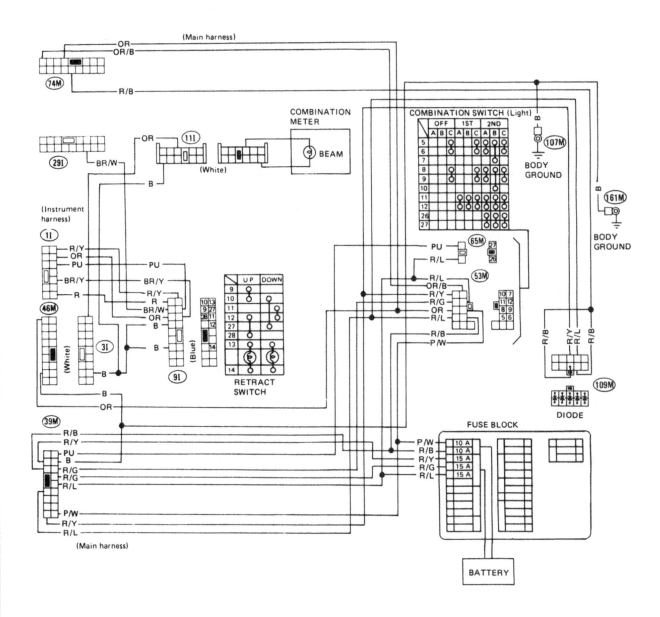

1985-86 Headlight diagram without headlamp sensor — SF model (Cont'd)

274

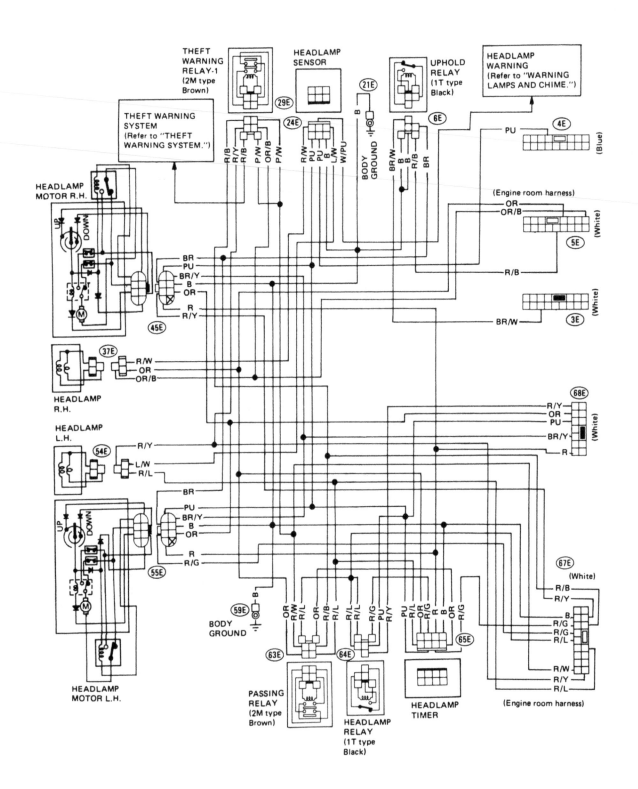

1985-86 Headlight diagram with headlight sensor — GL and GLL models

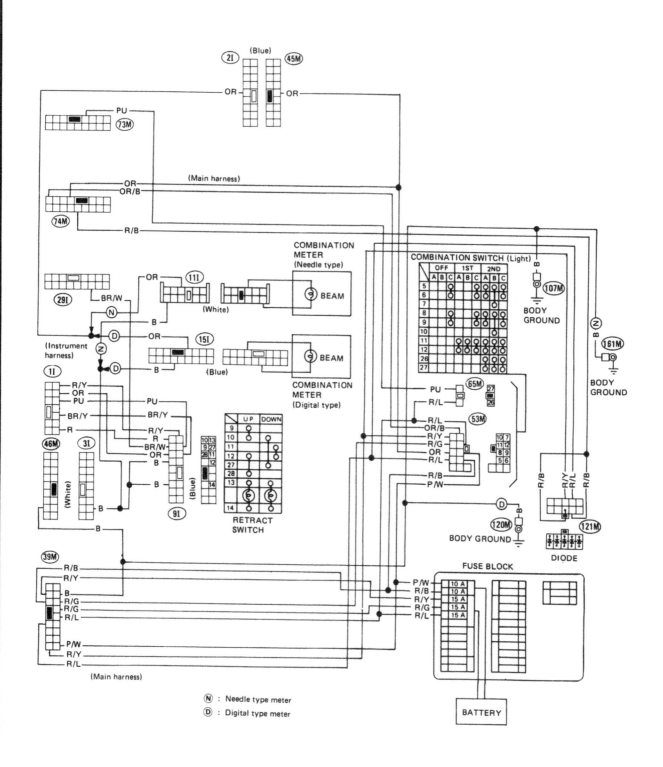

1985-86 Headlight diagram with headlight sensor — GL and GLL models (Cont'd)

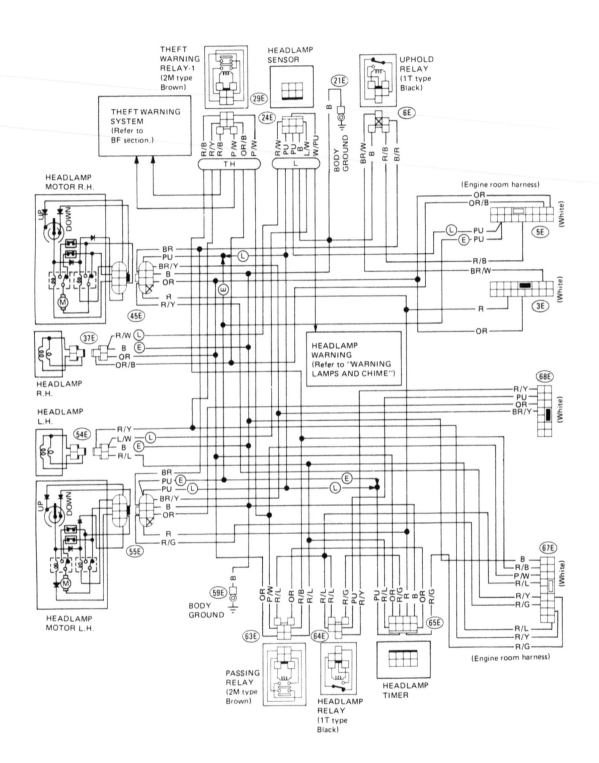

1984 Headlight diagram

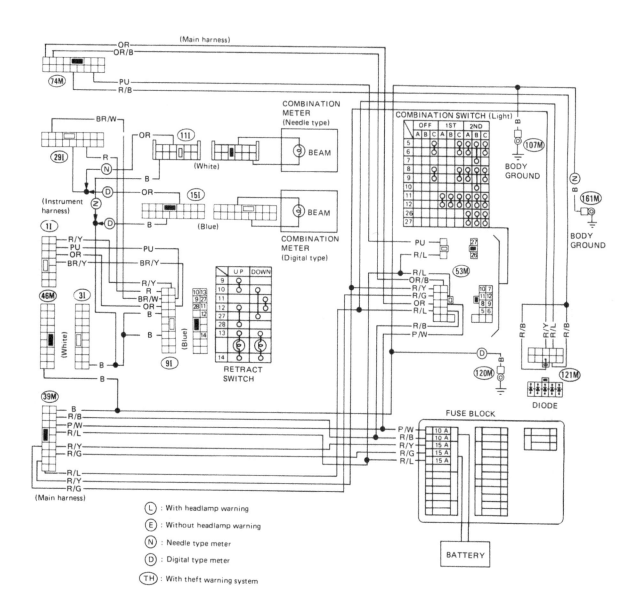

1984 Headlight diagram (Cont'd)

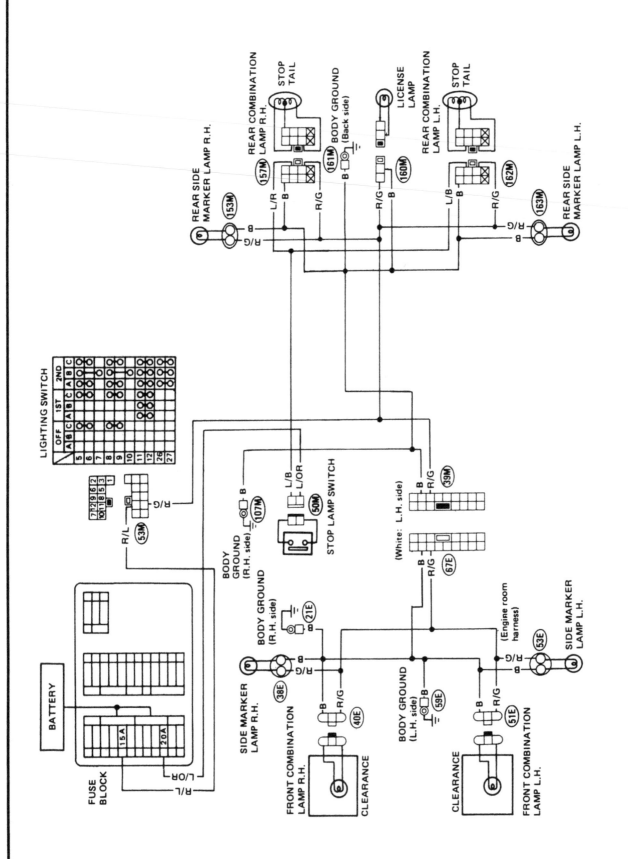

Exterior lamps without stop and tail lamp sensor (SF model) — clearance, license, tail and stop lamps

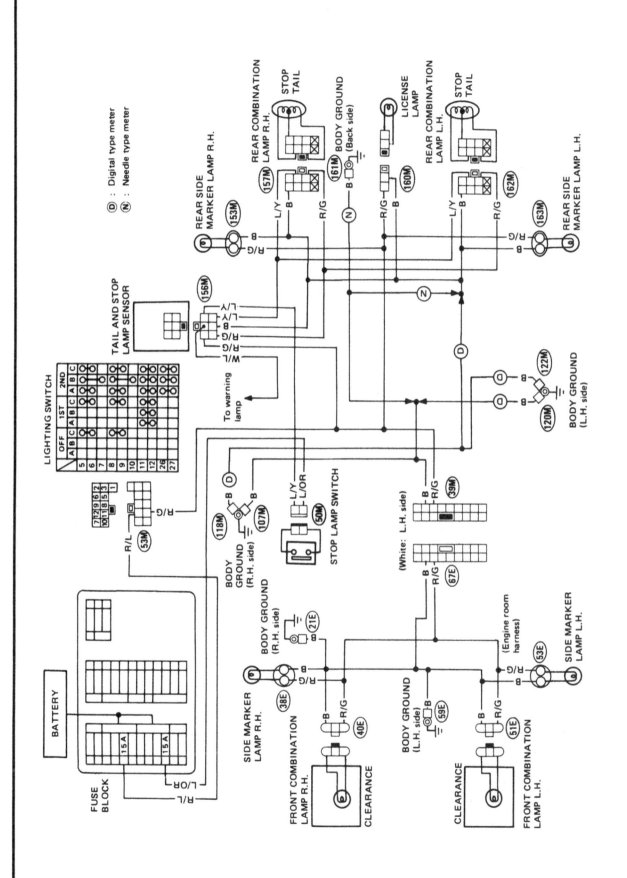

Exterior lamp with stop and tail lamp sensor (GL and GLL models)

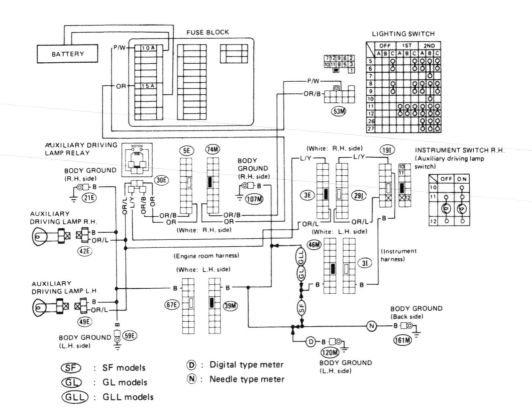

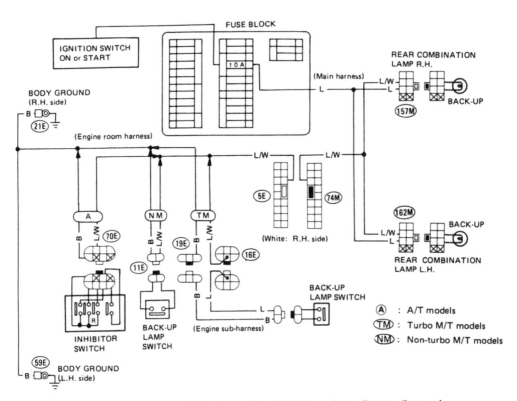

Auxiliary driving lamp (top) and back-up lamp diagram (bottom)

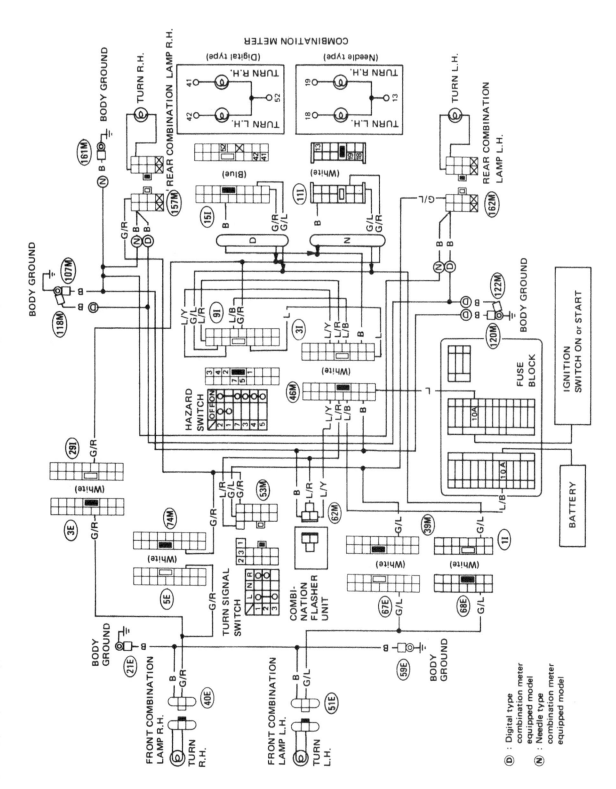

Turn signal and hazard warning lamps

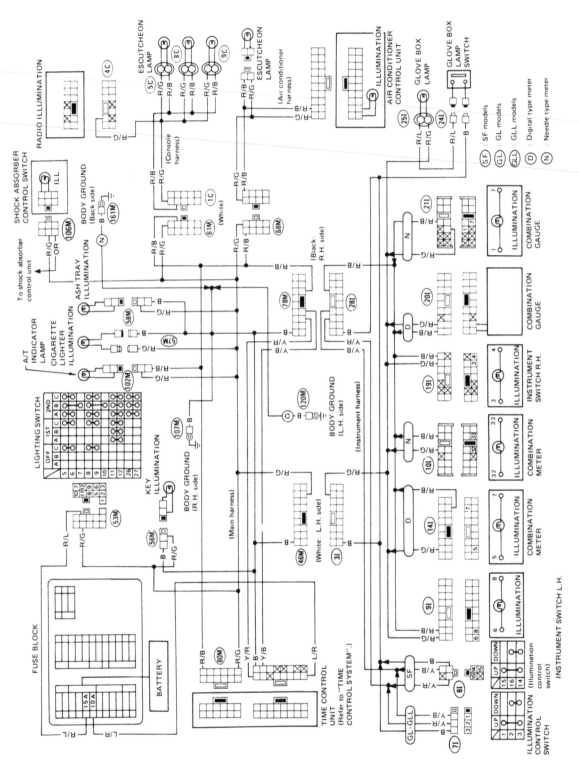

1984 Interior lamp illumination

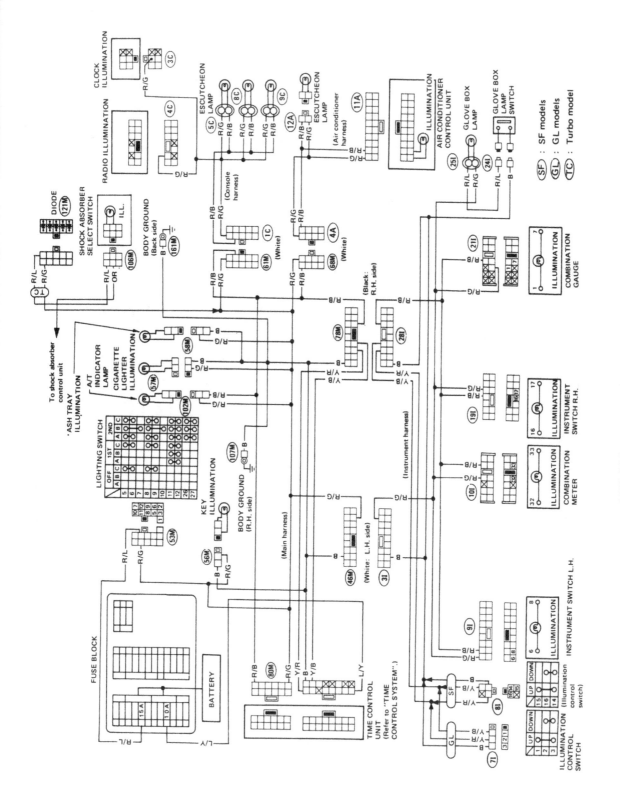

1985-86 Interior lamp illumination

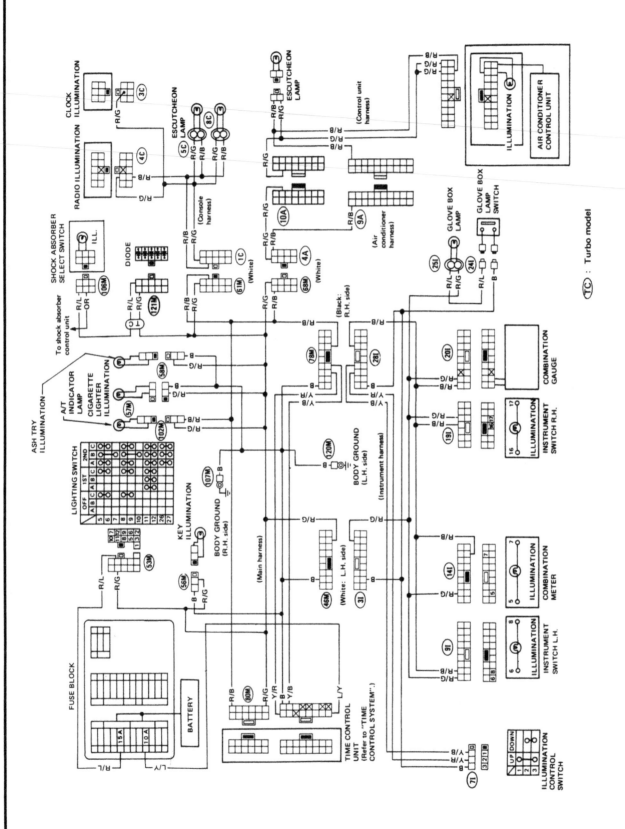

Interior lamp illumination if equipped with digital type combination meter (GLL model)

(TC) : Turbo model

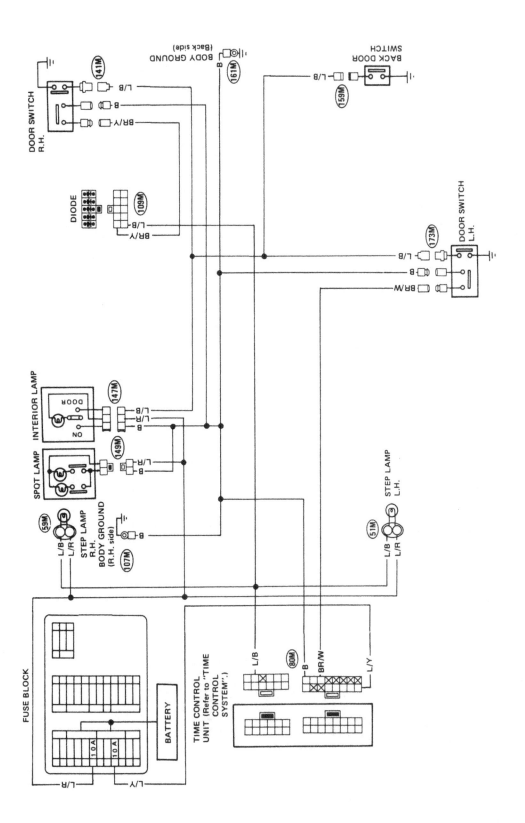

Interior, luggage and step lamps (SF model)

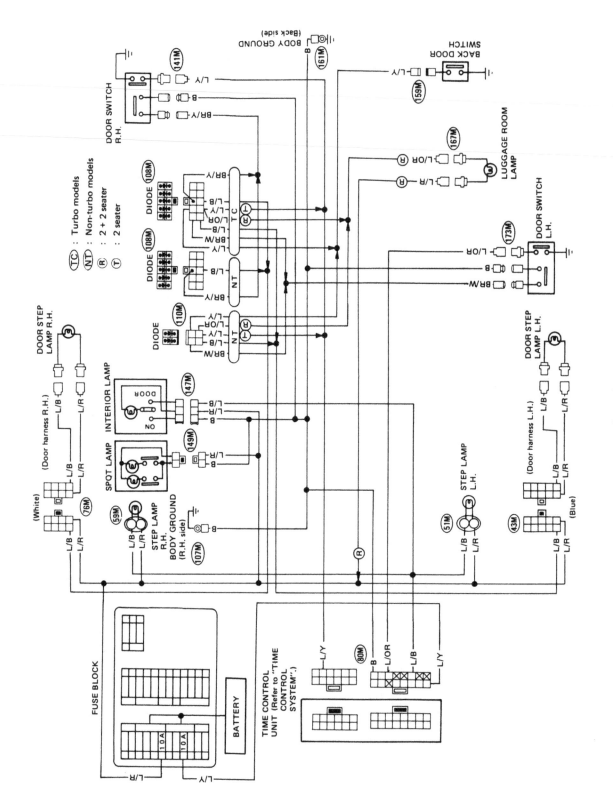

Interior, luggage and step lamps (GL model)

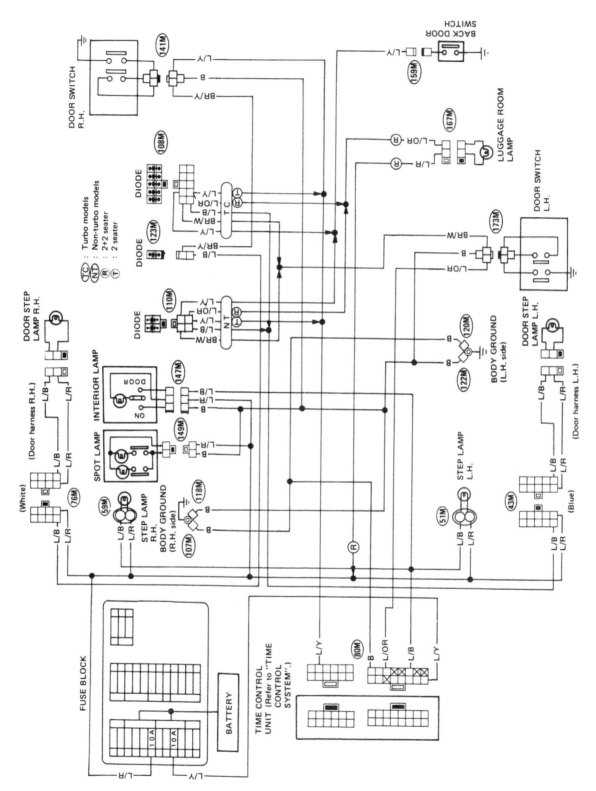

Interior, luggage and step lamps (GLL model)

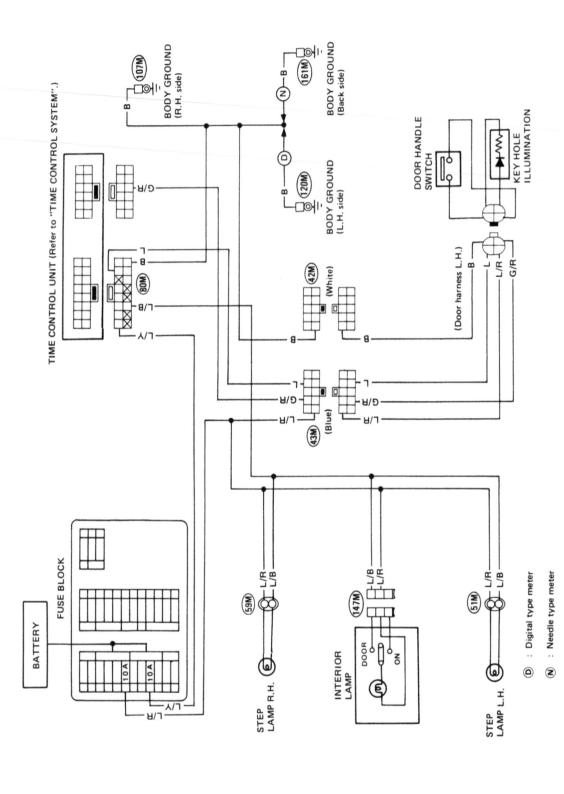

Door key and illuminated entry system

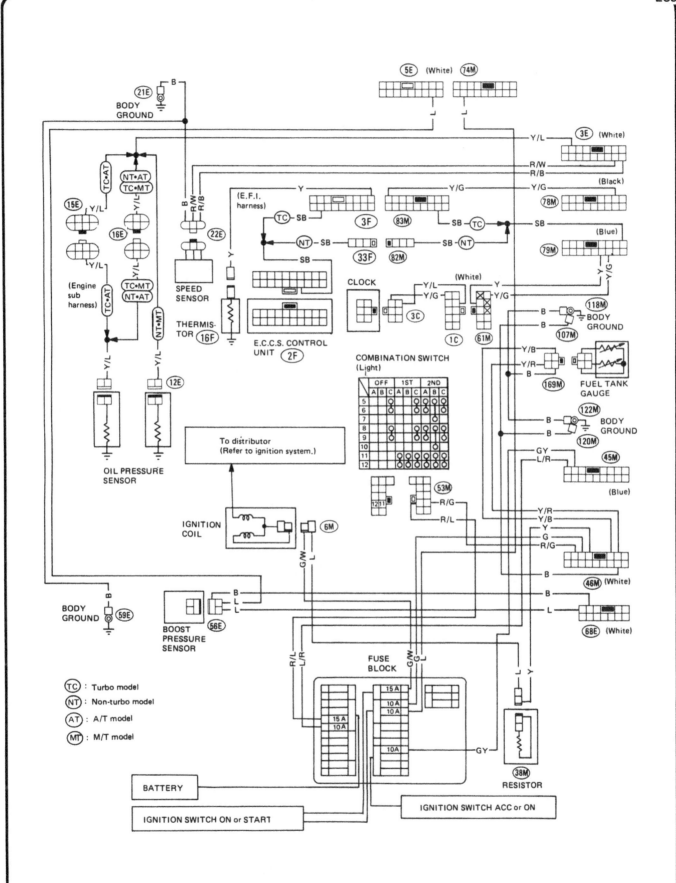

Digital type combination meter

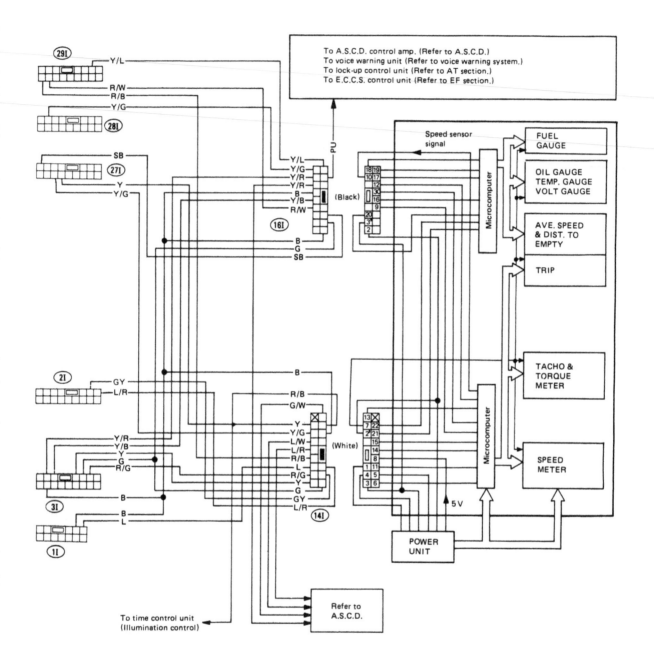

Digital type combination meter (Contd)

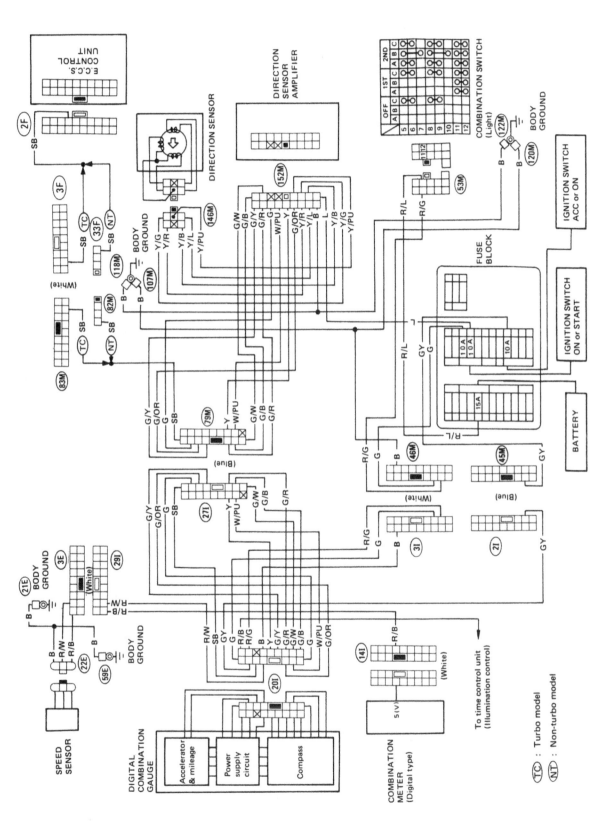

Digital type combination gauge

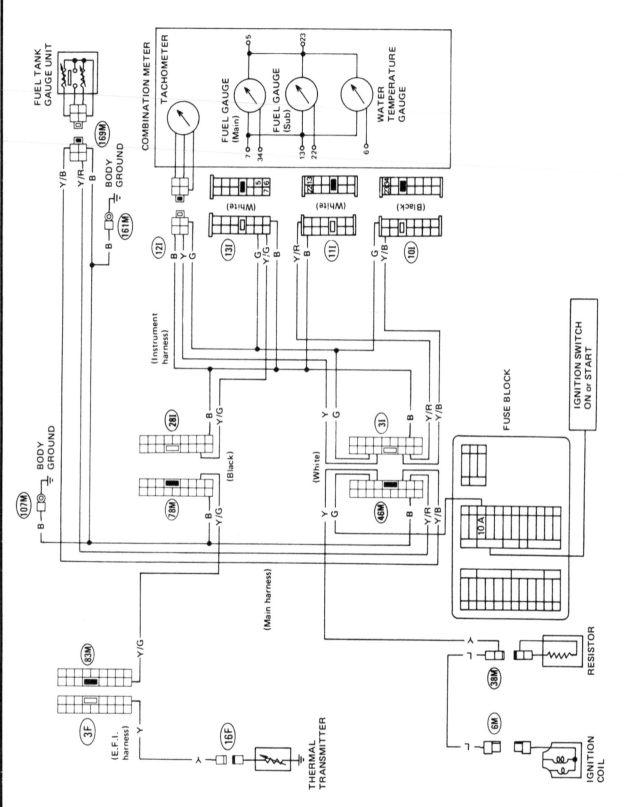

Needle type combination meter wiring diagram showing the tachometer, fuel and water temperature gauges

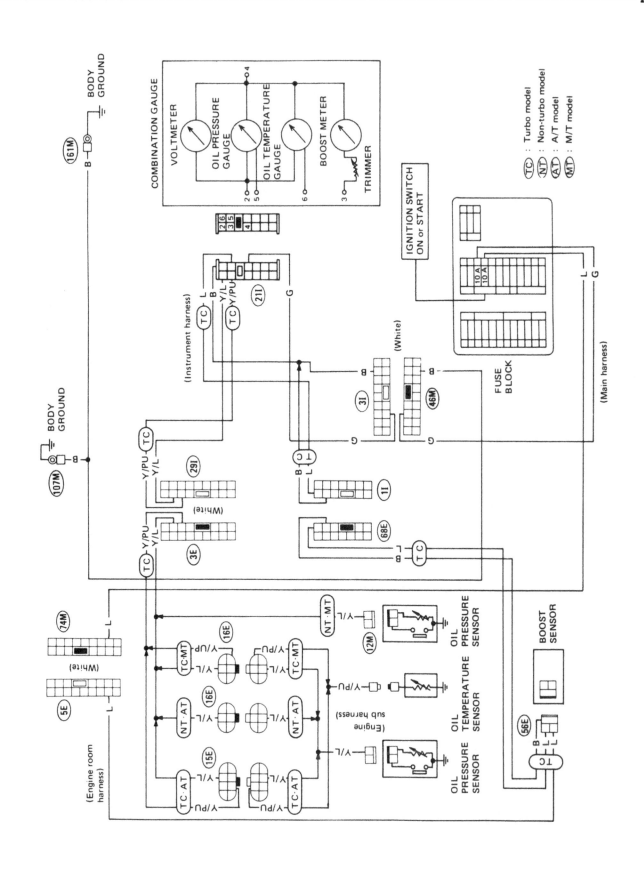

Needle type combination gauge — oil temperature, oil pressure, boost and volt gauges

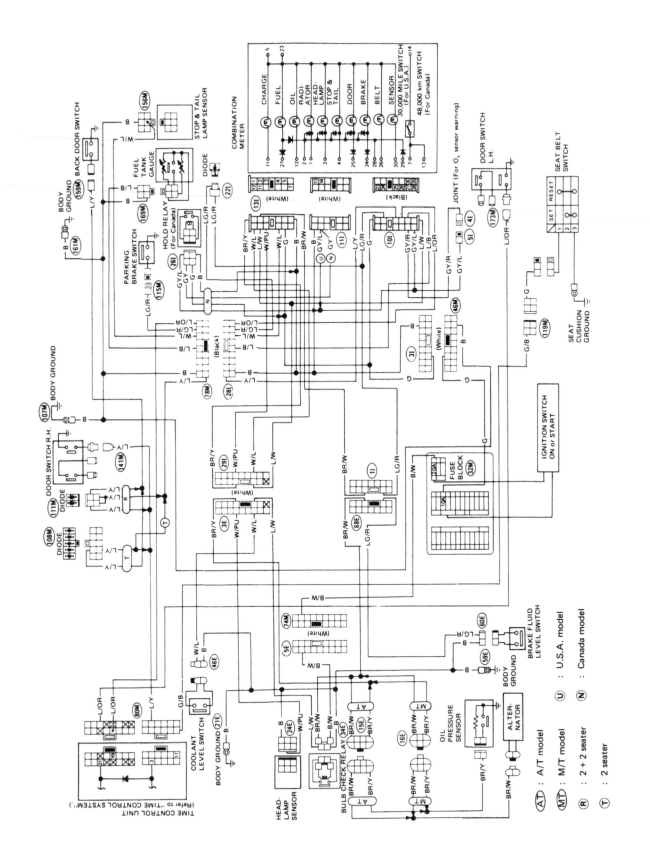

Turbo model (GL) with needle type combination meter warning lamps

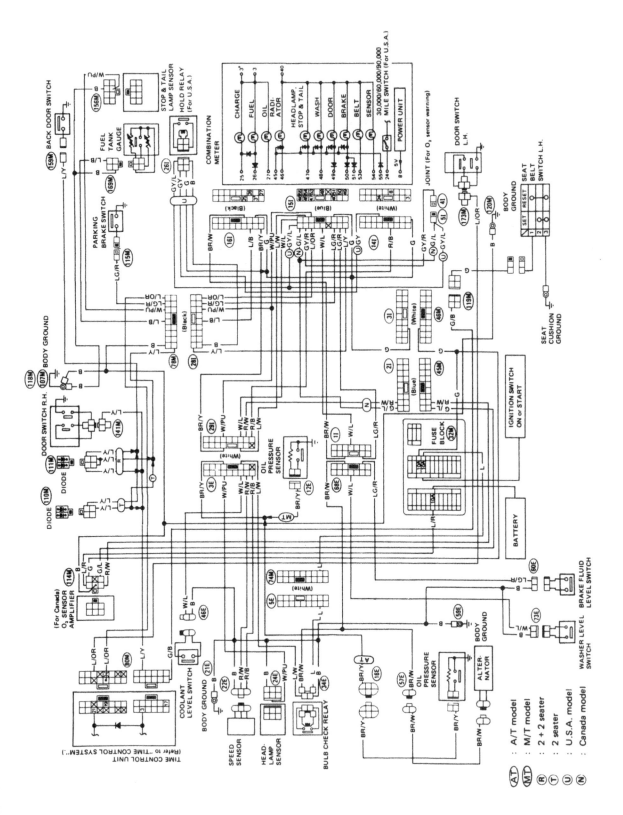

Non-turbo model (GLL) with digital type combination meter warning lamps

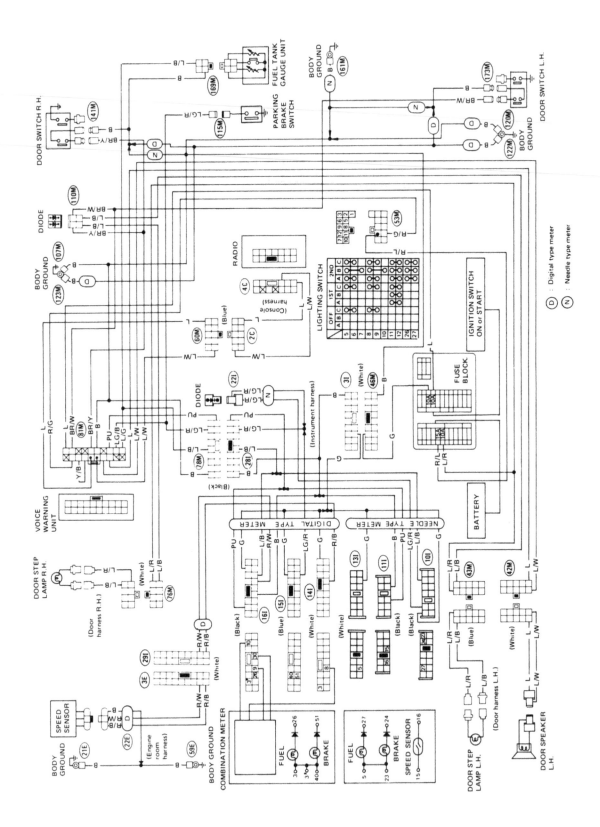

1984 voice warning system

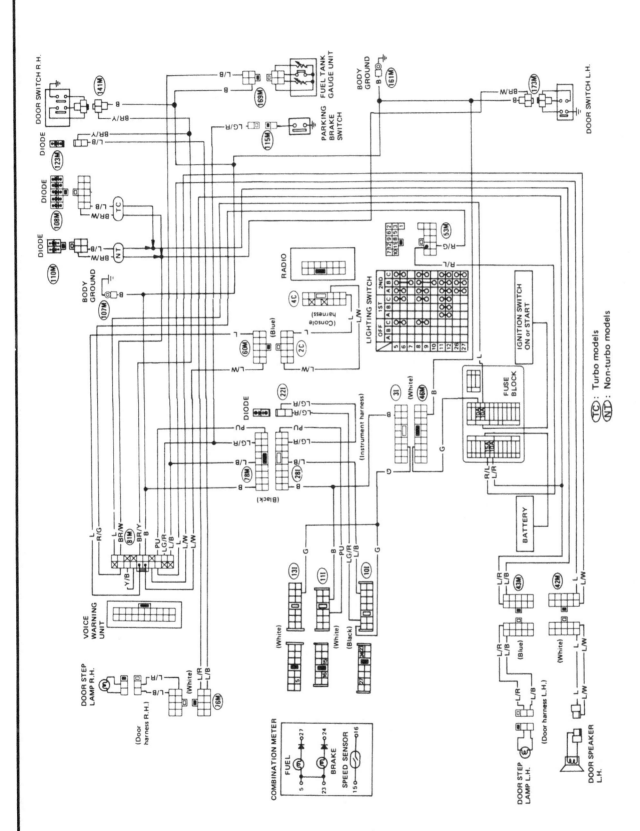

Typical 1985-86 needle type combination meter voice warning system

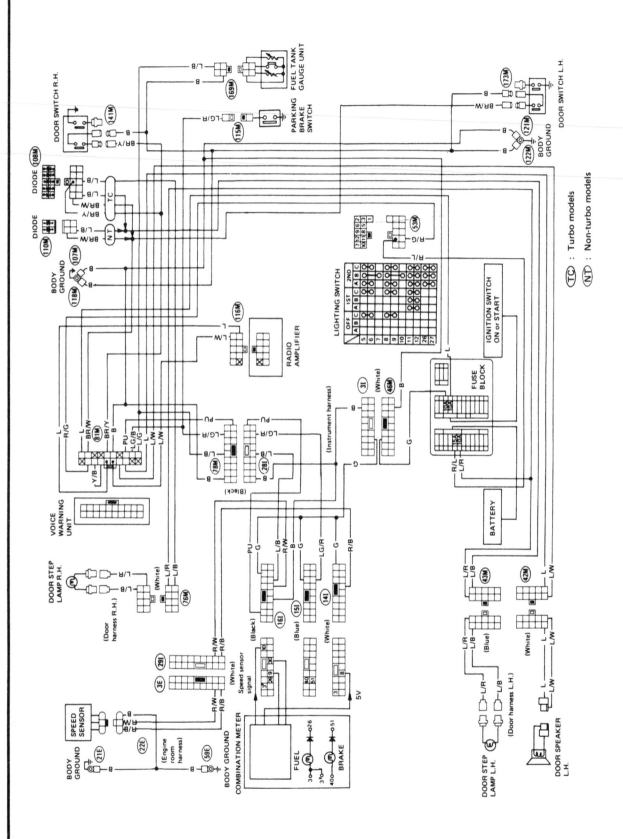

1985-86 Needle type combination meter voice warning system (Cont'd)

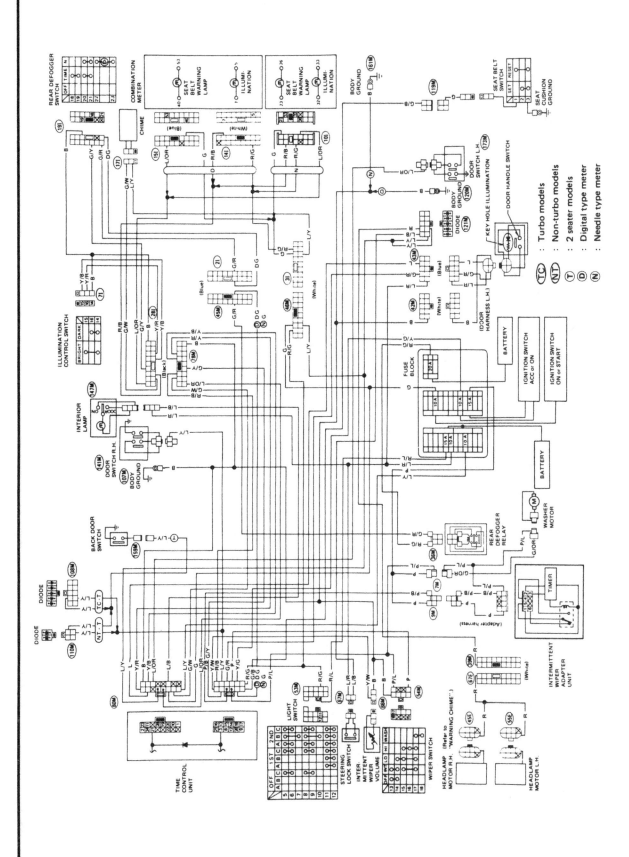

Time control system on GL and GLL models

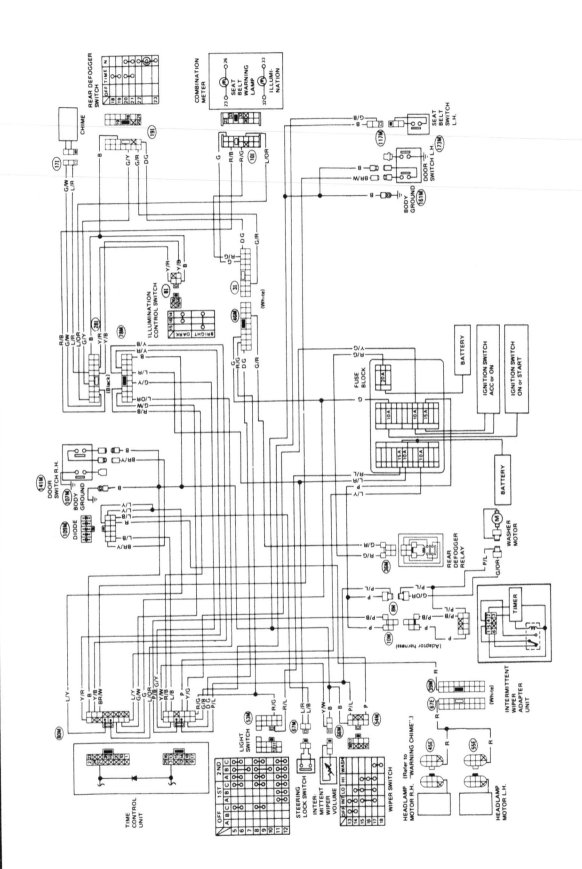

SF model time control diagram

WITHOUT HEADLAMP SENSOR

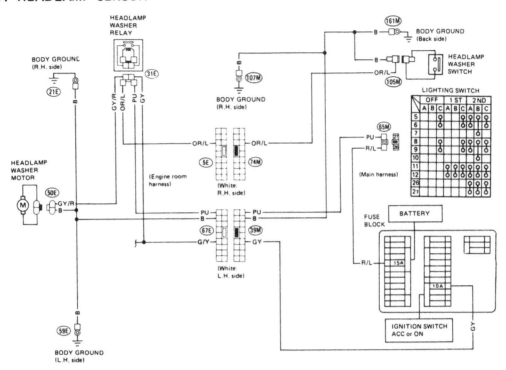

WITH HEADLAMP SENSOR

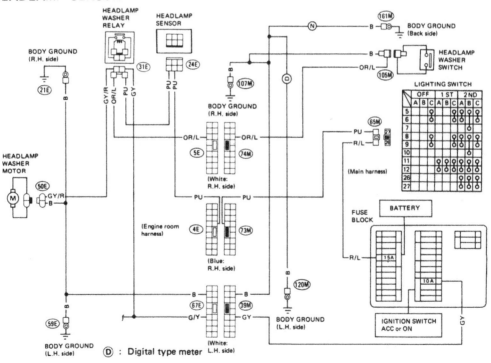

Ⓓ : Digital type meter

Ⓝ : Needle type meter

Front headlight washer system with or without a headlamp sensor

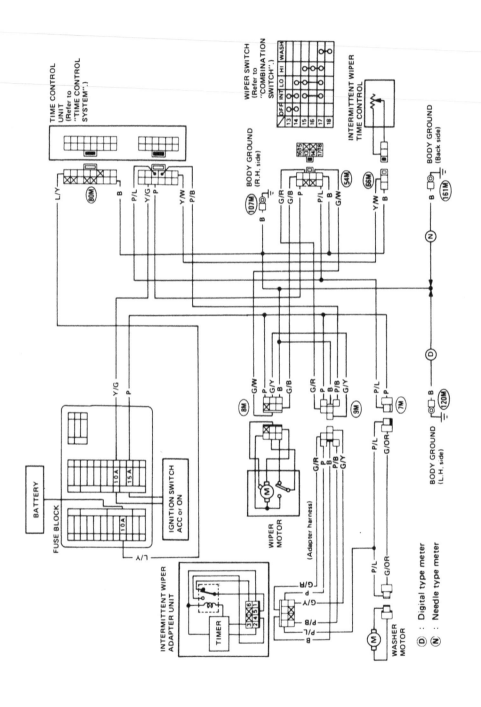

Windshield wiper and washer wiring diagram

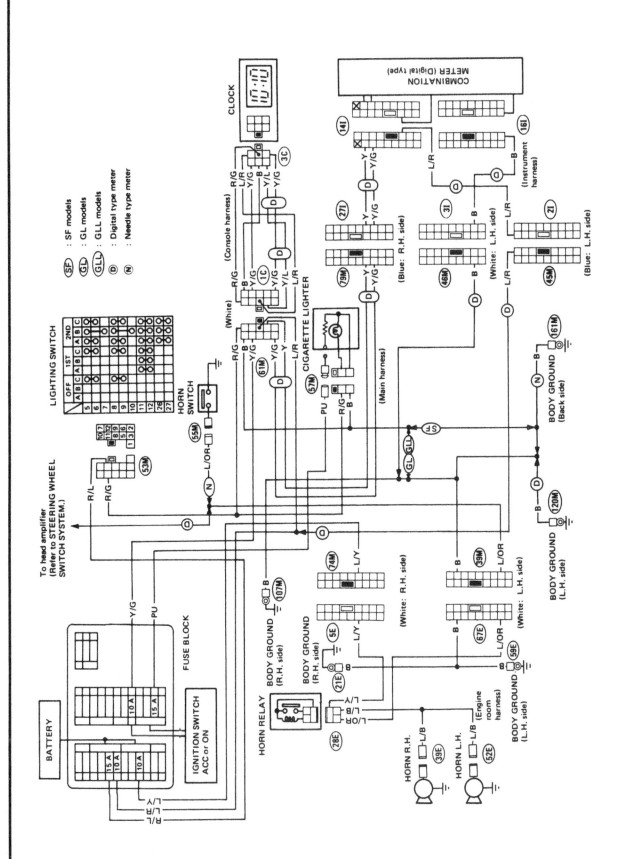

Horn, cigarette lighter and clock diagram

SF AND GL MODELS

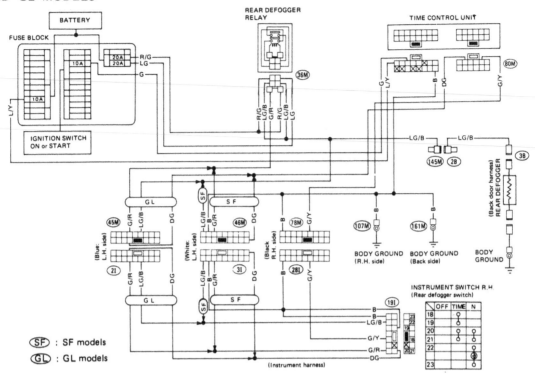

SF : SF models
GL : GL models

GLL MODEL

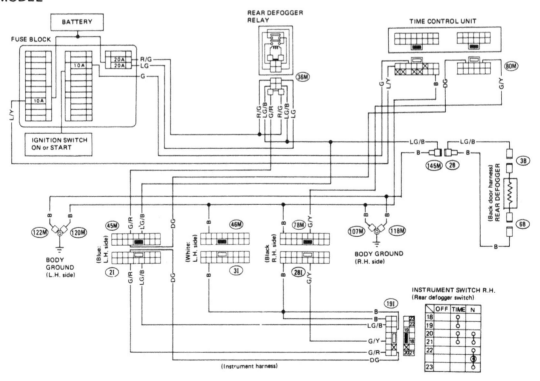

Rear window defogger; SF and GL models are shown together and the GLL model is separate

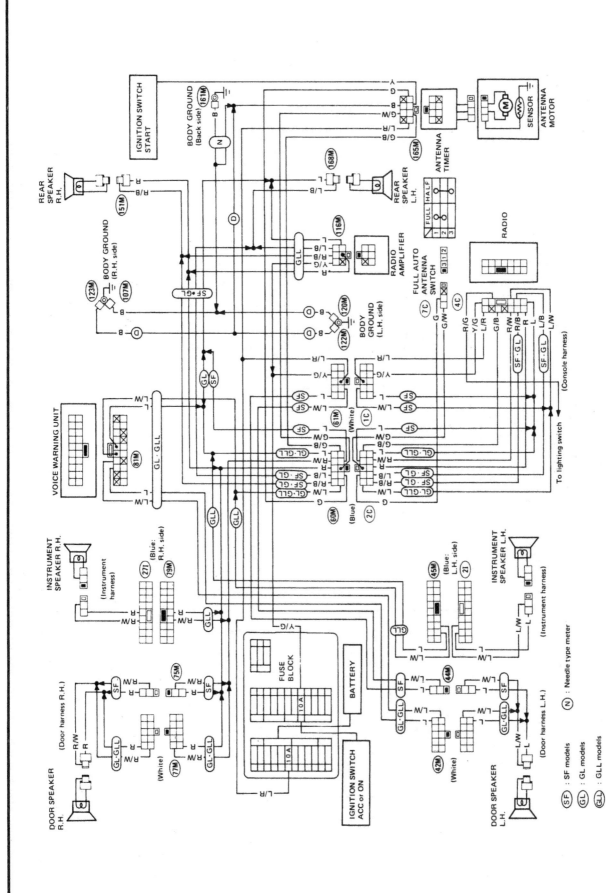

Audio and power antenna diagram

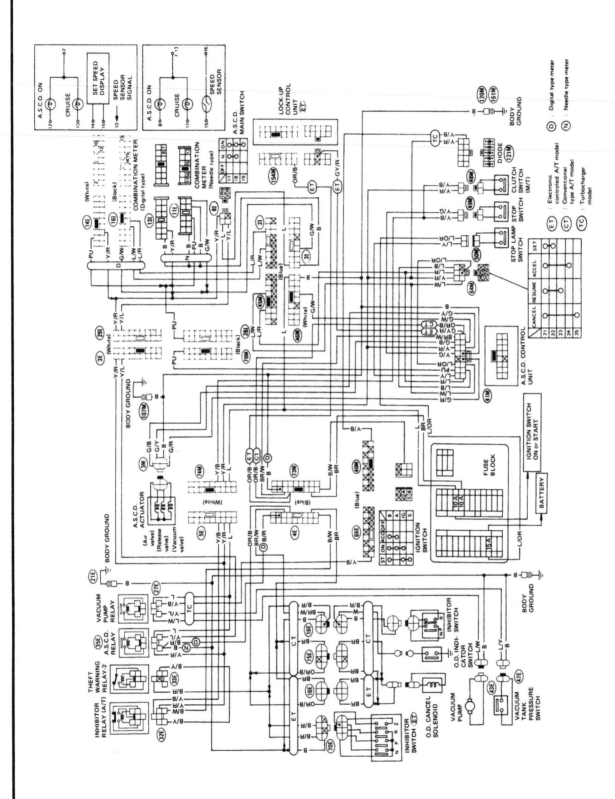

Automatic speed control device (ASCD) typical wiring diagram

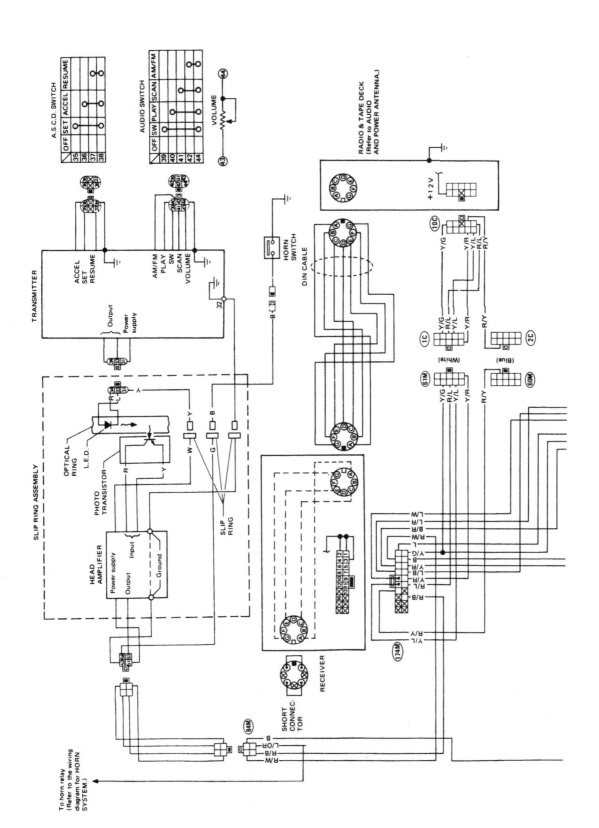

Steering wheel switch assembly

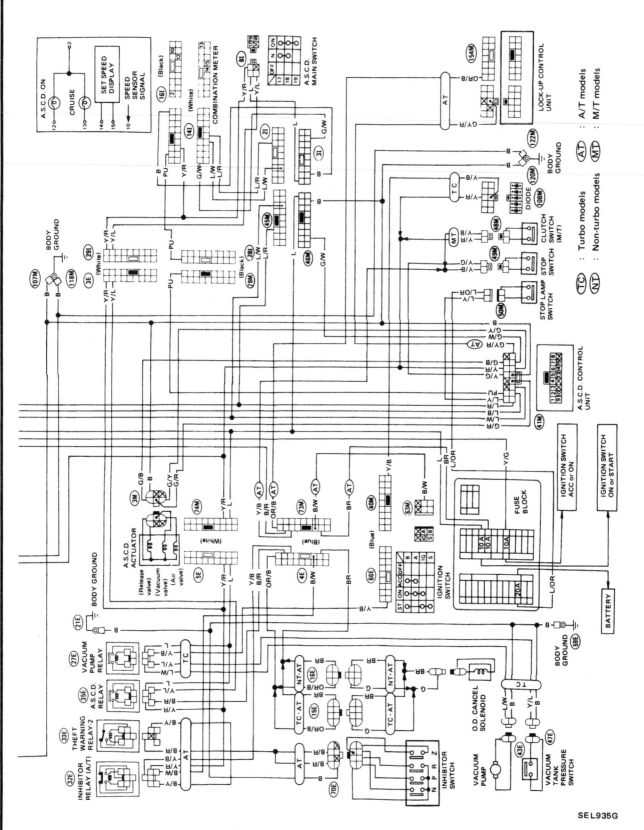

Steering wheel switch assembly (Cont'd)

SEL935G

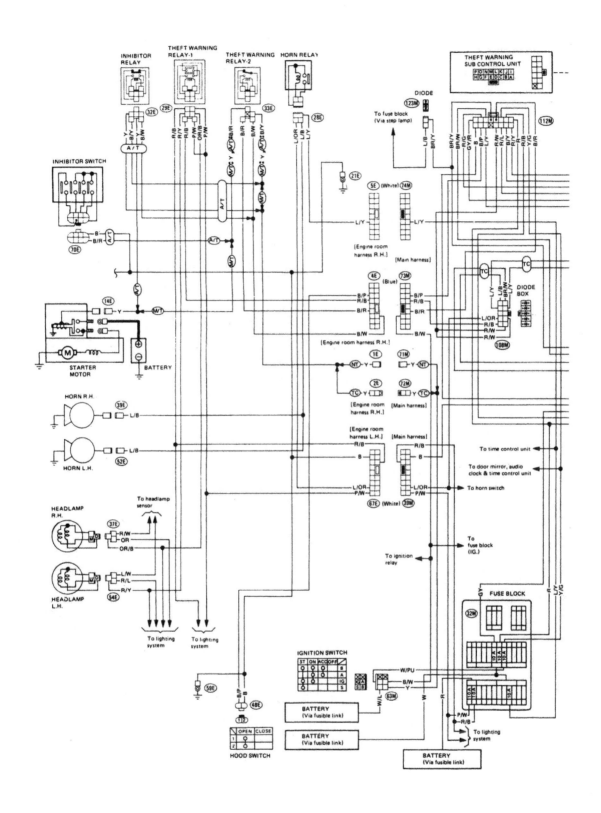

Theft warning system wiring diagram

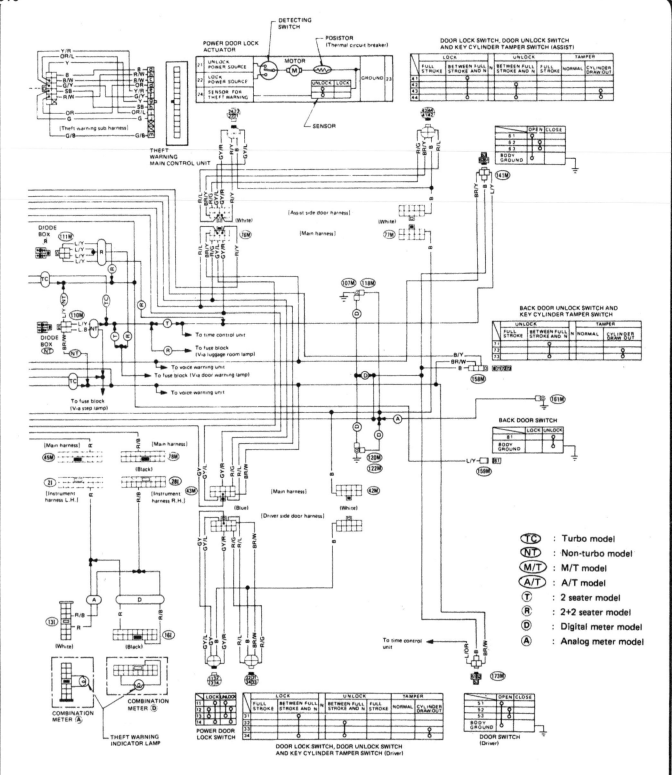

Theft warning system wiring diagram (Cont'd)

SEL936G

Index

Notes

Service record

Date	Mileage	Work performed

Service record

Date	Mileage	Work performed

Haynes Automotive Manuals

NOTE: If you do not see a listing for your vehicle, please visit **haynes.com** *for the latest product information and check out our* ***Online Manuals!***

ACURA
12020 Integra '86 thru '89 & Legend '86 thru '90
12021 Integra '90 thru '93 & Legend '91 thru '95
Integra '94 thru '00 - *see HONDA Civic (42025)*
MDX '01 thru '07 - *see HONDA Pilot (42037)*
12050 Acura TL all models '99 thru '08

AMC
14020 Mid-size models '70 thru '83
14025 (Renault) Alliance & Encore '83 thru '87

AUDI
15020 4000 all models '80 thru '87
15025 5000 all models '77 thru '83
15026 5000 all models '84 thru '88
Audi A4 '96 thru '01 - *see VW Passat (96023)*
15030 Audi A4 '02 thru '08

AUSTIN-HEALEY
Sprite - *see MG Midget (66015)*

BMW
18020 3/5 Series '82 thru '92
18021 3-Series incl. Z3 models '92 thru '98
18022 3-Series incl. Z4 models '99 thru '05
18023 3-Series '06 thru '14
18025 320i all 4-cylinder models '75 thru '83
18050 1500 thru 2002 except Turbo '59 thru '77

BUICK
19010 Buick Century '97 thru '05
Century (front-wheel drive) - *see GM (38005)*
19020 Buick, Oldsmobile & Pontiac Full-size (Front-wheel drive) '85 thru '05
Buick Electra, LeSabre and Park Avenue; Oldsmobile Delta 88 Royale, Ninety Eight and Regency; Pontiac Bonneville
19025 Buick, Oldsmobile & Pontiac Full-size (Rear wheel drive) '70 thru '90
Buick Estate, Electra, LeSabre, Limited, Oldsmobile Custom Cruiser, Delta 88, Ninety-eight, Pontiac Bonneville, Catalina, Grandville, Parisienne
19027 Buick LaCrosse '05 thru '13
Enclave - *see GENERAL MOTORS (38001)*
Rainier - *see CHEVROLET (24072)*
Regal - *see GENERAL MOTORS (38010)*
Riviera - *see GENERAL MOTORS (38030, 38031)*
Roadmaster - *see CHEVROLET (24046)*
Skyhawk - *see GENERAL MOTORS (38015)*
Skylark - *see GENERAL MOTORS (38020, 38025)*
Somerset - *see GENERAL MOTORS (38025)*

CADILLAC
21015 CTS & CTS-V '03 thru '14
21030 Cadillac Rear Wheel Drive '70 thru '93
Cimarron - *see GENERAL MOTORS (38015)*
DeVille - *see GENERAL MOTORS (38031 & 38032)*
Eldorado - *see GENERAL MOTORS (38030)*
Fleetwood - *see GENERAL MOTORS (38031)*
Seville - *see GM (38030, 38031 & 38032)*

CHEVROLET
10305 Chevrolet Engine Overhaul Manual
24010 Astro & GMC Safari Mini-vans '85 thru '05
24013 Aveo '04 thru '11
24015 Camaro V8 all models '70 thru '81
24016 Camaro all models '82 thru '92
24017 Camaro & Firebird '93 thru '02
Cavalier - *see GENERAL MOTORS (38016)*
Celebrity - *see GENERAL MOTORS (38005)*
24018 Camaro '10 thru '15
24020 Chevelle, Malibu & El Camino '69 thru '87
Cobalt - *see GENERAL MOTORS (38017)*
24024 Chevette & Pontiac T1000 '76 thru '87
Citation - *see GENERAL MOTORS (38020)*
24027 Colorado & GMC Canyon '04 thru '12
24032 Corsica & Beretta all models '87 thru '96
24040 Corvette all V8 models '68 thru '82
24041 Corvette all models '84 thru '96
24042 Corvette all models '97 thru '13
24044 Cruze '11 thru '19
24045 Full-size Sedans Caprice, Impala, Biscayne, Bel Air & Wagons '69 thru '90
24046 Impala SS & Caprice and Buick Roadmaster '91 thru '96
Impala '00 thru '05 - *see LUMINA (24048)*
24047 Impala & Monte Carlo all models '06 thru '11
Lumina '90 thru '94 - *see GM (38010)*
24048 Lumina & Monte Carlo '95 thru '05
Lumina APV - *see GM (38035)*
24050 Luv Pick-up all 2WD & 4WD '72 thru '82
24051 Malibu '13 thru '19
24055 Monte Carlo all models '70 thru '88
Monte Carlo '95 thru '01 - *see LUMINA (24048)*
24059 Nova all V8 models '69 thru '79
24060 Nova and Geo Prizm '85 thru '92
24064 Pick-ups '67 thru '87 - Chevrolet & GMC
24065 Pick-ups '88 thru '98 - Chevrolet & GMC
24066 Pick-ups '99 thru '06 - Chevrolet & GMC
24067 Chevrolet Silverado & GMC Sierra '07 thru '14
24068 Chevrolet Silverado & GMC Sierra '14 thru '19
24070 S-10 & S-15 Pick-ups '82 thru '93, Blazer & Jimmy '83 thru '94,
24071 S-10 & Sonoma Pick-ups '94 thru '04, including Blazer, Jimmy & Hombre
24072 Chevrolet TrailBlazer, GMC Envoy & Oldsmobile Bravada '02 thru '09
24075 Sprint '85 thru '88 & Geo Metro '89 thru '01
24080 Vans - Chevrolet & GMC '68 thru '96
24081 Chevrolet Express & GMC Savana Full-size Vans '96 thru '19

CHRYSLER
10310 Chrysler Engine Overhaul Manual
25015 Chrysler Cirrus, Dodge Stratus, Plymouth Breeze '95 thru '00
25020 Full-size Front-Wheel Drive '88 thru '93
K-Cars - *see DODGE Aries (30008)*
Laser - *see DODGE Daytona (30030)*
25025 Chrysler LHS, Concorde, New Yorker, Dodge Intrepid, Eagle Vision, '93 thru '97
25026 Chrysler LHS, Concorde, 300M, Dodge Intrepid, '98 thru '04
25027 Chrysler 300 '05 thru '18, Dodge Charger '06 thru '18, Magnum '05 thru '08 & Challenger '08 thru '18
25030 Chrysler & Plymouth Mid-size front wheel drive '82 thru '95
Rear-wheel Drive - *see Dodge (30050)*
25035 PT Cruiser all models '01 thru '10
25040 Chrysler Sebring '95 thru '06, Dodge Stratus '01 thru '06 & Dodge Avenger '95 thru '00
25041 Chrysler Sebring '07 thru '10, 200 '11 thru '17 Dodge Avenger '08 thru '14

DATSUN
28005 200SX all models '80 thru '83
28012 240Z, 260Z & 280Z Coupe '70 thru '78
28014 280ZX Coupe & 2+2 '79 thru '83
300ZX - *see NISSAN (72010)*
28018 510 & PL521 Pick-up '68 thru '73
28020 510 all models '78 thru '81
28022 620 Series Pick-up all models '73 thru '79
720 Series Pick-up - *see NISSAN (72030)*

DODGE
400 & 600 - *see CHRYSLER (25030)*
30008 Aries & Plymouth Reliant '81 thru '89
30010 Caravan & Plymouth Voyager '84 thru '95
30011 Caravan & Plymouth Voyager '96 thru '02
30012 Challenger & Plymouth Sapporro '78 thru '83
30013 Caravan, Chrysler Voyager & Town & Country '03 thru '07
30014 Grand Caravan & Chrysler Town & Country '08 thru '18
30016 Colt & Plymouth Champ '78 thru '87
30020 Dakota Pick-ups all models '87 thru '96
30021 Durango '98 & '99 & Dakota '97 thru '99
30022 Durango '00 thru '03 & Dakota '00 thru '04
30023 Durango '04 thru '09 & Dakota '05 thru '11
30025 Dart, Demon, Plymouth Barracuda, Duster & Valiant 6-cylinder models '67 thru '76
30030 Daytona & Chrysler Laser '84 thru '89
Intrepid - *see CHRYSLER (25025, 25026)*
30034 Neon all models '95 thru '99
30035 Omni & Plymouth Horizon '78 thru '90
30036 Dodge & Plymouth Neon '00 thru '05
30040 Pick-ups full-size models '74 thru '93
30042 Pick-ups full-size models '94 thru '08
30043 Pick-ups full-size models '09 thru '18
30045 Ram 50/D50 Pick-ups & Raider and Plymouth Arrow Pick-ups '79 thru '93
30050 Dodge/Plymouth/Chrysler RWD '71 thru '89
30055 Shadow & Plymouth Sundance '87 thru '94
30060 Spirit & Plymouth Acclaim '89 thru '95
30065 Vans - Dodge & Plymouth '71 thru '03

EAGLE
Talon - *see MITSUBISHI (68030, 68031)*
Vision - *see CHRYSLER (25025)*

FIAT
34010 124 Sport Coupe & Spider '68 thru '78
34025 X1/9 all models '74 thru '80

FORD
10320 Ford Engine Overhaul Manual
10355 Ford Automatic Transmission Overhaul
11500 Mustang '64-1/2 thru '70 Restoration Guide
36004 Aerostar Mini-vans all models '86 thru '97
36006 Contour & Mercury Mystique '95 thru '00
36008 Courier Pick-up all models '72 thru '82
36012 Crown Victoria & Mercury Grand Marquis '88 thru '11
36014 Edge '07 thru '19 & Lincoln MKX '07 thru '18
36016 Escort & Mercury Lynx all models '81 thru '90
36020 Escort & Mercury Tracer '91 thru '02
36022 Escape '01 thru '17, Mazda Tribute '01 thru '11, & Mercury Mariner '05 thru '11
36024 Explorer & Mazda Navajo '91 thru '01
36025 Explorer & Mercury Mountaineer '02 thru '10
36026 Explorer '11 thru '17
36028 Fairmont & Mercury Zephyr '78 thru '83
36030 Festiva & Aspire '88 thru '97
36032 Fiesta all models '77 thru '80
36034 Focus all models '00 thru '11
36035 Focus '12 thru '14
36045 Fusion '06 thru '14 & Mercury Milan '06 thru '11
36048 Mustang V8 all models '64-1/2 thru '73
36049 Mustang II 4-cylinder, V6 & V8 models '74 thru '78
36050 Mustang & Mercury Capri '79 thru '93
36051 Mustang all models '94 thru '04
36052 Mustang '05 thru '14
36054 Pick-ups & Bronco '73 thru '79
36058 Pick-ups & Bronco '80 thru '96
36059 F-150 '97 thru '03, Expedition '97 thru '17, F-250 '97 thru '99, F-150 Heritage '04 & Lincoln Navigator '98 thru '17
36060 Super Duty Pick-ups & Excursion '99 thru '10
36061 F-150 full-size '04 thru '14
36062 Pinto & Mercury Bobcat '75 thru '80
36063 F-150 full-size '15 thru '17
36064 Super Duty Pick-ups '11 thru '16
36066 Probe all models '89 thru '92
Probe '93 thru '97 - *see MAZDA 626 (61042)*
36070 Ranger & Bronco II gas models '83 thru '92
36071 Ranger '93 thru '11 & Mazda Pick-ups '94 thru '09
36074 Taurus & Mercury Sable '86 thru '95
36075 Taurus & Mercury Sable '96 thru '07
36076 Taurus '08 thru '14, Five Hundred '05 thru '07, Mercury Montego '05 thru '07 & Sable '08 thru '09
36078 Tempo & Mercury Topaz '84 thru '94
36082 Thunderbird & Mercury Cougar '83 thru '88
36086 Thunderbird & Mercury Cougar '89 thru '97
36090 Vans all V8 Econoline models '69 thru '91
36094 Vans full size '92 thru '14
36097 Windstar '95 thru '03, Freestar & Mercury Monterey Mini-van '04 thru '07

GENERAL MOTORS
10360 GM Automatic Transmission Overhaul
38001 GMC Acadia '07 thru '16, Buick Enclave '08 thru '17, Saturn Outlook '07 thru '10 & Chevrolet Traverse '09 thru '17
38005 Buick Century, Chevrolet Celebrity, Oldsmobile Cutlass Ciera & Pontiac 6000 all models '82 thru '96
38010 Buick Regal '88 thru '04, Chevrolet Lumina '88 thru '04, Oldsmobile Cutlass Supreme '88 thru '97 & Pontiac Grand Prix '88 thru '07
38015 Buick Skyhawk, Cadillac Cimarron, Chevrolet Cavalier, Oldsmobile Firenza, Pontiac J-2000 & Sunbird '82 thru '94
38016 Chevrolet Cavalier & Pontiac Sunfire '95 thru '05
38017 Chevrolet Cobalt '05 thru '10, HHR '06 thru '11, Pontiac G5 '07 thru '09, Pursuit '05 thru '06 & Saturn ION '03 thru '07
38020 Buick Skylark, Chevrolet Citation, Oldsmobile Omega, Pontiac Phoenix '80 thru '85
38025 Buick Skylark '86 thru '98, Somerset '85 thru '87, Oldsmobile Achieva '92 thru '98, Calais '85 thru '91, & Pontiac Grand Am all models '85 thru '98
38026 Chevrolet Malibu '97 thru '03, Classic '04 thru '05, Oldsmobile Alero '99 thru '03, Cutlass '97 thru '00, & Pontiac Grand Am '99 thru '03
38027 Chevrolet Malibu '04 thru '12, Pontiac G6 '05 thru '10 & Saturn Aura '07 thru '10
38030 Cadillac Eldorado, Seville, Oldsmobile Toronado & Buick Riviera '71 thru '85
38031 Cadillac Eldorado, Seville, DeVille, Fleetwood, Oldsmobile Toronado & Buick Riviera '86 thru '93
38032 Cadillac DeVille '94 thru '05, Seville '92 thru '04 & Cadillac DTS '06 thru '10
38035 Chevrolet Lumina APV, Oldsmobile Silhouette & Pontiac Trans Sport all models '90 thru '96
38036 Chevrolet Venture '97 thru '05, Oldsmobile Silhouette '97 thru '04, Pontiac Trans Sport '97 thru '98 & Montana '99 thru '05
38040 Chevrolet Equinox '05 thru '17, GMC Terrain '10 thru '17 & Pontiac Torrent '06 thru '09

GEO
Metro - *see CHEVROLET Sprint (24075)*
Prizm - *'85 thru '92 see CHEVY (24060), '93 thru '02 see TOYOTA Corolla (92036)*
40030 Storm all models '90 thru '93
Tracker - *see SUZUKI Samurai (90010)*

(Continued on other side)

Haynes Automotive Manuals (continued)

GMC
Acadia - see GENERAL MOTORS (38001)
Pick-ups - see CHEVROLET (24027, 24068)
Vans - see CHEVROLET (24081)

HONDA
42010 **Accord CVCC** all models '76 thru '83
42011 **Accord** all models '84 thru '89
42012 **Accord** all models '90 thru '93
42013 **Accord** all models '94 thru '97
42014 **Accord** all models '98 thru '02
42015 **Accord** '03 thru '12 **& Crosstour** '10 thru '14
42016 **Accord** '13 thru '17
42020 **Civic 1200** all models '73 thru '79
42021 **Civic 1300 & 1500 CVCC** '80 thru '83
42022 **Civic 1500 CVCC** all models '75 thru '79
42023 **Civic** all models '84 thru '91
42024 **Civic & del Sol** '92 thru '95
42025 **Civic** '96 thru '00, **CR-V** '97 thru '01 **& Acura Integra** '94 thru '00
42026 **Civic** '01 thru '11 **& CR-V** '02 thru '11
42027 **Civic** '12 thru '15 **& CR-V** '12 thru '16
42030 **Fit** '07 thru '13
42035 **Odyssey** all models '99 thru '10
Passport - see ISUZU Rodeo (47017)
42037 **Honda Pilot** '03 thru '08, **Ridgeline** '06 thru '14 **& Acura MDX** '01 thru '07
42040 **Prelude CVCC** all models '79 thru '89

HYUNDAI
43010 **Elantra** all models '96 thru '19
43015 **Excel & Accent** all models '86 thru '13
43050 **Santa Fe** all models '01 thru '12
43055 **Sonata** all models '99 thru '14

INFINITI
G35 '03 thru '08 - see NISSAN 350Z (72011)

ISUZU
Hombre - see CHEVROLET S-10 (24071)
47017 **Rodeo** '91 thru '02, **Amigo** '89 thru '94 & '98 thru '02 **& Honda Passport** '95 thru '02
47020 **Trooper** '84 thru '91 **& Pick-up** '81 thru '93

JAGUAR
49010 **XJ6** all 6-cylinder models '68 thru '86
49011 **XJ6** all models '88 thru '94
49015 **XJ12 & XJS** all 12-cylinder models '72 thru '85

JEEP
50010 **Cherokee, Comanche & Wagoneer Limited** all models '84 thru '01
50011 **Cherokee** '14 thru '19
50020 **CJ** all models '49 thru '86
50025 **Grand Cherokee** all models '93 thru '04
50026 **Grand Cherokee** '05 thru '19 **& Dodge Durango** '11 thru '19
50029 **Grand Wagoneer & Pick-up** '72 thru '91 Grand Wagoneer '84 thru '91, Cherokee & Wagoneer '72 thru '83, Pick-up '72 thru '88
50030 **Wrangler** all models '87 thru '17
50035 **Liberty** '02 thru '12 **& Dodge Nitro** '07 thru '11
50050 **Patriot & Compass** '07 thru '17

KIA
54050 **Optima** '01 thru '10
54060 **Sedona** '02 thru '14
54070 **Sephia** '94 thru '01, **Spectra** '00 thru '09, **Sportage** '05 thru '20
54077 **Sorento** '03 thru '13

LEXUS
ES 300/330 - see TOYOTA Camry (92007, 92008)
ES 350 - see TOYOTA Camry (92009)
RX 300/330/350 - see TOYOTA Highlander (92095)

LINCOLN
MKX - see FORD (36014)
Navigator - see FORD Pick-up (36059)
59010 **Rear-Wheel Drive Continental** '70 thru '87, **Mark Series** '70 thru '92 **& Town Car** '81 thru '10

MAZDA
61010 **GLC (rear-wheel drive)** '77 thru '83
61011 **GLC (front-wheel drive)** '81 thru '85
61012 **Mazda3** '04 thru '11
61015 **323 & Protegé** '90 thru '03
61016 **MX-5 Miata** '90 thru '14
61020 **MPV** all models '89 thru '98
Navajo - see Ford Explorer (36024)
61030 **Pick-ups** '72 thru '93
Pick-up '94 thru '09 - see Ford Ranger (36071)
61035 **RX-7** all models '79 thru '85
61036 **RX-7** all models '86 thru '91
61040 **626 (rear-wheel drive)** all models '79 thru '82
61041 **626 & MX-6 (front-wheel drive)** '83 thru '92
61042 **626** '93 thru '01 **& MX-6/Ford Probe** '93 thru '02
61043 **Mazda6** '03 thru '13

MERCEDES-BENZ
63012 **123 Series Diesel** '76 thru '85
63015 **190 Series** 4-cylinder gas models '84 thru '88
63020 **230/250/280** 6-cylinder SOHC models '68 thru '72
63025 **280 123 Series** gas models '77 thru '81
63030 **350 & 450** all models '71 thru '80
63040 **C-Class:** C230/C240/C280/C320/C350 '01 thru '07

MERCURY
64200 **Villager & Nissan Quest** '93 thru '01
All other titles, see FORD Listing.

MG
66010 **MGB** Roadster & GT Coupe '62 thru '80
66015 **MG Midget, Austin Healey Sprite** '58 thru '80

MINI
67020 **Mini** '02 thru '13

MITSUBISHI
68020 **Cordia, Tredia, Galant, Precis & Mirage** '83 thru '93
68030 **Eclipse, Eagle Talon & Plymouth Laser** '90 thru '94
68031 **Eclipse** '95 thru '05 **& Eagle Talon** '95 thru '98
68035 **Galant** '94 thru '12
68040 **Pick-up** '83 thru '96 **& Montero** '83 thru '93

NISSAN
72010 **300ZX** all models including Turbo '84 thru '89
72011 **350Z & Infiniti G35** all models '03 thru '08
72015 **Altima** all models '93 thru '06
72016 **Altima** '07 thru '12
72020 **Maxima** all models '85 thru '92
72021 **Maxima** all models '93 thru '08
72025 **Murano** '03 thru '14
72030 **Pick-ups** '80 thru '97 **& Pathfinder** '87 thru '95
72031 **Frontier** '98 thru '04, **Xterra** '00 thru '04, **& Pathfinder** '96 thru '04
72032 **Frontier & Xterra** '05 thru '14
72037 **Pathfinder** '05 thru '14
72040 **Pulsar** all models '83 thru '86
72042 **Roque** '08 thru '20
72050 **Sentra** all models '82 thru '94
72051 **Sentra & 200SX** all models '95 thru '06
72060 **Stanza** all models '82 thru '90
72070 **Titan pick-ups** '04 thru '10, **Armada** '05 thru '10 **& Pathfinder Armada** '04
72080 **Versa** all models '07 thru '19

OLDSMOBILE
73015 **Cutlass V6 & V8** gas models '74 thru '88
For other OLDSMOBILE titles, see BUICK, CHEVROLET or GENERAL MOTORS listings.

PLYMOUTH
For PLYMOUTH titles, see DODGE listing.

PONTIAC
79008 **Fiero** all models '84 thru '88
79018 **Firebird** V8 models except Turbo '70 thru '81
79019 **Firebird** all models '82 thru '92
79025 **G6** all models '05 thru '09
79040 **Mid-size Rear-wheel Drive** '70 thru '87
Vibe '03 thru '10 - see TOYOTA Corolla (92037)
For other PONTIAC titles, see BUICK, CHEVROLET or GENERAL MOTORS listings.

PORSCHE
80020 **911** Coupe & Targa models '65 thru '89
80025 **914** all 4-cylinder models '69 thru '76
80030 **924** all models including Turbo '76 thru '82
80035 **944** all models including Turbo '83 thru '89

RENAULT
Alliance & Encore - see AMC (14025)

SAAB
84010 **900** all models including Turbo '79 thru '88

SATURN
87010 **Saturn** all S-series models '91 thru '02
Saturn Ion '03 thru '07- see GM (38017)
Saturn Outlook - see GM (38001)
87020 **Saturn L-series** all models '00 thru '04
87040 **Saturn VUE** '02 thru '09

SUBARU
89002 **1100, 1300, 1400 & 1600** '71 thru '79
89003 **1600 & 1800** 2WD & 4WD '80 thru '94
89080 **Impreza** '02 thru '11, **WRX** '02 thru '14, **& WRX STI** '04 thru '14
89100 **Legacy** all models '90 thru '99
89101 **Legacy & Forester** '00 thru '09
89102 **Legacy** '10 thru '16 **& Forester** '12 thru '16

SUZUKI
90010 **Samurai/Sidekick & Geo Tracker** '86 thru '01

TOYOTA
92005 **Camry** all models '83 thru '91
92006 **Camry** '92 thru '96 **& Avalon** '95 thru '96
92007 **Camry, Avalon, Solara, Lexus ES 300** '97 thru '01

92008 **Camry, Avalon, Lexus ES 300/330** '02 thru '06 **& Solara** '02 thru '08
92009 **Camry, Avalon & Lexus ES 350** '07 thru '17
92015 **Celica Rear-wheel Drive** '71 thru '85
92020 **Celica Front-wheel Drive** '86 thru '99
92025 **Celica Supra** all models '79 thru '92
92030 **Corolla** all models '75 thru '79
92032 **Corolla** all rear-wheel drive models '80 thru '87
92035 **Corolla** all front-wheel drive models '84 thru '92
92036 **Corolla & Geo/Chevrolet Prizm** '93 thru '02
92037 **Corolla** '03 thru '19, **Matrix** '03 thru '14, **& Pontiac Vibe** '03 thru '10
92040 **Corolla Tercel** all models '80 thru '82
92045 **Corona** all models '74 thru '82
92050 **Cressida** all models '78 thru '82
92055 **Land Cruiser** FJ40, 43, 45, 55 '68 thru '82
92056 **Land Cruiser** FJ60, 62, 80, FZJ80 '80 thru '96
92060 **Matrix** '03 thru '11 **& Pontiac Vibe** '03 thru '10
92065 **MR2** all models '85 thru '87
92070 **Pick-up** all models '69 thru '78
92075 **Pick-up** all models '79 thru '95
92076 **Tacoma** '95 thru '04, **4Runner** '96 thru '02 **& T100** '93 thru '98
92077 **Tacoma** all models '05 thru '18
92078 **Tundra** '00 thru '06 **& Sequoia** '01 thru '07
92079 **4Runner** all models '03 thru '09
92080 **Previa** all models '91 thru '95
92081 **Prius** all models '01 thru '12
92082 **RAV4** all models '96 thru '12
92085 **Tercel** all models '87 thru '94
92090 **Sienna** all models '98 thru '10
92095 **Highlander** '01 thru '19 **& Lexus RX330/330/350** '99 thru '19
92179 **Tundra** '07 thru '19 **& Sequoia** '08 thru '19

TRIUMPH
94007 **Spitfire** all models '62 thru '81
94010 **TR7** all models '75 thru '81

VW
96008 **Beetle & Karmann Ghia** '54 thru '79
96009 **New Beetle** '98 thru '10
96016 **Rabbit, Jetta, Scirocco & Pick-up** gas models '75 thru '92 & Convertible '80 thru '92
96017 **Golf, GTI & Jetta** '93 thru '98, **Cabrio** '95 thru '02
96018 **Golf, GTI, Jetta** '99 thru '05
96019 **Jetta, Rabbit, GLI, GTI & Golf** '05 thru '11
96020 **Rabbit, Jetta & Pick-up** diesel '77 thru '84
96021 **Jetta** '11 thru '18 **& Golf** '15 thru '19
96023 **Passat** '98 thru '05 **& Audi A4** '96 thru '01
96030 **Transporter 1600** all models '68 thru '79
96035 **Transporter 1700, 1800 & 2000** '72 thru '79
96040 **Type 3 1500 & 1600** all models '63 thru '73
96045 **Vanagon Air-Cooled** all models '80 thru '83

VOLVO
97010 **120, 130 Series & 1800 Sports** '61 thru '73
97015 **140 Series** all models '66 thru '74
97020 **240 Series** all models '76 thru '93
97040 **740 & 760 Series** all models '82 thru '88
97050 **850 Series** all models '93 thru '97

TECHBOOK MANUALS
10205 **Automotive Computer Codes**
10206 **OBD-II & Electronic Engine Management**
10210 **Automotive Emissions Control Manual**
10215 **Fuel Injection Manual** '78 thru '85
10225 **Holley Carburetor Manual**
10230 **Rochester Carburetor Manual**
10305 **Chevrolet Engine Overhaul Manual**
10320 **Ford Engine Overhaul Manual**
10330 **GM and Ford Diesel Engine Repair Manual**
10331 **Duramax Diesel Engines** '01 thru '19
10332 **Cummins Diesel Engine Performance Manual**
10333 **GM, Ford & Chrysler Engine Performance Manual**
10334 **GM Engine Performance Manual**
10340 **Small Engine Repair Manual,** 5 HP & Less
10341 **Small Engine Repair Manual,** 5.5 thru 20 HP
10345 **Suspension, Steering & Driveline Manual**
10355 **Ford Automatic Transmission Overhaul**
10360 **GM Automatic Transmission Overhaul**
10405 **Automotive Body Repair & Painting**
10410 **Automotive Brake Manual**
10411 **Automotive Anti-lock Brake (ABS) Systems**
10420 **Automotive Electrical Manual**
10425 **Automotive Heating & Air Conditioning**
10435 **Automotive Tools Manual**
10445 **Welding Manual**
10450 **ATV Basics**

Over a 100 Haynes motorcycle manuals also available

10/22

Haynes North America, Inc. • (805) 498-6703 • www.haynes.com